Ecological Studies, Vol. 98

Analysis and Synthesis

Edited by

W. D. Billings, Durham, USA
F. Golley, Athens, USA
O. L. Lange, Würzburg, FRG
J. S. Olson, Oak Ridge, USA
H. Remmert, Marburg, FRG

Ecological Studies

Volume 81
Nitrogen in Terrestrial Ecosystems (1990)
By C.O. Tamm

Volume 82
Quantitative Methods in Landscape Ecology (1990)
Edited by M.G. Turner and R.H. Gardner

Volume 83
The Rivers of Florida (1990)
Edited by R.J. Livingston

Volume 84
Fire in the Tropical Biota: Ecosystem Processes and Global Challenges (1990)
Edited by J.G. Goldammer

Volume 85
The Mosaic-Cycle Concept of Ecosystems (1991)
Edited by H. Remmert

Volume 86
Ecological Heterogeneity (1991)
Edited by J. Kolasa and S.T.A. Pickett

Volume 87
Horses and Grasses: The Nutritional Ecology of Equids, and Their Impact on the Camargue (1991)
By P. Duncan

Volume 88
Pinnipeds and El Nino: Responses to Environmental Stress (1991)
Edited by F. Trillmich and K.A. Ono

Volume 89
Plantago: A Multidisciplinary Study (1992)
Edited by P.J.C. Kuiper and M. Bos

Volume 90
Biogeochemistry of a Subalpine Ecosystem: Loch Vale Watershed (1992)
Edited by Jill Baron

Volume 91
Atmospheric Deposition and Forest Nutrient Cycling (1992)
Edited by D.W. Johnson and S.E. Lindberg

Volume 92
Landscape Boundaries: Consequences for Biotic Diversity and Ecological Flows (1992)
Edited by A.J. Hansen and F. di Castri

Volume 93
Fire in South African Mountain Fynbos: Ecosystem, Community, and Species Response at Swartboskloof (1992)
Edited by B.W. van Wilgen, D.M. Richardson, F.J. Kruger, and H.J. van Hensbergen

Volume 94
The Ecology of Aquatic Hyphomycetes (1992)
Edited by F. Bärlocher

Volume 95
Palms in Forest Ecosystems of Amazonia (1992)
By F. Kahn and J.-J. DeGranville

Volume 96
Ecology and Decline of Red Spruce in the Eastern United States (1992)
Edited by C. Eagar and M.B. Adams

Volume 97
The Response of Western Forests to Air Pollution (1992)
Edited by R.K. Olson, D. Binkley, and M. Böhm

Volume 98
Plankton Regulation Dynamics (1993)
Edited by N. Walz

Volume 99
Biodiversity and Ecosystem Function
Edited by E.-D. Schulze and H.A. Mooney

Norbert Walz (Ed.)

Plankton Regulation Dynamics

Experiments and Models in Rotifer Continuous Cultures

With 110 Figures

Springer-Verlag
Berlin Heidelberg New York
London Paris Tokyo
Hong Kong Barcelona
Budapest

Dr. NORBERT WALZ
Institut für Gewässerökologie und Binnenfischerei
im Forschungsverbund Berlin e. V.
Abteilung Limnologie von Flußseen
Müggelseedamm 260
12561 Berlin
FRG

ISBN 3-540-55955-8 Springer-Verlag Berlin Heidelberg New York
ISBN 0-387-55955-8 Springer-Verlag New York Berlin Heidelberg

Library of Congress Cataloging-in-Publication Data.
Plankton regulation dynamics: experiments and models in Rotifer continuous cultures / Norbert Walz (ed.). p. cm. – (Ecological studies; vol. 98) Includes bibliographical references and index. 1. Rotifera – Ecology. 2. Plankton populations. 3. Rotifera – Cultures and culture media. 4. Continuous culture (Microbiology). 5. Chemostat. I. Walz, Norbert. II. Series: Ecological studies: v. 98. QL391.R8P58 1993 595.1'81045247–dc20 92-39672 CIP

Printed in Germany

Cover illustration drawn by Walter Koste
Production Editor: Renate Münzenmayer
Typesetting: Thomson Press Ltd., India
31/3145-5 4 3 2 1 0 – Printed on acid-free paper

Dedicated to Dr. Walter Koste
on the occasion of his 80th birthday

Preface

This volume considers principles of plankton regulation dynamics based on experimental results and models derived from continuous cultures. These kinds of cultures, and especially the chemostat, are regarded as model systems for more complicated natural communities. Apart from their different complexities, the connecting link between both is that the plankton community as well as the chemostat have similar system properties since they both operate over the same span of trophic levels: primary producers, consumers, predators. This will be discussed extensively.

At this point it seems to be of advantage to reveal the philosophy of the system theory on which the general parts of this volume are based. On each level of a system (e.g., on the organism, population, community, ecosystem level) "emergent properties" of the system arise which can not be predicted by knowledge of the separate parts alone. For example, on the population level, an important loss of energy is mortality, which is not considered in an energy balance of individuals. The succession of populations and defense strategies may not be predictable from population ecology. The adaptive value of a variable cladocera helmet size was only recognized through knowledge of predator-prey interactions (Havel 1987; Pijanowska 1990). On the ecosystem level, such properties as stability and diversity appear. On the upper levels of the system hierarchy, much more information is necessary than is obtainable from the separate, lower levels.

However, these emergent properties are not myths. In contrast to other views (e.g., Allen and Starr 1982), these properties are additional but not independent of the lower levels. The "emergent" properties of the higher level go beyond the separate properties of the lower level because different levels are in interaction. For this reason, interactions on each level and feedback between the levels and their parts (e.g., between populations) have to be studied. The three-dimensional conformation of a protein molecule cannot be predicted from its amino acid sequence (which is stored exclusively in its genetic information) without further knowledge of e.g. the hydrophobic interactions between the nonpolar rest groups of different acids (Lehninger 1975). The spatial configuration of the molecule is achieved by folding on the basis of these forces. Thus, this is understood as "self-organization", higher ecological levels also organize themselves, in contrast to a superorganism view of the system. These interactions result from the characteristics of the underlying parts in nonlinear systems and are expressed by feedback reactions between these parts. A short, but very interesting introduction to the ecological literature on this subject is given by Gross (1986).

Feedback reactions and interactions are arranged in regulation loops known from control engineering and do not need a superior control center. Therefore, continuous cultures are an excellent tool to study the basic elements of interactions and regulation principles. In this system no other effective factors except "natural selection" are assumed. Regardless of other contradictions, this corresponds to an evolutionary ecosystem ecology of Oksanen (1988) or to a "Darwinian ecology" of Sommer (1989d).

Spring 1993 NORBERT WALZ

Contents

Contributors

WILLIAM N. BENNETT
University of Wisconsin-Milwaukee
Center for Great Lakes Studies
600E Greenfield Avenue
P.O. Box 413
Milwaukee, Wisconsin 53204
USA

MARTIN E. BORAAS
University of Wisconsin-Milwaukee
Center for Great Lakes Studies
600E Greenfield Avenue
P.O. Box 413
Milwaukee, Wisconsin 53204
USA

J. BRIAN HORTON †
The Pennsylvania State University
University Park
Pennsylvania 16802
USA

CHARLES M. JAMES
Marine Sciences Research Center
State University of New York
Stony Brook, NY 11794
USA

KARL O. ROTHAUPT
Max-Planck-Institut für Limnologie
Abteilung Ökophysiologie
Postfach 165
24302 Plön
FRG

DIANNE B. SEALE
University of Wisconsin-Milwaukee
Center for Great Lakes Studies
600E Greenfield Avenue
P.O. Box 413
Milwaukee, Wisconsin 53204
USA

NORBERT WALZ
Institut für Gewässerökologie und Binnenfischerei im Forschungsverbund Berlin e.V.
Abteilung Limnologie von Flußseen
Müggelseedamm 260
12561 Berlin
FRG

1 Introduction

1.1 Systems Approach to Natural and Artificial Plankton Communities by Continuous Cultures

NORBERT WALZ

Shortly after the invention of the microscope (about 1590 by Z. Jansen) Antony van Leeuwenhoek (1632–1723) was the first to observed infusoria in a drop of water. Since that time and since the first use of plankton gauze nets by the physiologist, Johannes Müller (1801–1858), the organisms in the plankton of aquatic environments have fascinated generations of scientists and amateurs. Every child looking for the first time in a microscope is spellbound by this living world of crawling and hopping water fleas, rotating wheel and slipper animalcules, and bizarre by shaped algae. What a contradiction to the formerly prevailing conception of a monotonous and simple structure of this pelagic environment! This apparent simplicity, however, seemed to be a good starting point for many scientists to study structure, function, and regulation of this community as a model for more complicated ecosystems. Hutchinson (1961) with his formulation of the "paradox of the plankton" illuminated this inconsistency. Meanwhile, with the publication of volumes such as *Complex Interactions in Lake Communities* (Carpenter 1988) and *Plankton Ecology* (Sommer 1989a), the specific character of this system is approached more closely. It is not its simplicity which makes the plankton a good candidate for model studies but its applicability to quantitative and experimental studies. This community has lost none of its fascination.

For a considerable time, abiotic factors (e.g., temperature, weather, chemical factors) have been in the foreground of interest in physiological ecology as controlling factors, and the tolerance and optima of populations to these variables have been widely studied. But in recent decades plankton ecologists have considered interactive biotic factors to an increasing extent. This was already appreciated by Thienemann (1918) and others, who realized that biological communities change their environments either cyclically, e.g. annually, or in secular time scales. Today, we know unequivocally that organisms and populations of organisms interact in a myriad of ways and that they can change their environment and the living conditions of their successors dramatically or subtly.

In 1942 two publications appeared which greatly influenced the study of population and community ecology. Lindeman (1942), in his study of the trophic-dynamic aspects of ecosystems, established the basis for energetic studies of communities, which was further developed by Slobodkin (1962) and others.

At the same time, Monod (1942) formulated a kinetics of the growth of microorganisms which is dependent on substrate concentration. In ecological as well as in chemical processes, given several reactions that are energetically possible, only that one will take place that is kinetically the most favorable. Already in 1925, Lotka (1925) called attention to the narrow trade-off between kinetics and energetics in biological and ecological energy transformations.

Are these models also valid when there are many interacting populations such as in communities? The one-dimensional reaction, as the counteraction of the other populations, must obey these principles of kinetics and energetics. In the ecophysiology of the community, as in the physiology of a living cell, coupled reactions use free energy, but gain information (Stugren 1986) for structuring the community, since information and negative entropy correspond to each other (Brillouin 1956). The members of the community, therefore, are connected in a flux of information which cybernetically organizes the system (Patten 1959). For this reason, the laws of regulation techniques are indeed applicable to the populations in an ecosystem (Schwerdtfeger 1978), in direct analogy to their successful application to the physiology of organisms (e.g., Hassenstein 1966). A technical regulation loop, therefore, is a model for the interactions between populations in a community, allowing one to examine the regulation mechanisms underlying the processes (Siebeck 1986). However, the identity of the elements of this control loop, i.e., regulators, correcting elements, sensors, etc., require experimental analysis (Chap. 7.2).

The ecophysiological link between kinetic and energetic elements at the community level is the "self-regulation" (Haken 1982) of the community. Producers structure the community from the "bottom-up" and energetic aspects play the main role in food-limited communities. On the other hand, the community may also be controlled from the "top-down" by consumers and predators, such that the surviving populations are those with the necessary kinetic attributes and those protected by defense mechanisms. This regulation is achieved by the community itself, there is no superior control unit. Extending the approach of Lindeman (1942) and Monod (1942), according to regulatory points of view, the kinetics of many bioenergetic activities (i.e., the "functional response" of feeding rate, production rate, growth rate, etc.) may be identified formally as the characteristics of the correcting elements of a control circuit. The populations may differ considerably in these characteristic features. These ideas will be developed further in the following chapters (especially in Chap. 6.3).

Ruttner (1962) emphasized how the use of the fine gauze net by Müller was a necessary precondition for the study of the life-forms of plankton. The chemostat (Monod 1950; Novick and Szilard 1950), besides other continuous methods, is an additional instrument especially suitable to study the interactions of plankton organisms in a model ecosystem. The main attribute of these continuous cultures is the constant or regular dilution of the culture volume with new substrate for the organisms and simultaneous drawing off or washing out the same volume together with the organisms.

In the chemostat, a limiting substrate regulates the growth of the consumers but, at the same time, the substrate itself is regulated by a feedback reaction and just that amount is used to maintain this growth process. The chemostat, therefore, is per se a self-regulating system which allows one to study the regulating relationship inside a model plankton community.

The chemostat is only one of four types of continuous cultures. The others are semicontinuous cultures, fed-batch cultures, and the turbidostat. Semicontinuous cultures, as technically simple, continuous cultures (Chap. 2.1), have been used primarily for fast-growing organisms, e.g., for bacteria by Felton and Dougherty (1924) and for algae by Ketchum and Redfield (1938). Myers and Clark (1944) constructed a semicontinuous, self-operating turbidostat for *Chlorella*. Homocontinuous chemostat cultures have been used for bacteria (Monod 1950), for yeasts (Tempest and Herbert 1965), and bacteria-feeding protozoans which were cultivated in single-stage (e.g., Cook 1968) or two-stage chemostats (Curds and Cockburn 1971). Homocontinuous alga turbidostats were used by Phillips and Myers (1954). The fourth type, periodic or continuous fed-batch reactors (FBR), are closed batch reactors of variable volume with a substrate influent (Pirt 1975; Webster 1983). Organisms and medium are not withdrawn. This is a very interesting method for zooplankton studies (Chap. 2.1.2).

A highly refined mathematical theory was developed for homocontinuous cultivation in chemostats and turbidostats (Herbert et al. 1956; Williams 1971, 1972, 1973 and others) and for semicontinuous cultivation (Fencl et al. 1961). This was another strict precondition for further studies.

Undoubtedly, population dynamics of metazoans is more complicated than that of unicells at birth, death, and during developmental processes. Rotifers are the only metazoans which have been successfully cultivated in these devices. Rotifer chemostat cultures were first set up by Droop (1976) and used by Droop and Scott (1978), Boraas (1980, 1983), Walz (1983b), Rothhaupt (1985), and Schlüter et al. (1987). Hitherto, long generation periods hindered the growth of other plankton organisms, except protozoans, in such chemostats. Cladocerans have sufficiently high growth rates, but the time necessary to produce the first egg may result in residence times and time delays that are too long for a successful, continuous culture (Chaps. 3.3, 6.3). Culture vessels have to be very large and flow rates very small. In principle, these difficulties could be overcome by semicontinuous cultures (Chaps. 2.1, 6.1). A semicontinuous culture apparatus for daphnids was described by Harvey (1972), but experimental results obtained by this technique have not been published. Flow-through cultures for cladocerans have been developed to keep the food concentration constant (Sorgeloos and Persoone 1973; Lampert 1976; Vijverberg 1989), but these are not continuous cultures since the individuals are retained by a plankton gauze net. Such cultures are technically a special kind of "fed-batch culture" (Pirt 1975) in the sense of Chapter 2.1.1.

Rotifer chemostats have been extensively used for practical applications. This volume is primarily concerned with the use of chemostats to study regulation principles in the plankton community and examines the practical application in ecotoxicology and aquaculture only (Chaps. 7.3 and 7.4). The same regulation principles are the basis for the aquaculture of rotifers (e.g., as fish food) and for the processing of activated sludge in sewage treatment plants. Practical work can also stimulate theoretical science. It was exactly the occurrence of rotifers in sewage treatment plants that motivated me to evaluate the possibility of rotifer growth in chemostats.

Aquaculture of rotifers (Lubzens 1987; Lubzens et al. 1989) is mainly carried out in batch (e.g., Theilacker and McMaster 1971), semicontinuous (e.g., Mitchell 1986), and feedback cultures (Hirata et al. 1983) but also in true chemostats (James and Abu-Rezeq 1989b). Such cultures, mostly of the euryhaline *Brachionus plicatilis*, have been widely used for the extensive cultivation of marine fish larvae, but also of freshwater fish larvae (Widigdo 1988, with *B. calyciflorus*). Processing of activated sludge in sewage treatment plants, in which rotifers play an important role (Doohan 1975; Toman and Rejic 1985), acts in a feedback relationship with a small part of that sludge (= biomass).

Because of the self-regulating character of the chemostat and due to the fact that every community is "diluted" by predators, emigrants, or lake outflow, continuous cultures are a good model for the regulation of the plankton community. Chapter 7.2 will treat this subject in more detail.

As with any model (Stachowiak 1973), only those attributes of the community which are relevant for the study of regulation proesses need to be included, not every aspect of the natural system. In a further step of abstraction we have made computer simulations as models of chemostats (Chaps. 5.1, 5.2). Because of the rigorous formulation of hypotheses, computer programs are, or should be, testable statements of the respective theories (Patridge et al. 1984). Tests are easily done in continuous cultures, but have to be extended to the real communities in the lake. This produces many problems of scaling up (O'Neill 1989), but, as indicated above, pelagial habitats with many different kinds of plankton communities can also be adequately tested. This scaling up is best done stepwise in enclosures (limnocorrels) having increasing degrees of complexity (Chap. 7.2).

Apart from aquatic microbiology, where the work with continuous cultures is a well-established component of theory and practice, also in other ecological fields the idea of the "dilution" of the ecological community is not new. Nicholson (1954) and Slobodkin and Richman (1956) studied population dynamics of blowflies and daphnids by harvesting parts of the population. The theory of fish catchment (e.g., Gordon 1954, Beverton and Holt 1957, both according to Clark 1989; Cushing 1981) is based on harvesting parts of the fish populations without reducing the stock. This is nothing other than the dilution principle. Semicontinuous experiments in fields other than microbiology have been undertaken on rotifers by Dorazio (1986) and Seale and coworkers (Chap. 6.1). The regular harvesting of *Daphnia magna* large-volume tanks was carried out by Borgmann et al. (1988). Semicontinuous culture or harvesting imitates the action of a predator which will profit by maximizing the prey taken in, while ensuring that the prey does not become extinct (Begon and Mortimer 1989). This situation has often been treated by mathematical models (e.g., Conrad 1986).

The good testability of plankton communities in enclosures or by manipulating small lakes was mentioned above. This experimental treatment is the second main attribute of continuous cultures. Today, population and community ecology suffers from an overabundance of models which are not testable and from the lack of experimental systems to test models (Price et al. 1984). In continuous cultures all energy and matter fluxes and all abiotic conditions are readily controlled. This facilitates the testing of many hypotheses, as indicated above, especially the regulation of plankton communities.

The first step in this process is to clarify whether models of plankton communities, derived from chemostat theory and related experiments, are based a priori on equilibrium or nonequilibrium assumptions. Since continuous cultures tend to achieve steady states, such models are believed to be founded on equilibrium assumptions.

According to equilibrium hypothesis, populations and communities are in equilibrium with their resources, i.e., they are food-limited and population densities remain constant or oscillate around mean values as long as resources do not vary (Wiens 1984). Consumer populations reflect temporal changes in these resources and coexistence is maintained by resource partitioning. Equilibrium is achieved by density-dependent regulation. However, these models are not testable as they are based on the logistic and Lotka-Volterra equations, in which neither resources nor the mechanims of density dependence are explicitly defined. Therefore, we hold these models unsuitable for a mathematical application in population and community dynamics (Chaps. 3.2, 5.1, 5.2, 6.1).

Nonequilibrium models, on the other hand, maintain that changing conditions are responsible for stabilizing the existence of populations and especially their coexistence with other populations (DeAngelis and Waterhouse 1987). Environmental fluctuation, predation, competition, or interaction (DeMott 1989) may cause species successions. As with equilibrium assumptions, most of these models are not testable. For example, interspecific competition based on $r - K$ strategy is not testable since the $r - K$ model itself is derived from the Lotka–Volterra model (Boraas et al. 1990). For this reason, an r_{max}/K_s model was developed for rotifer life-history strategies which may be tested experimentally (Chap. 6.3).

The regulation model developed in this volume (Chaps. 3.3, 5.1, 5.2, 6.3, 7.2) does not assume equilibrium a priori. Because the resources are explicitly treated, it is a testable model. The steady state, not to be confounded with a more or less ill-defined equilibrium, is the result, not the starting point, of a regulation process, which is achieved by mass, energy, and information exchange between populations and their substrates, and held constant as long as the environment does not change. While this state is rare both in nature and in the laboratory (under special conditions, Perram 1973), because of time lags and changing environmental conditions, the inherent stability is both testable and satisfying.

The regulation model overcomes the contradiction between bottom-up and top-down control in plankton ecology (Chap. 7.2). It is not based a priori on one of these hypotheses. The mutually exclusive assumptions of both models have been falsified. The bottom-up model, which assumes only food limitation (energetically based models) is contradicted by the widely observed effects of predators on the community structure, as established by many manipulation experiments.

In the same way, the hypothesis of exclusive control of the community from the top-down by predators (kinetically based models) was falsified by the evidence of food limitations. This was more difficult to determine, especially in terrestrial systems (Wiens 1984), but aso in plankton communities (Tilman 1982). However, strictly speaking, this was demonstrated by applying the Monod-saturation relationship (e.g., Figs. 4.1.7, 4.1.8, 6.1.3, 6.2.9). This means that saturation not only occurs at high food concentrations, but also that limitation occurs at low food concentrations. Such low concentrations were often found in the field (see, e.g., several contributions in Lampert 1985). It follows that kinetic as well as energetic effects contribute to control community structures. The question remains as to how they precisely contribute? With this regulation model a testable hypothesis has been established. This model may be tested by tracing the reactions of the populations after perturbations. In controlled systems, e.g. in continuous cultures, the densities should return to steady states (Chap. 3.2). The continuous culture, therefore, is its own control.

After describing the techniques and practical methods of the three main types of continuous cultures (Chap. 2), their specific characteristics are elucidated in Chapter 3, beginning with the benefits of semicontinuous cultures in shift-up and -down experiments (Chap. 3.1). Chemostat cultures are carried out in single-stage chemostats with reactions to pulse additions of nutrients or algae (Chap. 3.2). In Chapter 3.3, on two-stage cultures, chemostat theory is introduced with regard to steady-state experiments and the behavior of a two-stage system. The turbidostat (Chap. 3.4) is most useful at maximum growth rates and may, therefore, be used to assess maximum population dynamic processes.

Two-stage chemostat cultures have been used to investigate the functional response and the bioenergetic processes of two different rotifer populations (Chap. 4.1) and to establish energy balances (Chap. 4.2). Chapter 4.3 compares the steady-state dynamics of two differently sized rotifer species, analyzing the role of body size in carbon metabolism.

Based on the relations found in chemostats, especially the Monod relationships between resources and population kinetics, two different models have been constructed and simulated (Chap. 5). These have been tested with chemostat transient-state experiments.

In Chapter 6.1 semicontinuous cultures have been used to test the competitive outcome between two rotifer species when grown on common food. Continuous cultures and related techniques with rotifers are highly suited to test integrating concepts of predation and resource-related processes. This was shown by examples of numerical and functional responses to food quantity, food quality, food thresholds, resource partitioning, and resource variance (Chap. 6.2). Using the relations between body size and Monod kinetics known from continuous cultures, the "r_{max}/K_s" concept of life-history strategies of rotifers was developed in Chapter 6.3.

In Chapter 7.1 the limits of continuous cultures are critically examined. By comparing other related techniques some errors may be avoided. In Chapter 7.2 the processes found in the chemostat, considered here as a model ecosystem, have been summarized and evaluated to formulate a common regulation model of the plankton community.

The application of continuous cultures is not limited to investigations of theoretical ecology, since knowledge of the regulation principles for the management of communities, e.g. in aquatic environments, may be important. Continuous culture methods may also be used in more practical applications.

These regulation features must also be valid in chemostats with industrial dimensions. The application of rotifer cultures is widespread in the aquacultural production of fish and crustcean larvae which, in many countries, hopefully contribute to human nutrition. Chemostats used for this purpose (Chap. 7.4) effectively employ the inherent regulation characteristics regarding the optimization of the operation variables. The productivity of rotifers and their quality as food are considerably higher in such chemostat systems than in other aquacultural plants.

Semicontinuous cultures and the turbidostat proved to be powerful instruments in studies of the toxic effects of heavy metals on algae and rotifers (Chap. 7.3). This ecotoxicological system approach, by no means limited to the study of heavy metals, looks for the integrated injury reaction of a model ecosystem with the least number of trophic levels. The rotifers, in their function as herbivores, are involved in a model interaction with other organisms (food, predators) under controlled conditions. This avoids the unequivocal but irrelevant monospecies investigations and the ambiguity of widely used but uncontrolled multispecies tests.

2 General Continuous Culture Methods

2.1 Semicontinuous Culture Methods

Martin E. Boraas

2.1.1 General Considerations

Semicontinuous cultures, herein referred to as SCs, are specialized forms of two-stage cultures, discussed in Chapter 2.2. SCs have been applied in zooplankton and microbial studies in a considerable variety of physical configurations, including both open and partially closed systems. Here, I will be primarily concerned with animal cultures. Open designs include those where a volume of the culture, including both residual food and animals, is replaced periodically with a fresh food suspension. This design approximates a chemostat culture both in theory and in operation. The methods for SCs in this more restricted sense are described in Chapter 2.1.4. The SC enters a fluctuating steady state or quasi-steady state. The degree of approximation to a chemostat increases as the frequency of dilution events increases.

Closed SCs are cultures where food is added periodically but animals are not removed. Such cultures may increase in volume if the food is a suspension or in a more concentrated form. Such fed-batch cultures or FBCs (see Pirt 1975) do not enter true steady states; some mass component is always changing. However, the biomass of the organisms can stabilize if the animal population is energy limited and metabolic wastes do not become excessively toxic. FBCs have been widely used in zooplankton studies, although the culture method has rarely, if ever, been identified as such.

SCs can be simple or eaborate. The primary requirements of rotifer SCs are (1) a repeatable, steady source of food suspension and (2) dilutions at exactly regular intervals. Most of those who have worked with rotifer SCs have used nutrient-limited algal chemostats as the source of food suspension. Other food sources, e.g., dried algae or yeast (e.g., James et al. 1987), have also been used successfully. When using live photosynthetic food, the rotifer stage is usually maintained in the dark to minimize algal growth on regenerated nutrients.

SCs have been used in a wide variety of laboratory and field studies. Bacterial applications have ranged from theoretical analyses of growth (Anagnostropoulos 1971) to clinical analyses (e.g., Manning et al. 1987). SCs have found wide application in phycological studies (e.g., Terry and Edmunds 1969; Holm and Armstrong 1981; Laing and Helm 1981; Suttle and Harrison

1986; Brezenzinski and Nelson 1988) and in studies of yeast (e.g., Broderick and Greenshields 1982).

Rotifers are grown in SCs both to examine the basic properties of their physiology and ecology (e.g., Boraas, Chap. 3.1, this Vol., Schlüter et al. 1987) and to produce biomass for aquacultural purposes (e.g., Hirayama and Ogawa 1972). When rotifers are grown to feed fish larvae, some care must be taken to regulate their chemical composition so that they are an adequate food (see James et al. 1983b for a review). In addition, SCs can be used for toxicity assays (e.g., Seale and Boraas, Chap. 7.3 this Vol.).

If an algal chemostat is used for the food source, it is operated continuously, with steady accumulation of outflow. Usually, this outflow is then used to dilute the rotifer culture on a daily basis. This means that some of the algal cells are new, i.e., fresh from the algal reactor vessel, while others are 24 h old. Care must be taken when extending data collected from dark-adapted algae to conditions under which the algae are growing in the light (see Fig. 5.2.1 and accompanying discussion). The rotifer reactor vessel can be very simple, possibly nothing more than an Erlenmeyer flask on a shaker table. The rotifer SC can then be operated by removing a fraction of the rotifer culture periodically and replacing it with fresh algal suspension. Every effort should be made to dilute the cultures within 15 min of the target time each day, a 1% error. Also, the volumetric measurements are best made gravimetrically, either with a triple beam balance or, ideally, with an electronic pan balance. With reasonable care, volumes can be determined to about $\pm 0.2\%$. The volume removed from the rotifer SC can then be used to determine populations of rotifers and residual algae and other variables as needed.

In contrast to SC cultures with no direct coupling between the algal source and the rotifer vessel, Schlüter et al. (1987) used a direct-coupled, intermittent-flow culture device. This device allowed the automatic delivery of food and monitoring of ingestion, filtration, and respiration rates over the 24-h cycle.

An advantage of inexpensive SCs is that multiple cultures can be maintained simultaneously, i.e., replication is easier than with chemostats or turbidostats. In general, the number of replicates is limited primarily by the volume of available food suspension. Other problems with chemostat culture can also be reduced with SCs, e.g., wall growth. Wall growth can dramatically alter the dynamics of cultures and even affect stability properties (e.g., Bonomi and Fredrickson 1976). However, removing wall growth inside vessels, while maintaining a sterile technique, is difficult. This problem can essentially be eliminated in SCs by changing rotifer culture vessels with every dilution. On the other hand, unless the dilutions are done using a sterile technique, especially in a laminar-flow bacteriological hood, the cultures are subject to biological contamination from bacteria, protozoans, etc.

SCs can also be applied on a large scale. For example, Groenweg and Schlüter (1981) and Mitchell (1986) used outdoor cultures for the mass production of rotifers. Although the principle of continuous and semicontinuous cultures apply to such mass cultures in general, specific solutions to technical problems must be found in each specific case.

2.1.2 Fed-Batch Cultures

Fed-batch cultures, or FBCs, have substrate inputs but generally do not have outputs of organism biomass. They may have a continuous or discontinuous input of food suspension, with consequent volume increase, or may be fed with concentrated food or a continuous flow of food suspension with constant volume but retention of organisms. Pirt (1975) analyzed the cases of continuously increasing volume and continuous input with a portion of the culture withdrawn (including organisms) at regular intervals. Webster (1983) demonstrated how growth parameters can be measured with FBCs, particularly low values of K_s.

Simulations of two fed-batch designs are shown in Fig. 2.1.1. A Monod growth function vs substrate concentration and a constant volume of the culture vessel are assumed. The model used in the simulations is essentially the model described in Chapter 5.2. The simulation shown in the left-hand panel of Fig. 2.1.1 assumes that the residual food (dashed line) and waste product (dotted line) are diluted by the incoming food suspension, but the entire animal population (solid line) is retained within the vessel. The simulation shown in the right-hand panel, in contrast, assumes a concentrated input of food into the growth vessel with no hydraulic flow through the culture. Both cultures

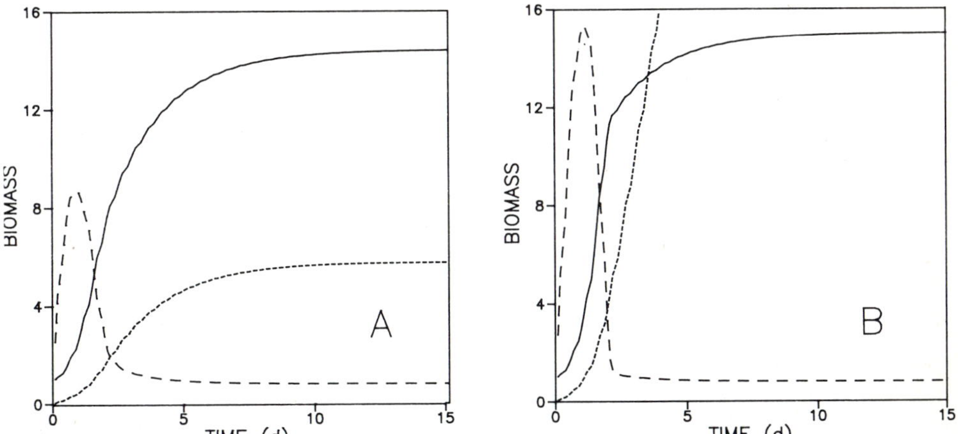

Fig. 2.1.1 Simulations of fed-batch cultures. The *left-hand panel* represents a system with no output. The *right-hand panel* represents a system with dilution of residual food and waste product only, not animals. The *solid line* represents animal biomass, the *dashed line* represents residual food biomass, and the *dotted line* represents metabolic waste products

attain the same steady-state population densities (ca. 15 μg ml^{-1}) at about the same time. The primary difference is retention of waste products in the system with no flow. This factor has almost certainly influenced some ecological studies.

While the simulations shown in Fig. 2.1.1 assume continuous flow, most applications of FBCs in animal studies have been semicontinuous. These simulations are intended to be illustrative of the method of FBCs. Careful application of the theory of FBC culture, as developed by Pirt (1975) and Webster (1983), could contribute greatly to the study of zooplankton physiology and ecology.

2.1.3 Derivations and Computations

The following derivations are largely due to the efforts of J. Brian Horton (deceased) when at the Pennsylvania State University and of D. Tilman (in Kilham and Kilham 1978).

In an SC, the system will enter into a quasi-steady state providing D the specific dilution rate, is less than r_{max} and dilution events are sufficiently frequent so that the physiological state of the organism is not disturbed (Pirt 1975; Kilham and Kilham 1978).

F is the fraction of the SC removed and replaced daily; N_1 is the number (or biomass) of organisms present just before the dilution event; N_0 is the organism concentration just after the dilution event. Then

$$N_1 = N_0 e^{rt}, \qquad (2.1.1)$$

where t is time and r is the specific growth rate. In the quasi-steady state, population densities do not change in time; therefore,

$$(1 - F)N_t = N_0, \qquad (2.1.2)$$

since the increase in population density equals the number removed. Hence, rearranging Eq. (2.1.2) and substituting with Eq. (2.1.1),

$$N_0/(1 - F) = N_0 e^{rt}.$$

It t = 1 day, then the above equation simplifies to

$$r = -\ln(1 - F). \qquad (2.1.3)$$

If the dilution events are more or less frequent than once per day, then a correction needs to be incorporated, as derived in Fencl et al. (1961),

$$r = n * \ln[n/(n - F)], \qquad (2.1.4)$$

where n = number of daily dilution events. As n increases, continuous dilution is approached and F approaches D as shown in Fig. 2.1.2. The numerical values in Fig. 2.1.2 indicate the number of daily dilutions.

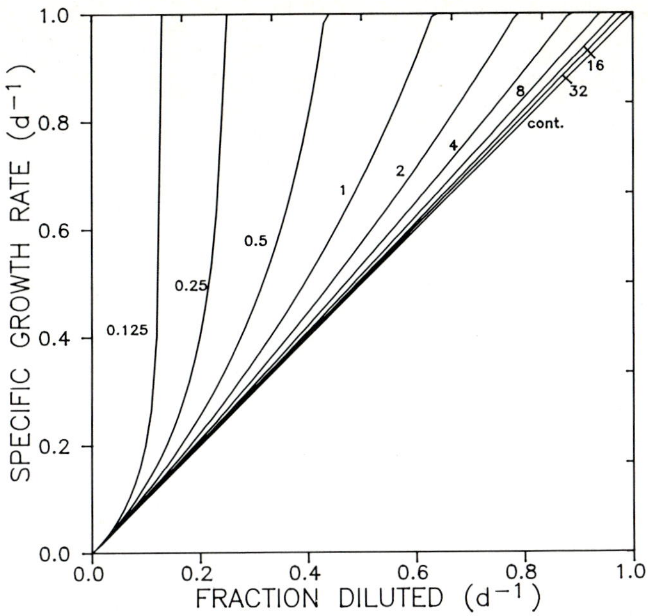

Fig. 2.1.2 Pseudosteady-state dilution rate (D) vs daily fraction removed (F). The values were calculated from Eq. (2.1.1), as derived from Fencl et al. (1961). The numerical values indicate the frequency of dilution events per day. The straight line marked *cont.* indicates continuous-flow dilution

As discussed by Fencl et al. (1961), an SC best approximates the chemostat at low values of F and with increasing dilution frequency. At low frequency of dilution or high values of F, the animals will experience increasingly variable environmental conditions. Such variation can be an experimental tool (e.g., Schlüter et al. 1987), particularly since the population returns to its same physiological state each day.

Growth rates can also be computed for the non-pseudosteady-state situation. F is set so that (1) the population washes out (e.g., $D > r_{max}$), and (2) C_1 (input concentration of the limiting nutrient) is sufficiently high to sustain nutrient-saturated growth of the dominant species at r_{max}. Then, r_{max} will be maintained, but the population will decline exponentially (Fig. 2.1.3). If the slope of the log(N) is g, then

$$r_{max} = g - \ln(1 - F), \tag{2.1.5}$$

assuming one dilution event per day.

When cultured in this manner, organisms can be maintined at r_{max} at useful concentrations for many generations. In addition, very accurate estimates of r_{max} can be obtained since the period of maximal exponential growth can be extended almost indefinitely.

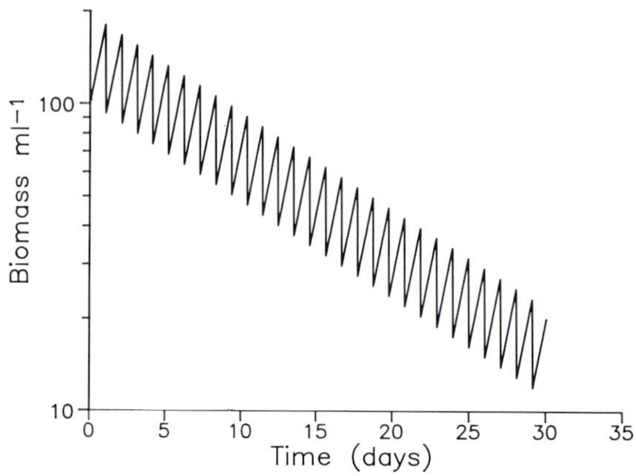

Fig. 2.1.3 Response of the biomass of an organism in an SC when $D > r_{max}$. Note the greatly extended time of sustained exponential growth relative to the time when nutrient limitation would come into play without dilution. However, FBCs may be operated continuously, as is usually the case in microbial studies (e.g., Webster 1983).

Finally, when one converts a growth rate obtained from a semicontinuous culture to other units, for example, a daily rate to an hourly rate, the discontinuity in sampling times must be taken into account. A simple division is not sufficient. In the derivation above, the interval of dilution is 1 day, but the hourly specific growth rate can only be approximated by dividing the daily rate by 24 h. Growth takes place continuously, but population densities are measured only at discrete daily intervals. The exact hourly rate is determined as follows. From the exponential growth equation, Eq. (2.1.1), or

$$N_t/N_0 = \text{EXP(daily rate/24 h)}. \tag{2.1.6}$$

For example, if a daily D of 0.916 day^{-1} (for *Brachionus rubens* during washout, see above) is set by one dilution per day, the hourly rate is N_t/N_0 $\text{EXP}(0.916\ \text{day}^{-1})(1\ \text{day}/24\ \text{h})$, and $\ln(N_t/N_0 = \text{EXP}(0.916/24) = 1.0389$, or a 3.89% increase hourly. In contrast, a simple division gives $0.916/24 = 0.0382$ or a 3.82% hourly increase, an error of 2%. While small, the error can affect precise work.

2.1.4 General Materials and Methods of Semicontinuous Culture

Semicontinuous rotifer chemostats can be operated as follows. The rotifer (*Brachionus rubens* and *B. calyciflorus*) cultures are maintained in polycarbonate Erlenmeyer screw-caped flasks on a reciprocal shaker table (80–120 oscillations min^{-1}) under low light ($< 100\ \mu$Einsteins $\text{m}^{-2}\,\text{s}^{-1}$) in an environmental chamber at $24 \pm 1\ °\text{C}$. The algae are grown in a culture vessel regulated at $25 \pm 0.2\ °\text{C}$ with an electronic temperature regulator on a mineral medium with limiting nitrate (0.5 mM) under continuous light (described in detail in Boraas 1980, 1983). These conditions provided *Chlorella pyrenoidosa* at $132\ \mu\text{g ml}^{-1}$

dry mass (SD = ± 0.99, n = 10), as determined with a Cahn electrobalance using Nuclepore filters dried at 60 °C for > 24 h. The overflow from the algal culture may be collected continuously (chemostat) or the culture volume may be allowed to increase until the cells are harvested. For the studies reported in Chapter 3.1, the rotifers and algae were grown in separate environmental chambers. The algae were grown in a vessel heated to 25 °C, and the cells were continuously collected at a chamber temperature of 15 °C.

Rotifer cultures are generally diluted daily. Immediately prior to the dilution of the rotifer SCs, the vessels are removed from the shaker table and transported to a room with an analytical balance. Each rotifer flask is thoroughly but gently mixed, and a predetermined volume is rapidly poured off. Because the rotifer suspension can rapidly become inhomogeneous, particularly when only a small volume is removed, no attempt is made to "fine-tune" the volume removed. With practice, the volumetric error is less than 5% of the removed volume. The actual volume removed was determined as mass with a balance. The algal suspension was then thoroughly remixed and the rotifer culture was returned to its former volume, again using the balance.

Both rotifer and algal populations can be routinely measured by counting individuals under the microscope. Estimates of mass, C, or nutrient content can be made using standard procedures. However, a particle counter, e.g., a Coulter counter, provides much more information on both populations and is about as fast or faster than microscope counts. Greater statistical precision can be obtained since several thousand individuals can be both counted and sized electronically as rapidly as a few hundred can be manually enumerated. With some species (see below), eggs are readily stripped from adults, providing a complete age-size structure of the rotifer population.

We have used a model ZB Coulter counter interfaced to a microcomputer using a custom program (written by J. B. Horton) and a Particle Data Celloscope. The rotifers were counted and sized with either a 560-μm (Coulter) or a 380-μm (Particle Data) aperture in an electrolyte consisting of 0.075% NaCl (w/v) in distilled water. The balance, rather than volumetric glassware, was used to calculate dilutions. Rotifers were still active after being passed through the aperture. Usually two rotifer subsamples and one algal subsample were counted. However, for some cultures there was only enough volume for one rotifer count (at least 10 ml). Both residual algae (algae in the culture vessel) and input algae were counted and sized in a custom electrolyte (see Boraas 1983, for a complete formula) with a 50-μm aperture. Rotifers were screened from the residual algal samples with a 45-μm Nitex screen in a 13-mm Swinnex Millipore filter holder. This screening greatly reduced the blockage rate of the Coulter aperture. In previous studies of *Brachionus calyciflorus* (e.g., Boraas 1983), the eggs of the rotifers were stripped from the females by passing the animals through a 25–26 gauge hypodermic needle. This method did not work with *B. rubens* since the force sufficient to strip the eggs also disrupted the animals. Therefore, unlike the *B. calyciflorus* studies, a size-age structure could not be obtained for the *B. rubens* populations.

2.1.5 Models of the Time Course of SCs

Models of SCs fall into two categories: the quasi-steady state and transient events or the time course. Only the transient dynamics will be considered here. Fencl (1966) derived the first model of semicontinuous cell growth. He assumed cells always grow exponentially between dilution events, e.g., as in sustained r_{max} shown in Fig. 2.1.2. He included no reference to environmental variability of limiting nutrients. In fact, however, the limiting resource may be depleted during the period between dilution events. Since most organisms show some type of saturation curve of r on limiting nutrient, this variability would inevitably affect the dynamics of the SC.

Recently, Quinian (1986) criticized the exponential assumption implicit in most descriptions of SCs [including Eqs. (2.1.3) and (2.1.4), above]. As an alternative, Quinian (1986, for bacteria) and Livansky (1979, for autotrophs) suggested that the assumption of exponential growth be replaced by an assumption of logistic growth between dilution events. That is, the population repeats a sigmoid time course between each dilution event. It is clear that under some conditions food depletion between dilution events occurs so that the growth of the population slows or stops. Under other conditions, e.g., during washout (e.g., Chap. 3.1), the animal population does increase exponentially between dilutions. It is unclear when populations would follow a logistic time course between dilutions. The critical assumption of the logistic: a negative linear dependence of r on N has been strongly criticized (e.g., Chaps. 5.1 and 5.2).

One solution is to model population growth in SCs as a function of food. However, as yet, no one appears to have published a model of SC growth using explicit relationships between the organisms and its limiting factor. Since Monod kinetics appear to describe the quasi-steady states of SCs (e.g., Boraas 1983; Schlüter et al. 1987), this approach might be a starting point for modeling the time course of SCs.

This is contribution no. 366 from the Center for Great Lakes Studies.

2.2 Methods of Single- and Two-Stage Chemostat Experiments

NORBERT WALZ

2.2.1 General Rotifer Culture Methods

Continuous rotifer cultures must conform to general rotifer cultivation methods. There is abundant information on rotifer culture in the literature. Hitherto, only *Brachionus* sp. and *Keratella cochlearis* have been used in continuous cultures due to their good cultivation attributes. Walz (1983a) described in more detail the culture of *B. angularis* and Walz (1983b) that of *K. cochlearis*. both species have been cultivated by the small unicellular green alga (Ulotrichales) *Stichococcus bacillaris*. Details of *B. calyciflorus* culture with *Chlorella pyrenoidosa* were given by Boraas (1983). Rothhaupt (Chap. 4.3, this Vol.) cultured *B. calyciflorus* and *B. rubens* with *Monoraphidium minutum*; details can be found in Rothhaupt (1985).

It would be most desirable to carry out continuous cultures of other rotifer species. Information on basic cultivation methods can be found in Nathan and Laderman (1959), Gilbert (1980a), Aloia and Moretti (1973), and Reed (1979). The choice of the most suitable food algae is very important for the cultivation of rotifers. Successful rotifer cultures have been reported by Pourriot (1965, 1980), Pejler (1977), Stemberger (1981), Vancil (1983) and May (1987). Culture media are discussed in Section 2.2.12.

2.2.2 Principle of Two-Stage Chemostat Cultures

A two-stage chemostat consists of two separate chemostats. The first, usually for the monoxenic culture of algae, is exposed to light. The second, for the monoxenic culture of rotifers, is in darkness to prohibit further growth of the algae. This arrangement allows a clear energetic relationship. In some cases a mixing reactor is inserted between these units to dilute the algal suspension (Fig. 2.2.1). Thus, the dilution rate in the rotifer chemostat can be varied independently of the algal chemostat. The algal chemostat should always be operated at a constant dilution rate. This is important because the biochemical composition of the algae is dependent on their growth rate which can affect the growth efficiency of the rotifers (Scott 1980, Schmid-Araya 1992).

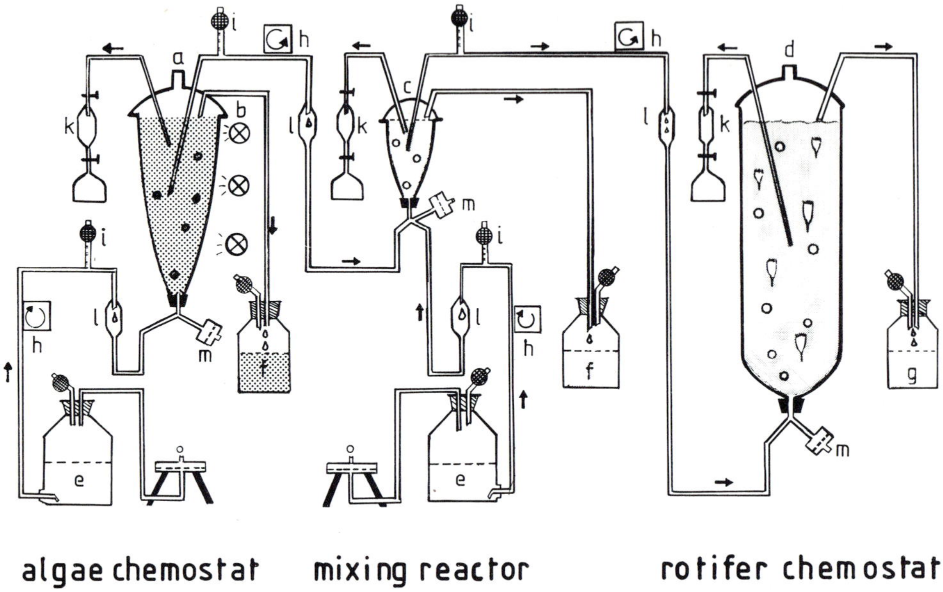

Fig. 2.2.1 A two-stage chemostat system. *a* Chemostat of algae; *b* fluorescent lights; *c* mixing reactor; *d* rotifer reactor; *e* culture medium reservoir; *f* reception vessel for excess algae; *g* reception vessel for produced rotifers; *h* peristaltic pumps; *i* flowmeters (control of flow rate of media and suspensions); *k* sterile sampling ports; *l* traps for bacteria; *m* sterile air filters; *o* sterile pressure filters for culture medium. (After Walz 1983b)

In most cases the mixing reactor and its connecting vessels are omitted and the algal suspension coming from the first stage is added directly to the rotifer chemostat. This can be accomplished without influencing the operation of the algal chemostat.

While the principle of construction follows the basic design for bacterial chemostat cultures (e.g., Rićica 1966; Evans et al. 1970), it may vary according to the individual investigators. Many variations for single-stage chemostats (Baker 1968; Solomons 1972; Calcott 1981) and for two-stage chemostats (Curds and Cockburn 1971; Schlüter et al. 1987) have been described. An example of a flexible apparatus, constructed by Walz (1983b, 1987b), is presented here. Useful data on design and instrumentation of biological reactors are given in Bailey and Ollis (1986), on operation and kinetic analyses in Moser (1981).

2.2.3 Principle of Single-Stage Chemostat Cultures

Single-stage chemostat cultures are mixed cultures containing both algae and rotifers growing in a single vessel, which is illuminated to promote algal growth. The construction is essentially identical to an algal chemostat. Boraas (1980;

Chap. 3.2, this Vol.) presents experiments with single-stage chemostats culturing *Brachionus calyciflorus*.

2.2.4 Chemostat for Algae

The chemostat presented here (Fig. 2.2.1a) is composed of a flat flange reactor vessel (Schott) with a bottom screw thread. Usually, the cylindrical vessel has the following dimensions: 460 mm length, 100 mm i.d., 2800 ml volume. The vessel is tapered conically at the base which reduces the danger of wall growth. The flat flange lid with five threads is closed with a quick-release clamp. The medium is pumped into the chemostat from a 20-l storage bottle (e) with a peristaltic pump (h). The storage bottle can be filled through a sterile pressure filter (o). The sterilized air (m), regulated to 85 l/h, is directed through the same bottom entrance to mix the medium rapidly. The algal chemostat is irradiated laterally by eight fluorescent tubes (b) (Phillips 18 W, four cool white, four warm white tubes) under constant light.

Stichococcus bacillaris cultures were grown at a dilution rate of $D = 2\ day^{-1}$, corresponding to a doubling time of 8.3 h (Chaps. 3.3, 4.1, 4.2). *Monoraphidium minutum* was grown at a dilution rate of $0.75\ day^{-1}$ which represents a doubling time of 22.3 h (Rothhaupt, Chap. 4.3, this Vol.).

2.2.5 Mixing Reactor (Fig. 2.2.1c)

The mixing reactor (Fig. 2.2.1d) has the same construction as the algal chemostat. In has a volume of 300 ml, a length of 150 mm, and an inner diameter of 70 mm. The residence time is very low. The food concentration in the mixing stage is regulated by the simultaneous regulation of the input of food suspension from the algal chemostat and new medium.

2.2.6 Rotifer Chemostat (Fig. 2.2.1d)

The food suspension (from the mixing reactor or directly from the algal chemostat) is pumped by a peristaltic pump (Fig. 2.2.1h) into the rotifer chemostat at the desired flow rate. The rotifer reactor is identical to the algal reactor. In some cases the reactors had larger volumes: up to 4800 ml, length: 500 mm, inner diameter: 160 mm; and the base was more rounded (see Fig. 2.2.3d). To prevent rotifers from being trapped in the air bubbles, air was blown in very coarse bubbles at a low intensity (10 l/h). The rotifer overflow was collected in its own vessel (g). Boraas (1983; Chap. 3.2, this Vol.) used rotifer vessels with a volume of only 570 ml. All vessels were regularly scraped using a magnet outside the flask and a small magnetic Teflon-encased stir bar inside the flask before autoclaving.

2.2.7 Transport and Dosage of Media and Suspensions

All connections are made with silicon tubing. Tubings between different system units are coupled by Corning tube couplings tightened by Teflon washers. To prevent wall growth and settling of algae the tubing, in which suspensions of algae are transported, should have a very narrow lumen (ca. 0.3 mm) to accelerated transport velocity; the pump may also be operated intermittently (pulsatile). Tubings of this kind are connected by narrow, stainless steel tubes.

The secure and constant transport of media and suspensions is essential for chemostat operation. This can only be accomplished by high-quality peristaltic pumps. Rotifer chemostats require longer residence times than bacteria and algae, therefore, culture vessels must be large or the flow rate must be low. The low dilution rate of 0.05 day^{-1} (residence time = 20 days) in a 2800-ml vessel requires a flow rate of 5.8 $ml\,h^{-1}$. The precise operation of peristaltic pumps at this low flow rate can be problematic. This low rate was achieved with a timer which switched off the pump for 40 s in every minute.

Other devices for pumping, besides peristaltic pumps, e.g. Mariotte flasks, air lifts and needle valves (Rićica 1966), are not recommended for the transport of particle suspensions due to the danger of wall growth and settling.

The flow rate of liquid can be measured with a burette protected by a cotton-wool filter inserted in vetical position into the tubing over a T-piece. In normal operation it is shut off by a clamp. For measuring it is filled up to the upper mark and the inflowing tube is shut off. Then the decrease in liquid in the burette with in a given time interval is determined (Rićica 1966; Fig. 18 in Evans et al. 1970). At low flow rates one often forgets to open the hose again. Therefore, the following construction is preferred (Fig. 2.2.2). A 1-ml pipette is inserted into the hose. Near the entrance into the pipette a small glass fiber filter (Sartorius, 25 mm diam.) is connected over a T-piece, which is normally closed by a clamp. To measure the flow rate the clamp is opened for a moment and the other hose closed just long enough to suck in a small air bubble. Movement of this bubble between two marks in the pipette is then measured. An alternative method is to determine the volume of water pumped into the waste flask on a daily basis, usually with a balance.

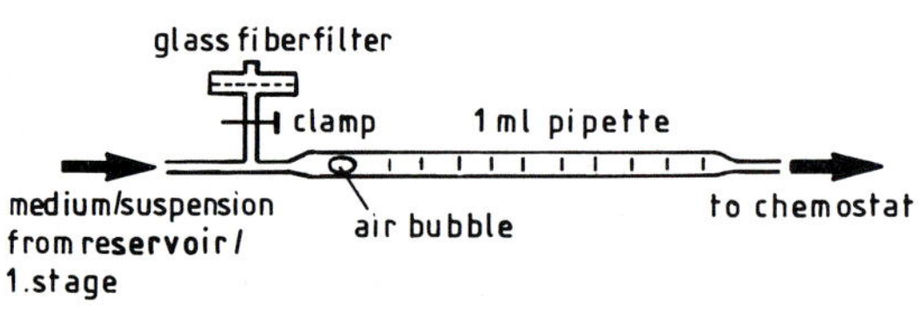

Fig. 2.2.2 Flowmeter for controlling flowrates of media and suspensions

2.2.8 Mixing, Aeration, and Sterilization

In order to prevent damage of algae and rotifers suspensions should not be agitated with motor-driven or magnetic stirrers. In contrast to bacteria cultures, organism concentrations are not as high and oxygen demand is lower in algal and rotifer cultures. Therefore, agitation by air bubbles is sufficient in most cases. Air bubbles in rotifer chemostats should be large to minimize the danger of trapping rotifers on the phase boundaries between water and air. Many rotifer species (*Polyarthra*, *Filinia*) are very vulnerable to bubbling. This fact has prevented up to now the chemostat culture of many delicate species.

Air is sterilized through two glass fiber filters and regulated by flowmeters. Schlüter et al. (1987) enriched the air for the algal chemostats by 1% CO_2, but this depends on the mineral medium and the desired algal concentration. Several authors hydrated the air which is very dry coming out of the compressor, with distilled water. For slow flow rates, evaporation can represent a large fraction of the total flow. Moreover, sterilizing filters have to be heated to prevent condensation. Hydration after sterilization presents problems regarding the eventual contamination of the water. As an alternative to air sterilization, Gelman ACRO 50 filters, 0.22 μm pore size, are autoclavable, hydrophobic, and do not air-lock when wet.

The different parts of the system are autoclaved separately; Boraas (1980) autoclaved the entire system. All permanent openings are sealed with filters of glass wool or glass fiber. All temporary openings are covered with aluminum foil and combined after flaming the tube couplings. The media are sterilized by pressure (at least 4 bar) through membrane filters or membrane laminates (Schleicher and Schüll, No. 1121, 0.2 μm). Boraas found 0.1 μm necessary for organic media.

2.2.9 Sampling and Counting Rotifers

Samples can be withdrawn directly from the rotifer chemostat and all other reactors through special devices (Fig. 2.2.1k). The sample is first pressed out of the chemostat by air pressure into a 100-ml bacteria trap, while the outflow of this trap is closed by a clamp. After closing the connection between reactor and trap, the clamp of the outflow is released and, under a sterile bell, the sample is filled into a sterile vessel. The bell is flamed after use and sealed with aluminum foil. This device is very similar to that described by Rićica (1966).

Boraas (1983) took samples with a 25-gauge needle attached to a 30–50 ml plastic syringe through a silicon stopper on the sample port. Ninety-five percent of the *Brachionus calyciflorus* egg were stripped from the females with no apparent damage to adults or eggs.

Algae are measured for absorbance in 5-cm cuvettes at 720 nm in a photometer. Ten to 100 ml of algae is filtered through glass fiber filters (Whatman

GF/F) to analyze chlorophyll a and pheopigments (Marker et al. 1980) and through precombusted glass fiber filters (Whatman GF/F) to analyze carbon by combustion. Algae from the rotifer chemostat are first prefiltered through a 30-μm gauge.

Rotifers and eggs are counted at least in duplicate, alive or fixed in 4% formalin, in screened plastic petridishes (5 cm diam.). Schlüter et al. (1987) counted rotifers in 1-ml pipettes directly under a binocular microscope. To determine the carbon content rotifers are filtered through a 30-μm gauze, washed, and rinsed in precombusted glass fiber filters (Whatman GF/C).

Boraas (1983) determined the total number, 100 channel-size distributions, and the biovolume (sum of channel frequency times the calibrated size of each channel) of alga and rotifer populations with a Coulter counter. Lang Ltd. (Munich) produces a suitable rotifer counter which counts the rotifers when they pass the beam in a laser barrier sensor.

2.2.10 Control of Culture Temperature

The temperature of the cultures can be controlled in a temperature-controlled water bath or in a room or in an incubator having a constant temperature. The culture vessels may be equipped with a water jacket connected to an external thermostat, through which the water is circulated. Boraas (1983) controlled the temperature of his algal chemostat by a fan, and of the rotifer stage by a heater which was controlled by a relay driven by an amplified thermistor bridge. Temperature was measured directly inside the cultures. A 1–2 s delay was programmed between the amplifier and the relay to prevent chatter (Boraas 1980).

2.2.11 Control of Suspension Volume

To maintain a constant volume of the suspension a continuous overflow is produced, which is blown into recipient vessels (Fig. 2.2.1f,g). Every stage, and even the mixing reactor, if present, has its own overflow. Schlüter et al. (1987) and Boraas (1983) removed the overflow with solenoid valves. The fluctuations in culture volume were only $< 3\%$ of the total volume (Boraas 1983) but Schlüter et al. (1987) grew their cultures semicontinuously with a one-third difference in total volume.

2.2.12 Culture Media

Culture media in single- and two-stage chemostats must be suitable for the growth of both algae and bacteria. Both stages should be identical. Most workers used common alga media such as Chu 12 (Chu 1942, modified by Müller 1972; Table 2.2.1: Walz, Chaps. 4.1, 4.2, this Vol.; Rothhaupt, Chap. 4.3, this Vol.) and

Table 2.2.1 Culture media in single- and two-stage chemostats

Culture media	Final concentration
Chu 12 mineral medium (modified by miller 1972):	
$Ca(NO_3)*4H_2O$:	42.9×10^{-3} g/l
$MgSO_4*7H_2O$:	75.0×10^{-3} g/l
K_2HPO_4:	5.0×10^{-3} g/l
Na_2CO_3:	20.0×10^{-3} g/l
KCl:	5.0×10^{-3} g/l
Trace element solution P IV (Provasoli and Pintner 1960):	
Na-EDTA:	1000.0×10^{-3} g/l
$FeCl_3*6H_2O$:	193.6×10^{-3} g/l
$MnCl_2*4H_2O$:	36.1×10^{-3} g/l
$ZnSO_4*7H_2O$:	4.9×10^{-3} g/l
$(NH_4)_6Mo_7O_{24}*4H_2O$:	9.2×10^{-3} g/l
Vitamin solution (Provasoli et al. 1957):	
Thiamine-HCl (B_1)	2.0×10^{-8} g/l
Pyridoxine (B_6)	0.2×10^{-8} g/l
Cyanocobalamin (B_{12})	0.5×10^{-8} g/l
Nicotinic acid amine	0.5×10^{-8} g/l
Ca-panthothenic acid	0.5×10^{-8} g/l
Biotin	0.1×10^{-8} g/l
Thymine	0.5×10^{-8} g/l
Riboflavin	0.1×10^{-8} g/l
p-Aminobenzoic acid	0.2×10^{-8} g/l
Folic acid	0.01×10^{-8} g/l

Chu 10 (modified by Williams 1971; Boraas 1983). Other culture media for rotifers are given by King and Snell (1978) and Reed (1979).

Trace elements are necessary, but must be used with caution. For example, the trace element solution P IV of Provasoli and Pintner (1960; Table 2.2.1) is too concentrated for *Keratella cochlearis*. Therefore, only half of this concentration was used (Walz 1983a), but a steady state in a *Stichococcus* chemostat could not be achieved even with that concentration (Walz 1983b).

Rotifer culture usually requires free vitamins. Table 2.2.1 gives the concentrations of the vitamin solution of Provasoli et al. (1957) used by Walz (1983a, b; Chap. 4.2, this Vol. Rothhaupt (Chap. 4.3, this Vol.) used vitamin B_{12}, biotin, and thiamine. Droop and Scott (1978) and Scott (1980, 1988) added thiamine and vitamin B_{12}. Vitamin B_{12} seems to be the most decisive factor (Scott 1981), e.g. Boraas (1983) added this vitamin only.

2.2.13 Status of Algae and Bacteria in the Chemostats

Continuous axenic rotifer cultures have not been achieved, although, in principle, such cultures are possible (Dougherty et al. 1961). Attempts to make rotifers bacteria-free with NaOCl (according to Gilbert 1970) or with antibiotic mixtures

Table 2.2.2 Bacterial density and proportion of bacterial carbon and algal carbon (percentage) in different culture vessels ($\pm 95\%$ confidence limits of the means, n = 19–21)

Culture	Bacteria $\times 10^6$ ml^{-1}	Bacteria-C/alga-C (%)
Algal chemostat	1.82 ± 0.52	0.90 ± 0.31
Rotifer chemostat	1.25 ± 0.69	5.34 ± 1.94

(Droop 1976) did not lead to sterile cultures unless the concentrations of the effective substances (D'Agostino 1975) were so high that the growth of the rotifers in the chemostats was hindered. Effective concentrations of penicillin to obtain bacteria-free cultures are 625 mg/l for *Brachionus angularis* and 1250 mg/l for *Keratella cochlearis*. After discontinuing antibiotics the same growth rates of the rotifers are reached as before but, sooner or later, cultures became contaminated. The same observation was made by Boraas (1983).

Bacteria in different chemostats (*Stichococcus bacillaris* and *Brachionus angularis*, Chaps. 4.2, 4.3) were counted by the DAPI method (Porter and Feig 1979). To calculate the bacterial carbon content the factor 1.5×10^{-13} g C cell^{-1} was accepted (Bratbak 1985), which is in the lower range given by Bjørnsen (1986) since the bacteria are not cultured with organic substrates.

The difference in bacterial density was, at most, only weakly significant (t = 1.76) in algal and rotifer chemostats (Table 2.2.2). The density in the rotifer chemostat is usually lower than in the alga vessel. In contrast, the carbon content of bacteria is higher in the rotifer chemostat (t = 9.17) because the C_2-algal density is lower. If the C_2-algal density is very low in steady-state phases, the carbon content of bacteria increases up to 40%. Rotifers may ingest bacteria. The percentage increase in the carbon content of the bacteria at unchanged, absolute density shows the selective feeding on algae by *B. angularis*. Rotifers of the genus *Brachionus* need a higher bacterial density than is found in these chemostats (Starkweather et al. 1979). Bacterial density is as low as in natural

Table 2.2.3 Active chlorophyll and pheopigments in chemostat systems with *Stichococcus bacillaris* as food, depending on the residence time ($\pm 95\%$ confidence intervals of the means, n = 9). (After Walz 1983b)

	Cumulated residence time (days)[a]					
	0.5*	0.5**	0.8**	1.1*	5.1*	7.4**
	Algale reactor (light)		Mixing reactor (dark)		Rotifer reactor (dark)	
Active chl a/ carbon (%)	1.21 ± 0.28	1.26 ± 0.42	1.00 ± 0.26	1.06 ± 0.19	1.16 ± 0.63	0.88 ± 0.88
Pheopigments/ carbon (%)	0.07 ± 0.07	0.03 ± 0.02	0.03 ± 0.03	0.04 ± 0.03	2.59 ± 2.01	1.89 ± 1.14

[a] * System with *Brachionus angularis*, ** system with *Keratella cochlearis*.

environments. In spring, maxima $>10 \times 10^6$ bacteria/ml are found in Lake Constance (Güde et al. 1985).

It is rather surprising that the amount of bacteria in the rotifer chemostat does not increase, since feces are abundant. This is shown by the increase in pheopigments per algal carbon with increasing residence time in the rotifer chemostats (Table 2.2.3). The chlorophyll a content per carbon decreases only very slightly with longer residence time in the system, but the decrease cannot be verified statistically ($R = -0.16$, $n = 54$, $P > 0.1$).

2.2.14 Calculation of Michaelis-Menten Parameters

The Michaelis-Menten- (Michaelis and Menten 1983) and the Monod- (1942)-saturation functions for enzyme kinetics and the growth of microorganisms may be presented as a rectilinear hyperbola:

$$V = \frac{V_{max} * S}{K_s + S} \quad \text{two-parameter model} \qquad (2.2.1)$$

or

$$V = y_0 + \frac{V_{max} * S}{K_s + S} \quad \text{three-parameter model} \qquad (2.2.2)$$

where V = reaction velocity, V_{max} = maximum reaction velocity, S = substrate concentration, K_s = half-maximum saturation substrate concentration for $S = 1/2\ V_{max}$, and y_0 = intersect on the y-axis.

The precision of the parameters depends to a great extent on the "fitting technique" used (Currie 1982). Especially linear transformations, e.g. the double-reciprocal Lineweaver-Burk plot of 1/V versus 1/S, produce a strong bias of the true values, often simulating better fits (Dowd and Riggs 1965).

Only nonlinear fitting techniques, e.g. those of Bliss and James (1966), give unbiased parameters. The approximation to the hyperbola is optimized by the method of least-square deviations. This was done according to an MS-Quick BASIC program (Walz, unpubl.), which could be operated optionally by applying either the two- or three-parameter model. For negative y_0 there are intersections with the x-axis, which represent a threshold for the substrate concentration. Every regression was calculated with two and three parameters, but the threshold solution was only accepted when y_0 was significantly ($P < 0.05$) different from zero. In other cases, the two-parameter solution was preferred according to the "rule of simplicity". The goodness of fit was tested by the Kolmogorov-Smirnov and the χ^2 tests (Sachs 1992).

Acknowledgment. This chapter was substantially improved by contributions of Martin Boraas. The chemostat work was supported by the Deutsche Forschungsgemeinschaft.

2.3 Rotifer Culture in the Turbidostat

WILLIAM N. BENNETT and MARTIN E. BORAAS

2.3.1 Introduction

Population dynamics is most conveniently analyzed in a steady state (Boraas 1983). In a steady state, the population is time-independent so it is at a constant physiological state, thereby providing a convenient starting point for experiments and measurements. Secondly, the mathematics describing the biological processes are greatly simplified. Finally, greater sampling precision is possible since multiple samples can be averaged together.

Because such steady states are difficult to find in natural populations, these analyses are usually restricted to the laboratory. The laboratory apparatus necessary to obtain steady-state population growth is either some form of the chemostat, including semicontinuous cultures, or the turbidostat. In the steady state, population growth is determined by the rate at which the limiting nutrient is delivered to the culture vessel. Essentially, the chemostat operates by providing the organisms a constant environment where their specific growth rate can be controlled by the dilution rate (flow rate/culture volume or D), with their population density as a dependent variable. In contrast, in the turbidostat the environment of the organisms (e.g. dilution rate) can be adjusted so that the population remains at a constant density, again in a steady state.

The turbidostat requires continuous monitoring of either the population density directly or some variable that is proportional to population density. These variables are usually an optical property of the culture. Therefore, it is necessarily a more complex instrument than most other culture devices. Although based on somewhat different operating principles, both chemostats and turbidostats can be described by the same mathematical models.

The turbidostat was originally developed by Myers and Clark (1944) to provide inocula in a constant physiological state for batch-culture studies of algal physiology. The subsequent development of the chemostat (Monod 1950; Novick and Szilard 1950) led to a vast literature, dealing primarily with unicells (see Chap. 3.3, this Vol. for a brief review). The turbidostat has not enjoyed this popularity for two major reasons. First, until the development of inexpensive microcomputers, considerable electronic sophistication was necessary to build and operate the sensing and control apparatus. Second, the turbidostat was perceived by many as simply a complicated means to maintain a constant "log-phase" of growth (sustained growth at r_{max}), such as is obtained in batch cultures. Unfortunately, this has resulted in an incomplete understanding of the physiological and demographic properties of organisms growing near or at r_{max} for many generations.

After Myers and Clark (1944), workers used turbidostat cultures to explore the population dynamics and physiology of bacteria (e.g., Munson 1970), algae (e.g., Hill et al. 1985), and other unicells. Turbidostat culture is particularly well suited to examine toxic effects on algae and other organisms, particularly regarding physiological adaptations to toxic stress (Bennett 1988, 1990; Bennett and Brooks 1989). Until recently, this type of continuous culture had not been used to study metazoan populations. The turbidostat has application in both basic and applied research. For example, in many agricultural practices, including the growing field of rotifer aquaculture (e.g., Theilacker and McMaster 1971; Oka et al. 1980; Schlüter 1980; Snell et al. 1987), attempts were made to maintain populations at a constant density at high r in order to maximize total production.

Population density in the turbidostat can be detected either turbidometrically or nephelometrically. In either case an optical signal is produced. The optical signal is converted to an electronic signal with a transducer which then, via control circuitry (e.g., Boraas and Bennett 1988), activates a pump to deliver medium when the signal exceeds a set point. The set point may be detected with either a "hard-wired" electronic regulator or a dedicated microcomputer. Thus, practical turbidostats are diluted at frequent intervals rather than continuously. The organisms, however, grow continuously. It is usually not difficult to operate the system with sufficient sensitivity so that the interval between dilutions is a small fraction of the doubling time of the population, effectively resulting in continuous operation.

Turbidostats and chemostats are complimentary systems with regard to measurements of a population's response to its limiting nutrient and the dilution or per capita growth rate (Fig. 2.3.1). The curves in Fig. 2.3.1 represent the steady-state relationship between the population mass and limiting nutrient at a given steady-state, specific growth rate, r. A rotifer turbidostat is operated by maintaining the population at a constant density (here defined as organisms per volume) by diluting the population with fresh algal suspension in response to increases in the rotifer density. As in a chemostat, the specific growth rate ($r = dX/Xdt$) (where X is population density) of the organisms in the steady state is numerically equal to the flow rate of the medium divided by the culture volume when the system is in a steady state. Thus, the regulated population density is an independent variable and r is a dependent variable.

The chemostat is most effectively used for steady-state measurements at specific growth rates which are between ca. 5 and 80% of the population's r_{max} (indicated by the right-hand arrow in Fig. 2.3.1) (Herbert et al. 1956). This is because small variations in the pumping rate or culture volume, and thus r, have little effect on the steady-state concentrations in this portion of the curve.

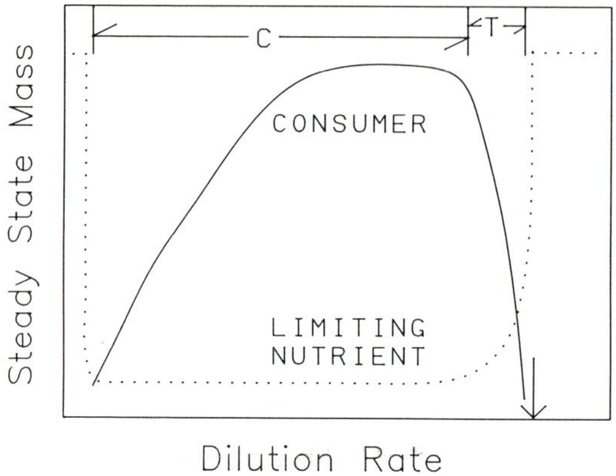

Fig. 2.3.1 Steady-state functional relationship between consumer (rotifer) and limiting nutrient (algal food) at different specific growth rates (r). *C* indicates range of r (= D in the steady state) in which the chemostat culture functions optimally. *T* indicates range where turbidostat culture functions optimally. *Arrow* defines relative position of r_{max}

That is, small errors in D do not result in large changes in X or C_2. At specific growth rates near r_{max}, however, any variations in pumping rate greatly affect the values of X and C_2. Hence, the chemostat is inherently inaccurate for regulating population growth at high values of D. In contrast, the turbidostat functions well at higher specific growth rates since it actively regulates the population at a given density on the vertical portion of the curve. Small variations in density regulation result in minor changes in the specific growth rate. However, at lower specific growth rates the turbidostat may become erratic because small variations in density regulation can result in large changes in r.

The turbidostat can be potentially operated at low values of r for carbon-limited populations (Fig. 2.3.1). Because these populations necessarily suffer respiration losses, their biomass, and frequently their population density, decline as r approaches 0 (Herbert 1958; Boraas 1983). Thus, depending on the initial conditions, there is a potential for operating turbidostatically controlled cultures over virtually the entire physiological range of r. However, this mode of operation has not been used to date.

2.3.2 Setup and Operation of the Turbidostat

Here, we outline a method to grow the rotifer, *Brachionus calyciflorus* Pallas, in a turbidostatically controlled, continuous culture system on the alga, *Chlorella pyrenoidosa* Chick (UTEX No. 26), the limiting factor for rotifer growth. The methods are described in more detail in Boraas and Bennett (1988). Unlike most turbidostat systems, where the population density of the organism of concern is regulated, this system was used to measure and regulate the concentration of the limiting nutrient. That is, the residual algal food was actively regulated, which indirectly regulated the rotifer population (Fig. 2.3.1). This indirect regulation was necessary since rotifers in turbidostat culture are in relatively low concentration and their comparatively large size gave no measurable optical signal. In contrast, the algal populations were closely correlated with optical density (Boraas and Bennett 1988).

The rotifers were isolated from the Milwaukee Harbor in Wisconsin, USA. When grown on *Chlorella*, this rotifer strain required a growth factor found in both tap water from Lake Michigan or in water from the Milwaukee Harbor. Despite extensive attempts using rotifer bioassays, the growth factor was not identified. It was characterized as water-soluble, about 800 molecular weight, and not belonging to the families of essential amino acids or vitamins. Therefore, we mixed filter-sterilized dechlorinated tap water with the *Chlorella* suspension in a 1:1 ratio to support growth of the rotifer population. The culture apparatus was a two-stage system consisting of a first-stage algal chemostat, a second-stage rotifer turbidostat, a sensor, and a computer (Fig. 2.3.2). The first stage can be considered a separate culture, from which algae were periodically harvested to feed the rotifers in the second stage. The algae in stage 1 were maintained at a dilution rate of 0.82 day^{-1}. All wetted surfaces were glass, silicone rubber, or Teflon. The inorganic medium was that of Sorokin and Krauss (1958), diluted 1:50 and with a final concentration of 1.5 mM nitrate. The medium was supplied

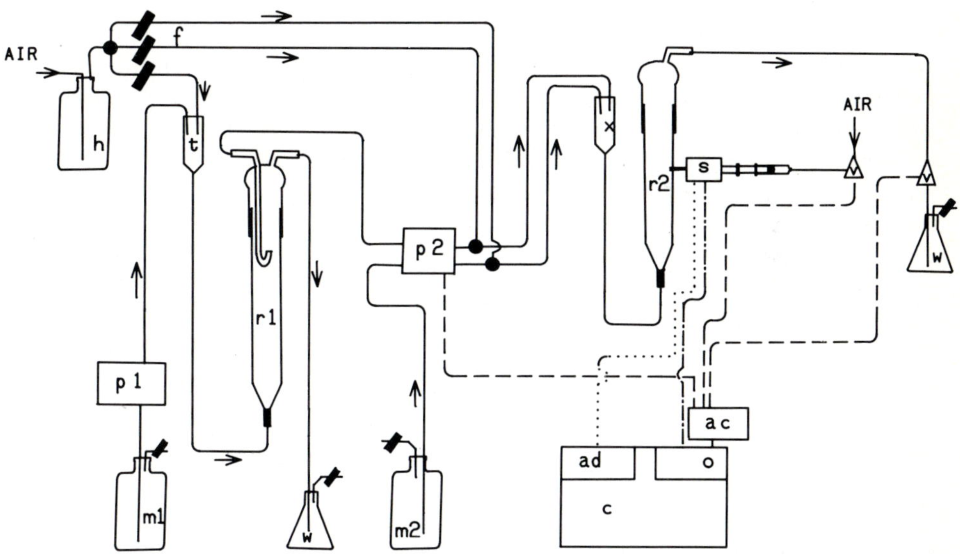

Fig. 2.3.2 Schematic of the turbidostatic culture system. From *left* to *right*: *h* air hydrator/filter containing water, charcoal, and ion-exchange resins; *f* 0.22-μm membrane air filterns; *p1* peristaltic pump; *m1* medium reservoir for algal chemostat; *t* trap; *r1* algal chemostat vessel (stage 1); *w* waste outflow; *m2* sterile-filtered lake water containing unknown rotifer growth factor; *p2* computer-controlled peristaltic pump; *x* mixing reservoir/trap; *r2* rotifer turbidostat culture; *s* turbidity sensor; *v* solenoid valves; *c* computer; *ad* analog-to-digital converter; *o* output port; *ac* 5 V to 110 V converter; *solid lines* fluid (air and liquid) lines; *dotted line* sensor input to ad; *dash-dot line* 5 V line to power the LED; *dashed line* 110 V to power solenoid valves and p2. (After Boraas and Bennett 1988)

to the algal chemostat by a Harvard model 1201 peristaltic pump. The algal culture was illuminated by a bank of "Cool-White" fluorescent lamps providing a mean luminous flux of 10^{16} quanta $s^{-1} cm^{-2}$ PAR on one side of the culture vessel. The algal culture was mixed using charcoal-filtered, hydrated, filter-sterilized air.

The second-stage rotifer turbidostat, in darkness, consisted of a mixing reservoir, a 250-ml culture vessel, a sensor unit, and an outflow flask. The algal suspension from the first stage and the rotifer growth factor medium were held in 4 separate medium reservoirs. Both were delivered to the mixing reservoir by a second, computer-controlled, peristaltic pump (Fig. 2.3.2). The growth factor medium was tap water enriched with the same nutrients as the algal growth medium. The algal suspension was drawn from the algal stage using a J-shaped 16-gauge Teflon tube, which prevented bubbles from entering the delivery tube and minimized variance in the flow rate of algal suspension to the rotifer stage. Sterile-filtered air was supplied to the rotifer stage, immediately downstream from the pump, minimizing settling of algae in the pump lines and mixing the rotifer stage.

The sensor used to measure the residual algal (limiting nutrient) concentration was adapted from Blachére and Jamart (1969) as described in Boraas and Bennett (1988; Fig. 2.3.3). The sensor consisted of the following components. The "super-bright" light-emitting diode (LED) (Stanley Electric Co., 2000 mcd output at 50 milliamps current) and the silicon photovoltaic light sensors (PIN photodiodes) were attached to an opaque polyvinylchloride housing. The light sensors were positioned at 0, 23 and 45° from the central axis of the LED. A glass sample chamber and a pressure-operated, solenoid-actuated plunger system were attached directly to the culture vessel (see Boraas and Bennett 1988 for details of operation). The plunger system, controlled by a microcomputer, operated as follows. The culture suspension was first drawn into the glass sample chamber with a rubber-tipped plunger. The turbidity was then determined as described in the next paragraph. The sample suspension was then returned to the culture vessel, which also cleaned

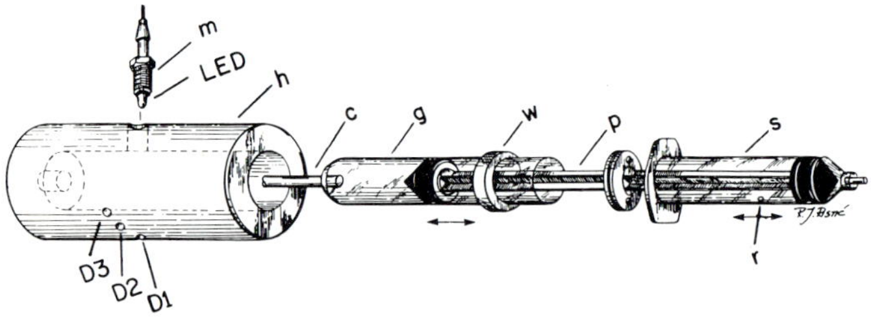

Fig. 2.3.3 Diagram of the turbidity sensor. *m* LED support; *h* PVC housing; *c* connecting tube to rotifer vessel; *g* glass sample chamber; *w* silicone washer; *p* plunger; *s* 20-ml plastic syringe; *r* air release aperture; *D1–D3* attachment sites for photovoltaic light sensors at 0, 23 and 45°, respectively, from the axis of the LED. (After Boraas and Bennett 1988)

the glass tubing of the plunger system of any attached material. The primary advatages of this type of sensor construction are that the light path window is cleaned at each measurement and no culture volume is lost.

Turbidity measurements were taken every 15 min by the computer. The signals from the light sensors were amplified, digitized with an analog-to-digital converter, and compared with a set point stored in the program. When this comparison indicated that the set point was exceeded, the computer activated the pump delivering mixed algae and growth factor medium to the rotifer stage. After each turbidity measurement, data were saved on a floppy disk file, including time, volume pumped per interval, pump status, and some values calculated from these data. Since the volume delivered could change due to aging of the tubing in the peristaltic pump, the total outflow from the turbidostat was measured daily. This volume was used to correct for variations in the pumping rate. Algal population densities from both the algal growth vessel and the turbidostat vessel were measured daily with an electronic particle counter, a Particle Data Celloscope. Live rotifers were either enumerated microscopically or with the particle counter within 0.5 h of sampling.

A computer program (Bennett, unpubl.) was written in Pascal to calculate D from the total volume (V) of algal medium pumped through the system during a given interval, obtained from the record of the pumping status for each sample. All pump volumes associated with pump events were summed to give a total volume (V). The following equation was then used to calculate D:

$$D = V/(I \times 248) \qquad (2.3.1)$$

where I is the calculation interval (in days), and 248 is the volume of the turbidostat culture (in ml). In the steady state, $r = D$. During periods of nonsteady state operation, r was corrected for changes in X using the exponential growth equation.

2.3.3 Examples of Turbidostat Runs

In this section we present typical data obtained with our rotifer turbidostat system. Immediately after the culture vessel is filled with algal suspension and the rotifers are inoculated, usually about 0.2 rotifers ml^{-1}, the residual algal concentration (C_2) is almost equal to the input algal concentration (C_1). At this time, the dilution rate of the culture (D) will remain zero for several intervals, although the rotifers are increasing exponentially at r_{max}. This is because the density of C_2 is high, and the set point is not "tripped" after a density measurement. A "trip", and subsequent pump activation, does not occur until C_2 is reduced through rotifer ingestion of algae.

A time course illustrating the change in culture light transmittance (= inverse of the relative concentration of C_2) and rotifer r from initial rotifer inoculation to steady-state conditions is shown in Fig. 2.3.4. The growth rates

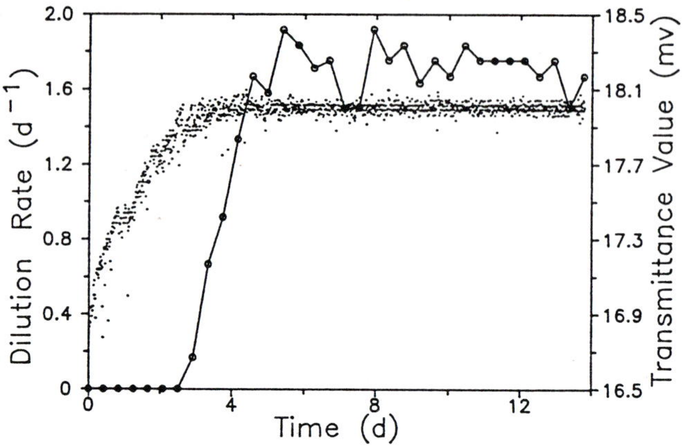

Fig. 2.3.4 Time course of specific growth rate D (*circles/solid line*) and culture light transmittance (*dots*), which is approximately the reciprocal of C_2, for a turbidostat culture inoculated with rotifers at time zero. D was calculated for 10-h intervals using pump status data (see text). The density set point for this culture was 18.0 mV

plotted in Fig. 2.3.4 were calculated from the volume pumped over 10-h intervals (see Sect. 2.3.2).

To illustrate how the turbidostat system responds to a perturbation, algae were added abruptly to a steady-state culture, and D was then calculated at 10-h intervals before, during, and after the perturbation (Fig. 2.3.5A). Algae were added at 5.2 days, which resulted in a decrease in culture transmittance. D declined because the optical density was above the set point. Although the actual D of the culture was reduced to zero at the time of algal addition, the calculated D (from the number of pump events per 10-h interval) shows a positive value after the addition (Fig. 2.3.5A). This is because the calculation interval contained the time of algal addition and thus averaged data from before and after the addition. As algae were consumed by the rotifer population, transmittance increased until the set point was reached at 6.3 days, resulting in an increase and stabilization of D to the levels which existed prior to the perturbation. It is important to note that r and D are equal in the turbidostat when the culture is near the set point and the biomass of both X and C_2 has stabilized. Immediately after the perturbation, r was no longer equivalent to D because X and C_2 were no longer in a steady state, resulting in a culture that had a population with a positive r but a D of zero. Some other types of perturbations, such as temperature fluctuations or the introduction of toxins (Bennett 1988, 1990; Bennett and Brooks 1989), would not affect the relationship between D and r because the steady-state biomass of both X and C_2 remains stable.

The accuracy of r (or D), as calculated from turbidostat pump data, is dependent on the sampling interval and the calculation interval as well as on the accuracy of the optical measurements. To illustrate this point, the same

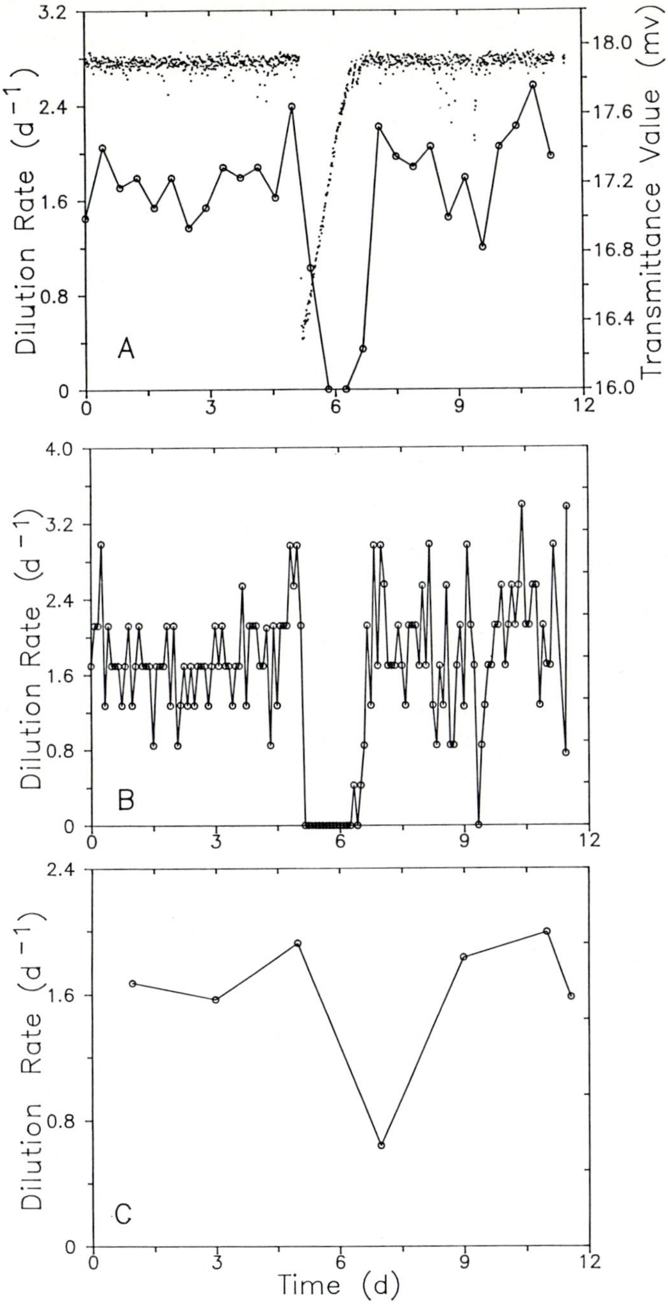

Fig. 2.3.5 **A** Time course of r and culture transmittance before, during, and after a perturbation in turbidostat culture. Definitions as in Fig. 2.3.4. The perturbation was the addition of algae at 5.2 days D was calculated at 10-h intervals. **B** shows the time course of D (as in **A**), calculating D at 2-h intervals. **C** shows the time course of D (as in **A**), calculating D at 48-h intervals

data set used in the calculation of D at a 10-h calculation interval (Fig. 2.3.5A) was used to calculate D at intervals of 2 and 48 h (Fig. 2.3.5B, C). The resolution of D increases as the calculation interval decreases, but artificial grouping of the calculated data occurs as the calculation interval approaches the sample interval. Using a calculation interval of 2 h (Fig. 2.3.5B) results in an artificial pattern which is related to the 15-min sample interval. Over a 2-h period, there can be a maximum of 8 pump events, i.e., 9 possible calculated values (based on 0–8 events). As the calculation interval approaches the sample interval, the number of possible calculated values decreases until, at a calculation interval of 15 min, only 2 possible values can result, based on 0 or 1 pump events. At a calculation interval of 48 h, the resolution of D is decreased (Fig. 2.3.5C). Calculations of D during steady-state periods are comparable to those calculated from 2- and 10-h intervals; however, resolution of the perturbation is lost since transient responses are averaged with steady-state growth rates.

The 15-min sample interval was deemed sufficiently precise. A 10-h calculation interval for r was subjectively determined to provide sufficiently high resolution without introducing artificial grouping of the calculated data (e.g., Fig. 2.3.5B). The sample interval, however, could have been reduced to a minimum of 6 min if a higher resolution of r was required.

The turbidostatically controlled rotifer culture functioned well with very little maintenance. Only two major types of problems occurred over a period of 4 years: failure of solenoid valves and contamination of the rotifer culture by *Vorticella*. This contaminant, possibly resulting from growth upstream through the delivery tube to the waste reservoir, caused clumping of the residual algae, which altered the light transmittance and scattering properties, thus interfering with the measurement of algal density. These events were rare.

This is contribution no. 375 from the Center for Great Lakes Studies.

3 Characteristics of Rotifer Continuous Culture Systems

3.1 The Growth of *Brachionus rubens* in Semicontinuous Culture

MARTIN E. BORAAS

3.1.1 Introduction

The factors regulating single-species growth in the simplest experimental systems must be clearly defined before the dynamics of multispecies systems can be understood. In many areas of ecology, progress in understanding single-species growth has been accelerated by combining complementary approaches in model building and model testing. The model-building process defines the important material transfer processes which can limit population growth, gives them mathematical formulations, and incorporates empirical estimates for coefficients. Complementary experiments can be performed in culture devices that satisfy the major physical assumptions and constraints of the models: a well-mixed, open system in which the substrate (food) can be supplied continuously to a population. Limitations in the models are revealed by comparisons with experimental results, suggesting refinements in the assumptions and mathematical formulations (Williams 1980). This stepwise approach is, perhaps, the best available method for revealing the fundamental mechanisms governing the population growth of both unicells and metazoans.

In this chapter, I describe the growth of the rotifer *Brachionus rubens* in semicontinuous cultures (SCs) with 24 h dilutions. Dilution rates, D, were calculated as presented in Chap. 2.1. SCs are approximations to a true chemostat culture as the rotifer and its algal food both enter quasi-steady states in which the environmental conditions and population structure vary but return to the same state just prior to the next dilution event (see Chap. 2.1). Data showing "transient events" or deviations from the quasi-steady state are also presented. These transients include the estimation of the r_{max} (the maximum specific growth rate) of *B. rubens* and food conversion efficiency when the dilution of the culture exceeds r_{max}, and the effects of step changes in the dilution rate of the cultures, or "shift-up" and "shift-down" experiments.

The dependent variables include concentrations of rotifers (X) and residual algae (uneaten algae in the culture vessel or C_2) as well as physiological (metabolic rate, conversion efficiency, and biomass output) variables. In addition, a parameter of a model of steady-state chemostat growth, the "half-saturation constant of growth," K_s (Herbert 1958), is calculated from the quasi-steady-state concentrations. Variations in K_s provide insight into factors controlling food-limited rotifer growth.

3.1.2 Special Methods

General semicontinuous methods are presented in Chap. 2.1. Only specific methods are given here. The rotifers were kindly supplied from the laboratory of U. Halbach. They were maintained on the chlorophyte *Chlorella pyrenoidosa* (Chick) (UTEX No. 26) on a mineral medium supplemented with vitamin B_{12} as described in detail in Boraas (1980, 1983).

Algal biovolume estimated by a particle counter (see Chap. 2.1) as $\mu m^3\ ml^{-1}$ was converted to dry mass using the conversion factor 0.57 g dry mass cm^{-3} live biovolume for *Chlorella pyrenoidosa* (Boraas 1983). The *B. rubens* biovolume-to-biomass conversion factor was obtained by first holding rotifers overnight in the absence of algae to clear their guts. The animals were then filtered onto 13 mm diameter, 12 μm poresize Nuclepore filters in triplicate, dried at 65 °C for 72 h, and weighed to 1 μg on a Cahn electrobalance. A biovolume in triplicate was taken for the same sample. The conversion factor thus obtained was 0.272 g cm^{-3} (SD = $\pm$ 0.016 g cm^{-3}).

3.1.3 Transient States

The results are separated into transient and quasi-steady state. For all studies, the symbols used for biomass concentrations are: rotifers, X; residual algae, C_2; and input algae, C_1. A superscripted $*$, e.g., X^*, here indicates a quasi-steady state.

Transient results are presented as "shift-down" (D is lowered; Fig. 3.1.1A, B) and "shift-up" (D is increased; Fig. 3.1.1C) experiments, and as an analysis of exponential washout when $D > r_{max}$ (Fig. 3.1.2). In this study, the shift-up transitions did not attain a steady state, either because r_{max} was exceeded or because the culture was shortly terminated thereafter.

A typical time course of a semicontinuous rotifer culture with a shift-down transition is shown in Fig. 3.1.1A. This culture was maintained as a batch culture until the algal concentration was depleted. On initiation of daily dilutions at $D = 0.29\ day^{-1}$, the rotifer concentration declined to a low value and then recovered, entering into a quasi-steady state. When D was stepped from 0.29 to 0.19 day^{-1} (indicated by the vertical arrow in Fig. 3.1.1A), there was a brief overshoot in rotifer mass followed by a new quasi-steady state. In another culture, shown in Fig. 3.1.1B, dilution was initiated at 0.13 day^{-1} after initial batch growth of the rotifers to 34 $\mu g\ ml^{-1}$. When the rotifer population stabilized, residual algal data were collected. Fourteen days after dilution began,

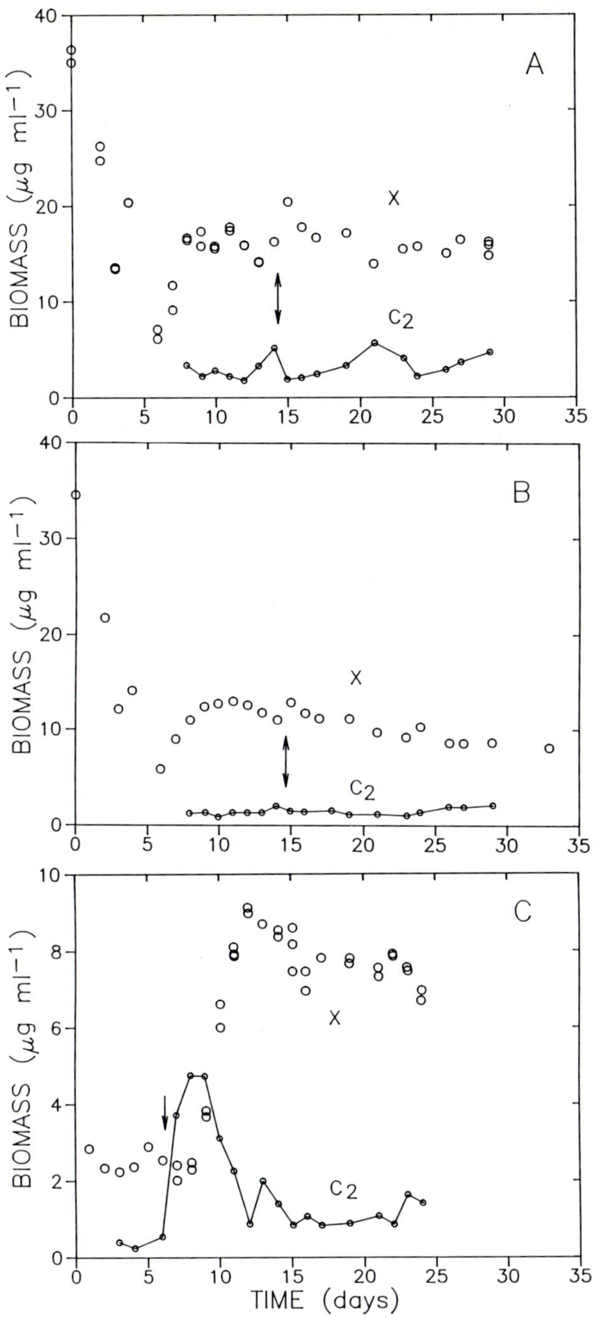

Fig. 3.1.1 A–C Step changes in dilution rate of two-stage semicontinuous cultures: **A** and **B** show decreases, or shift-down changes; **C** shows an increase or shift-up change

D was stepped from 0.13 to 0.078 day^{-1} (arrow in Fig. 3.1.1B). Immediately after the shift down, the rotifer population increased slightly, and C_2 decreased slightly. Thereafter, X declined, eventually reaching a new quasi-steady state slightly lower than the previous value.

A shift-up perturbation is shown in Fig. 3.1.1C. The mean D was increased from 0.051 to 0.47 day^{-1} (arrow in Fig. 3.1.1C). The rotifer population first declined slightly and then increased severalfold, subsequently stabilizing at a value about four times higher than at the slower D. In contrast, C_2 increased dramatically after the step change and then declined to a quasi-steady state that was higher than the earlier value at 0.051 d^{-1}.

The data in Fig. 3.1.1C indicate a lag in the rotifer response to the perturbation. After the step change, the algal concentration increased simply because more algae were being added to the culture. However, the rotifer population actually declined slightly, in part because a larger proportion of the rotifer population had been washed out. However, without any lags, the potential growth rate of *B. rubens* should have overwhelmed the loss due to harvesting. This recovery did not occur for at least 2 days. The size-frequency distributions of the rotifer population are consistent with a demographic lag, in that 2 days elapsed before large individuals, presumably with developing or external eggs, appeared (data not shown).

Shift-up and shift-down perturbations are very powerful methods (Maaløe and Kjeldgaard 1966) since they start with a stable state, enter a transient period, and then reach a new stable state. Thus, the state before the shift may be used as a control. This method can reveal lags, qualitative changes in physiological state and other biological phenomena. In addition, shift experiments can provide very precise data for the evaluation of models.

The shift-down data shown in Fig. 3.1.1A, B are consistent with the data and a population biomass model discussed in Boraas (1983) and elsewhere (e.g., Chap. 5.2). There was an initial increase in X and a decrease in C_2. One explanation for these data is that when D was decreased, the rotifer population continued to feed and metabolize for a short time at a rate consistent with the higher, earlier value of D. That is, there was a lag before the population attained a constant physiological state. Also, algae were being added to the culture vessel at a slower rate. These two effects resulted in a short-term decline in C_2 linked with a rise in X. Since C_2 was now too low to sustain growth at the new value of D, the rotifer population declined by washout while adjusting to the new value of D, finally entering into a new quasi-steady state. The final results of such experiments depend on the relative positions of the starting and ending values of D vs. X*. Here, changes in D were small and either near the peak of the curve of X* vs. D or near the origin. Washout occurs when D exceeds r_{max}, implying that the rotifer population is diluted faster than it can grow. When D was set to 0.92 day^{-1}, the rotifer population declined exponentially (Fig. 3.1.2A) while C_2 increased linearly in time (Fig. 3.1.2B). In addition, the biomass conversion efficiency, Y (ingested algal biomass/rotifer biomass), also declined linearly with time (Fig. 3.1.2C). The r_{max} was computed as D minus

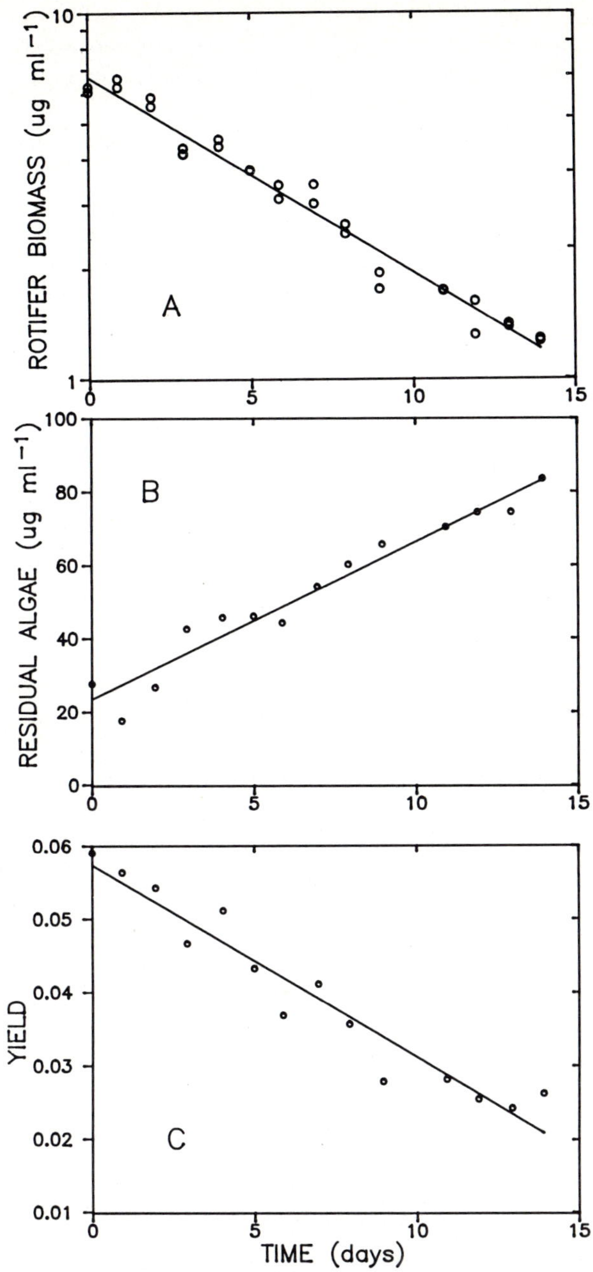

Fig. 3.1.2 A–C Washout of *Brachionus rubens* when D exceeds r_{max}: **A** shows exponential washout of *B. rubens*, **B** shows the concomitant increase in C_2, *Chlorella pyrenoidosa*, with decreasing rotifer concentration, **C** shows the yield, Y = algal biomass consumed/rotifer biomass produced, during the washout phase of the culture

the sloped of ln X vs. time. This method permits a much more accurate measurement of r_{max} than the exponential increase in simple batch culture since many more values can be collected, minimizing experimental error (Yee et al. 1981).

During washout, as X declines, C_2 increases. When r is constant, a constant physiological state is frequently assumed. However, the efficiency of conversion of algal biomass to rotifer biomass (yield, or rotifer mass produced per algal mass consumed) is not constant but steadily declines as C_2 increases (Fig. 3.1.2C). During this washout of X, the rotifers were growing at their physiological and demographic r_{max}, but C_2 was steadily increasing. Either the digestion efficiency must have declined or the respiration rate increased; one or both phenomena would result in a declining Y over time. Thus, the state of "balanced growth," as defined for prokaryotes and other osmotrophs (e.g., Pirt 1975), may not exist for holotrophs during growth at r_{max} under conditions of a changing food concentration.

3.1.4 Pseudo-steady States

In the time-independent cultures (Fig. 3.1.3), the rotifer response to D is a unimodal curve that appears to pass through the origin. The algal response is a monotonically increasing curve with D, with a positive intercept on the ordinate. The combined plots of C_2 and X* (Fig. 3.4.3) show patterns which can be compared with available theoretical models for chemostat culture (Chap. 5.2). The most conspicuous feature of the plot is the inverted "U" (humped) shape for steady-state rotifer density. This pattern is reasonably well explained by the Monod (1942) growth model, as modified by Herbert (1958) to account for metabolic loss (see Appendix).

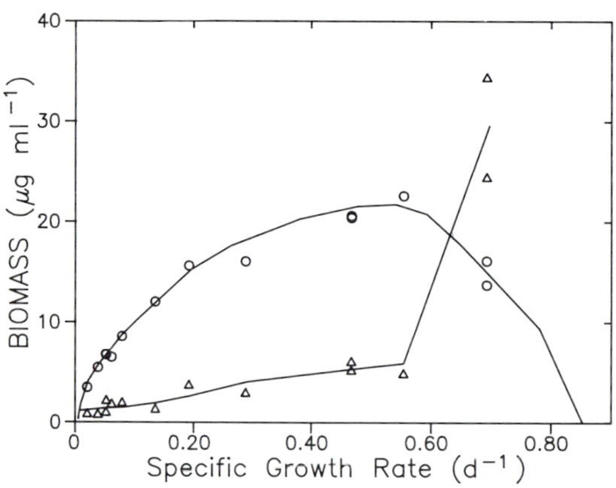

Fig. 3.1.3 Pseudo-steady state rotifer (*circles*) and residual algal (*triangles*) biomass vs. dilution rate, D, for semicontinuous *B. rubens* two-stage cultures. The lines are fitted visually

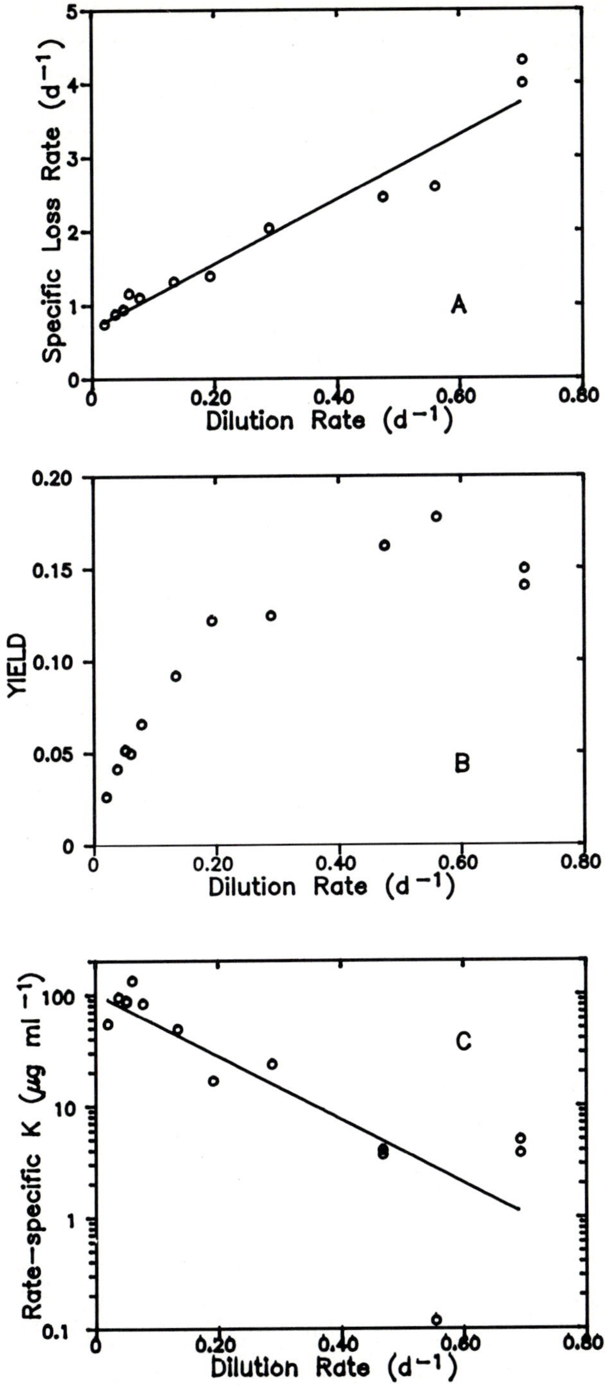

Fig. 3.1.4 A–C Characterization of quasi-steady-state culture of *B. rubens*. **A** shows the loss rate calculated from (algal biomass ingested – residual algal biomass)/rotifer biomass for each D. **B** shows yield, calculated as in the legend of Fig. 3.1.2C, for each D. **C** shows K_s values for each D, calculated according to the formula given in Boraas (1983)

Other dependent variables were calculated from formulae derived in Droop and Scott (1978) and summarized in Boraas (1983). These values include specific metabolic loss rates, L; a conversion efficiency or yield, Y; and output, $(D*X)$. With respect to D $(=r)$, L is linear with a positive intercept on the ordinate, Y is approximately hyperbolic, and $(D*X)$ is a roughly sigmoid unimodal function with an evident intercept at the origin.

The responses of X^* and C_2 vs. D and the physiological variables are consistent with the Monod-Herbert (M–H) microbial biomass growth model discussed in Boraas (1983). This model provides an alternative viewpoint to traditional animal population models, in that birth and death are not explicitly incorporated. Rather, the model treats biomass as if it were a chemical concentration and is, therefore, most appropriate for large populations of small organisms.

The predicted hyperbolic relationship of Y to D is due to the cost of maintenance, μ_c. At low D, a large fraction of the standing crop of biomass is lost by respiration, resulting in a low value of Y. The maximum value of Y, Y_{max}, occurs at r_{max}.

The predicted positive linear relationship of L vs. D is largely due to increasing metabolic loss rates from growth with increasing r, over and above a constant specific loss rate from maintenance over all D (the intercept of L on the ordinate at $D = 0$ is a measure of μ_c before the conversion of food to biomass, i.e., μ_c/Y_{max}). In this study, despite an increase in loss rate with D (as L), the loss of biomass due to maintenance (μ_c*X) resulted in a declining standing crop of rotifer biomass as D approached zero (Fig. 3.1.3).

The steady-state patterns observed in this study are very similar to *B. calyciflorus* data collected in chemostat culture (Boraas 1983; Bennett and Boraas 1990). The predicted responses between Y, L, and D were observed in both studies, as shown in Fig. 3.1.4 for this study. Therefore, it seems reasonable to conclude that the *B. rubens* populations in this study were also biomass limited by the carbon energy content of their diet. It is not clear whether or not mortality played a major role in biomass loss as indicated by Walz (Sect. 3.3.7), particularly as the mean residence time (1/D) approached the rotifer life span. A deviation in the linearity of L vs. D would be expected as D approaches the origin. This deviation was not observed (Fig. 3.1.4A). Indeed, empty loricas, indicating dead rotifers, were seldom seen in this study.

3.1.5 Trace Growth Factor vs. Carbon Energy Limitation

Most animal populations are considered to be energy (or carbon-energy) limited. However, there are other possibilities. Many human diseases are caused by deficiencies of vitamins and other trace growth factors (TGFs). It seem reasonable that some animal populations, particularly those raised in the laboratory on algal monocultures, might be limited by a TGF. Can evidence of TGF limitation be found in the rotifer literature?

I argued above that my *B. rubens* data are consistent with carbon-energy limitation, based on calculations derived from the Monod-Herbert model. The lack of evident TGF limitation in this study and in that of Boraas (1983) may have been because bacteria were not eliminated from the cultures. Prokaryotes could have provided the necessary TGF requirements, similar to that hypothesized for mixotrophic algae (phagotrophic phototrophs) under natural conditions (Boraas et al. 1990). However, *B. calyciflorus* growth on algae held in the dark (two-stage chemostats) and algae growing in the light (single-stage chemostats) appears to differ. As shown in Fig. 2.1.1, *B. calyciflorus* have higher growth rates at lower algal concentrations when feeding on the algae growing in the light. This may indicate that even if rotifers are carbon-energy limited, their ability to process the digested material may vary with different types of food. Such variations in processing ability could represent a different form of TGF limitation. That is, TGF limitation might take the form of biomass limitation or rate limitation of carbon-energy processing. An example of TGF biomass limitation is discussed below.

All animals require a panoply of vitamins, essential amino and fatty acids, and other TGFs. A priori, food species may be deficient in at least one TGF. If the population is biomass-limited by a TGF, then Y should not be hyperbolic. If carbon-energy is in excess, then there appear to be three metabolic "solutions" available to the animals: The carbon may be stored as lipid, the assimilation efficiency could decline, or the surplus carbon could be "burned" by respiration. Elevating the respiration rate or reducing the assimilation efficiency will tend to produce a flat Y vs. D, while increasing storage should result in a linear increase of Y with D. Y has been reported to be independent of D for bacteria grown on non-carbon energy limitation (e.g., Herbert 1976).

Since L increases linearly with D under carbon-energy limitation, under TGF limitation an increased respiration rate or reduced assimilation efficiency should produce an increasingly nonlinear L with D. Increased lipid storage would have little effect on L.

These predictions are supported by data from both rotifers and bacteria. Droop and Scott (1978) reported that Y of *B. plicatilis* was virtually constant against D while the specific respiration rate increased nonlinearly with the dilution rate. These observations are consistent with two hypotheses. First, their rotifers were TGF limited. Second, the rotifers digested and, perhaps, absorbed excess organic carbon, which was released by elevating the respiration. In a later study, Scott (1980) was unable to find a correlation between the macromolecular algal composition and rotifer Y. Such a correlation would not be expected if a TGF were the actual limiting factor.

Similar responses have been observed in the microbial literature. Under sulfate, nitrate, phosphate, or ammonia limitation, *Klebsiella aerogenes* demonstrate "overflow metabolism" (e.g., Neijssel and Tempest 1976). More organic carbon was taken up than was required for growth and was subsequently released from the cells either as organic acids or as CO_2. Thus, free-living prokaryotes and rotifers do not appear to respond to a surplus of organic carbon until the material is intracellular.

3.1.6 Variation in K_s with Dilution Rate

Figure 3.1.4C shows the value of the half-saturation constant of growth, K_s, calculated for each steady-state rotifer population. This calculation, described in Chap. 5.2 (Eq. 5.2.6), is based on a steady-state solution of the Monod-Herbert growth model (Herbert 1958). There was an apparent log-linear relationship between K_s and D (Fig. 3.1.4C).

Most workers assume that K_s is constant compared with D. However, this is not a biological necessity. The derivation of K_s is purely empirical (Monod 1942), and no direct analogy to the enzymatic definition can be assumed. When K_s is calculated at each D for these *B. rubens* data, there is a clear negative, approximately log-linear relationship (Fig. 3.1.4C).

The decline of K_s vs. D is also consistent with limiting carbon energy in the *B. rubens* cultures. Rotifers have a high maintenance coefficient, i.e., μ_c. That is, the carbon-energy expenditure necessary to maintain biomass is a significant fraction of r_{max} (e.g., Boraas 1983). Under severe energy limitation, so little energy may be left over from maintenance and reproduction that locomotion, expressed as swimming speed, and hence food-gathering ability could decline.

Acknowledgements. The bulk of the data in this chapter were collected while I was a student in the Department of Biological Sciences at The Pennsylvania State University. This study was funded in part by NSF grants BMS 75–18749 and DEB 77-29404 to D.B. Seale and F.M. Williams at The Pennsylvania State University and in part by the University of Wisconsin Sea Grant Institute under grants from the Office of Sea Grant, NA80AA-D-00086. I wish to thank Richard Back and Russel Cuhel for critical reviews of the manuscript.

This is contribution no. 348 from the Center for Great Lakes Studies.

3.2 Single-Stage Predator-Prey Algal-Rotifer Chemostat Culture

MARTIN E. BORAAS

3.2.1 Introduction

The primary biological processes that structure freshwater aquatic communities are usually considered to be competition and predation (e.g., Sommer 1989). Competition has been extensively investigated using steady-state methods (e.g., Chaps. 6.1 and 6.2, this Vol.). However, predation has not received comparable attention in this context. Herbivory, as a form of predation, is unquestionably a critical factor in the dynamics of aquatic systems. While the predator-prey interaction can be defined in many ways (Taylor 1984), I here define it as the predator (rotifer) consuming whole algal prey, with sustained growth of both predator and prey populations. The chemostat can be readily adapted to study this form of multiple-species interaction.

Unlike the two-stage culture systems described in Chaps. 3.3, 4.1–4.3, a single-stage system allows for growth of both the rotifers and the algal food in a predator-prey interaction. The system input is inorganic medium, with a nutrient limiting to algal growth, at least in the absence of the predator. The predator receives the algal prey as its sole food source. This system can be used to analyze major historical themes. For example, what governs the stability of predator-prey interactions? Do they oscillate? What are the effects of perturbations?

The analysis of mixed-species predator-prey continuous cultures, in which both the predator and the prey populations are actively growing, has been largely restricted to bacterial-protozoan studies (e.g., Bungay and Bungay 1968). Such studies are represented by glucose-bacterial-ciliate interactions (e.g., Jost et al. 1973a, b) and glucose-bacterial-amoeba interactions (e.g., Tsuchiya et al. 1972; Dent et al. 1976). Taub and McKenzie (1973) and Bader et al. (1976a, b) studied the continuous culture of a protozoan predator on an algal prey.

Tsuchiya et al. (1972) modeled the predator-prey interaction of the slime mold (*Dictyostelium discoideum*) amoebae and *E. coli* using the assumption of Monod growth (see Chap. 5.2) for both species. Estimating the model parameters independently, their simulations formed an almost exact correspondence with experimental observations. Subsequent work with ciliates failed to repeat this correspondence (e.g., Drake and Tsuchiya 1976). The most likely explanation for the correspondence observed in the bacteria-amoeba study was that the amoebae tend to adhere to surfaces, thereby removing bacteria growing on the walls of the culture vessel (Fredrickson 1977), which can dramatically affect the stability of mixed-species cultures (Bonomi and Fredrickson 1976). Earlier studies, e.g., Gause (1934) and Huffaker (1958), often concluded that the predator-prey interaction is unstable, with one or both of the species going extinct. In contrast, the continuous-flow studies usually exhibited either sustained oscillations or steady states for tens to hundreds of generations (e.g., Boraas 1980).

With few exceptions, chemostat predator-prey studies have explicitly incorporated three random variables: the predator population, the prey population, and they prey's resource. Operationally, the addition of the predator is simply an elaboration of a unispecific algal culture, with a precisely defined nutrient environment. The requirement for an explicit statement of the prey resource is obvious when a prey and its predator are cultured together in a single-stage chemostat. This plant-herbivore interaction cannot be completely described and analyzed without the inclusion of the plant resource, which permits mass-balance estimates to be made for the entire system. The growth rate of the algal prey, a variable unique to this system, is here defined as μ; the maximum algal growth rate is termed μ_{max}.

3.2.2 Special Techniques

The culture conditions were essentially as given in Boraas (1983). A single culture vessel was used rather than a two-stage arrangement (Chap. 2.2). The inorganic medium consisted of a modified Chu-10 with 0.5 mM nitrate as the factor in shortest supply for the algae. The medium, flow rate, and light levels were held constant; and temperature was maintained at $25 \pm 0.05\,°C$. *Chlorella pyrenoidosa* (UTEX No. 26) was the prey alga and *Brachionus calyciflorus*, the predator. The dilution rate, D, was 0.66 day^{-1}.

Both the algae and rotifers were counted and sized with a Model ZBI Coulter Counter and Model 100 Channelyzer to determine the biovolumes. This method was typically reproducible to $\pm$ 1–2% for the algae and to $\pm$ 2–3% for the rotifers (Boraas 1983), although the errors tend to increase as the population levels decline. The biomass was calculated from biovolume-to-biomass (dry weight) conversion factors of 1.01 and 0.57 $g\,cm^{-3}$ for rotifers and algae, respectively (Boraas 1983). Rotifer age categories were determined from the particle size distributions as described in Boraas (1983). Nitrate was determined from frozen samples by the cadmium reduction method (Wood et al. 1967).

The experimental design was as follows. Algae were inoculated into the culture vessel and allowed to come to a steady state at a concentration of 110–114 $\mu g\,ml^{-1}$, which reduced the input nitrate concentration of 7.0 $\mu g\,N\,ml^{-1}$ to undetectable levels. Rotifers were then inoculated into the culture vessel.

The system went through initial oscillations of all three variables and then entered a steady state (data not shown). The data presented in this chapter extend from the time the three-component system (nitrate, algae, rotifers) had been stable for ca. 15 days. At this time there was a low algal concentration of 3–4 $\mu g\, ml^{-1}$, a rotifer density of 6–8 $\mu g\, ml^{-1}$, and a high nitrate concentration of ca. 5 μg-N ml^{-1}, such that the bulk of nitrogen in the system was inorganic.

The system achieved a steady state with respect to the rotifers, algae, and nitrate. The rotifer-specific growth rate, r, was calculated from the changes in the rotifer and algal concentrations during the transient dynamics after inoculation of the rotifers by assuming exponential growth (or decline) between samples plus the dilution loss, D (see Fig. 6.1.1 and associated text).

The system stability was experimentally challenged by two pulse experiments, to test the predictions of the Lotka-Volterra predator-prey models (described below). The system was tested for nutrient-saturated algal growth by pulsing it with nitrate. Subsequent to the nitrate pulse, the system was perturbed with a pulse of algae. The nitrate pulse was produced by injecting filter-sterilized concentrated calcium nitrate directly into the culture vessel. The algal pulse was produced by injecting 100 ml of algal suspension, which was growing in a parallel unispecific algal chemostat maintained under environmental conditions identical to the mixed-species chemostat. This resulted in approximately an eight fold increase in the algal concentration in the mixed-species chemostat and a dilution of the nitrate and rotifer concentrations.

In these experiments, each pulse represented a new set of initial conditions for the culture. If the Lotka-Volterra model were even approximately valid, then resetting the initial conditions should, as a minimum effect, result in a change in the predator-prey dynamics. A return of the system to the steady state existing before the perturbation would effectively falsify the Lotka-Volterra model.

3.2.3 Rotifer Growth Dynamics in Single-Stage Chemostats

The rotifer-specific growth rate is shown as a function of algal density in Fig. 3.2.1. The entire range of algal densities observed in the study reported here are indicated on the figure. The line plotted in Fig. 3.2.1 is a regression fit of a rectangular hyperbola using the Lineweaver-Burke linearization. Over the range of algal biomass during the algal perturbation, 0.5–25 $\mu g\, ml^{-1}$, the relationship with r was approximately linear.

3.2.4 Single-Stage Dynamics After a Pulse of Nitrate

As an experimental system, the nitrate-algal-rotifer interaction was remarkably well-behaved. The culture demonstrated very little long-term drift or short-term noise. When the system was in a steady state, the noise was roughly within the

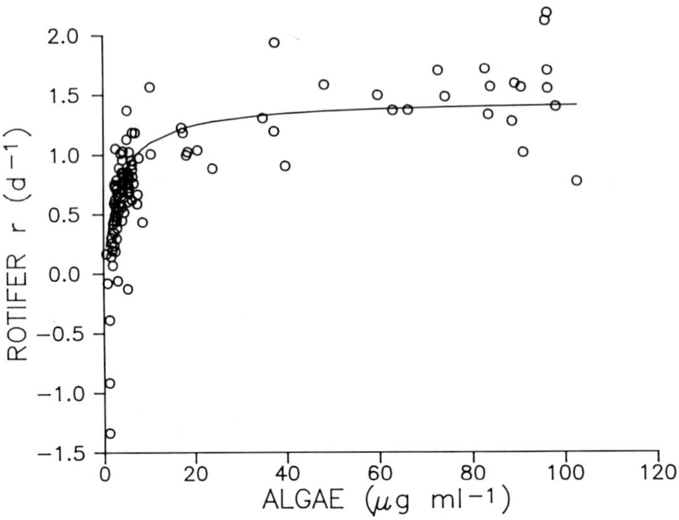

Fig. 3.2.1 Rotifer-specific growth rate vs. algal concentration. These data were calculated from changes in rotifer and algal concentrations during transient dynamics of the predator-prey culture after the rotifers were inoculated

limits of measurement error. Therefore, the system could be used as its own control before and after the pulse perturbations.

The time course of the mixed-species culture before and during the pulse of nitrate is shown in Fig. 3.2.2. There was no response by either the algal or rotifer populations that can be unequivocally attributed to the nitrate pulse, indicating that algal growth was saturated with respect to nitrate.

Since neither the algae nor the rotifer populations responded to the nitrate pulse, algal growth must have been saturated with respect to nitrate. There is no evidence that another factor was growth limiting to the algae. Hence, it seems reasonable to assume that the algae were growing at μ_{max}. Based on previous evidence, the nitrate concentration was sufficient throughout the study to maintain this strain of *C. pyrenoidosa* at μ_{max} (Williams 1971).

3.2.5 Single-Stage Dynamics After a Pulse of Algae

The time course of the mixed-species culture during and after the pulse of algae is shown in Fig. 3.2.3. These data were collected from the same culture immediately after the nitrate pulse shown in Fig. 3.2.2. The rotifer and algal populations and the nitrate concentration all followed smoothly damped oscillations with a decreasing amplitude and increasing period that eventually

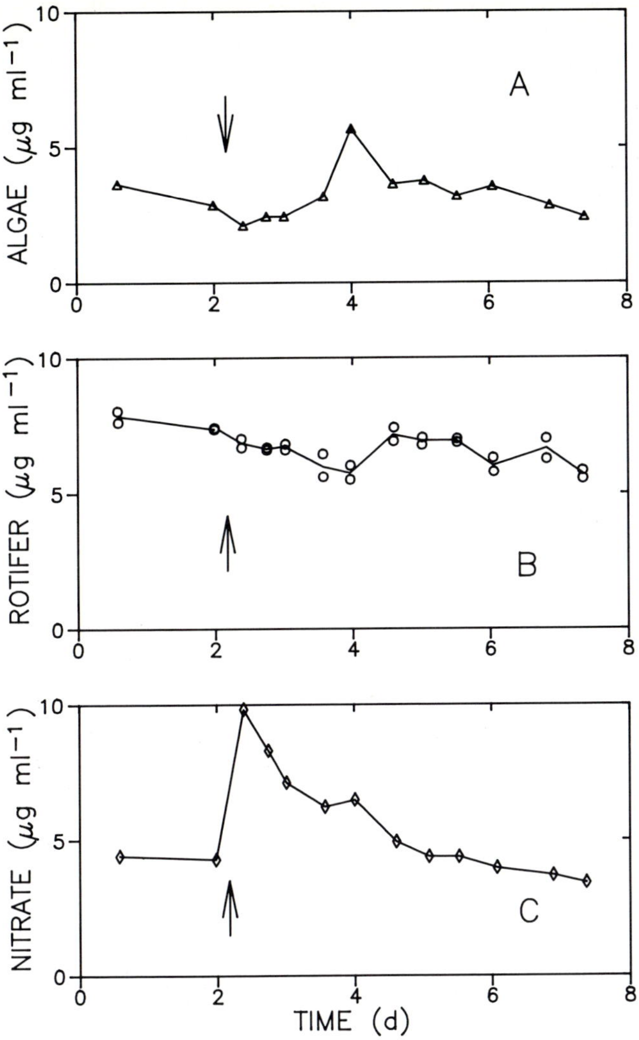

Fig. 3.2.2 A-C Time course of mixed-species culture with a nitrate pulse. Sterile-filtered $Ca(NO_3)_2$ was added to the culture at the *arrow*

returned to a steady state. These data indicate (1) that r was not saturated with respect to algal biomass and (2), that the system was characterized by an extremely stable steady state.

The nitrate-algal-rotifer responses to the algal pulse are shown as three two-by-two phase planes in Fig. 3.2.4. The algal-nitrate and algae-rotifer plots return (approximately) to their original state with little evidence of oscillation. The rotifer-nitrate plot has one loop before the former steady state is regained.

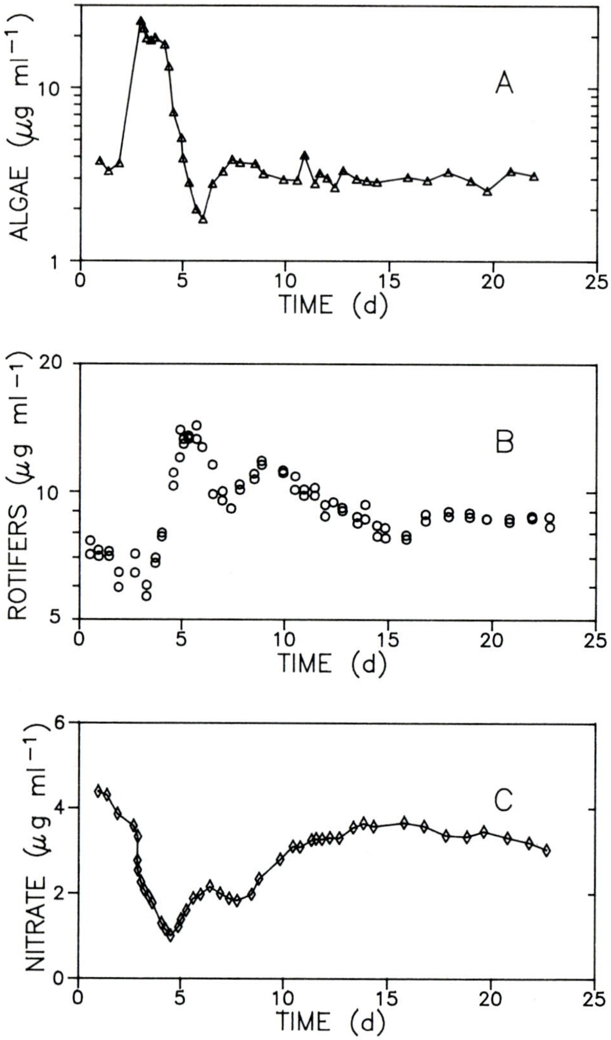

Fig. 3.2.3 Time course of the mixed-species culture following a pulse perturbation of the algal concentration algal were added at the *arrow*. Further explanation in text

The age structure of the rotifer population is shown in Fig. 3.2.5. Following the algal pulse, the egg numbers increased rapidly. This was followed by increases in juveniles and then adults. All of the age components then showed damped oscillations, eventually returning to the steady state. Numerical (Fig. 3.2.5) and biomass (Fig. 3.2.3) oscillations followed virtually the same time course.

Despite the increase in algal density after the pulse (Fig. 3.2.3A), the algal population immediately began to decline, indicating that the algal growth rate did not exceed the rotifer grazing rate. This may have been because the newly

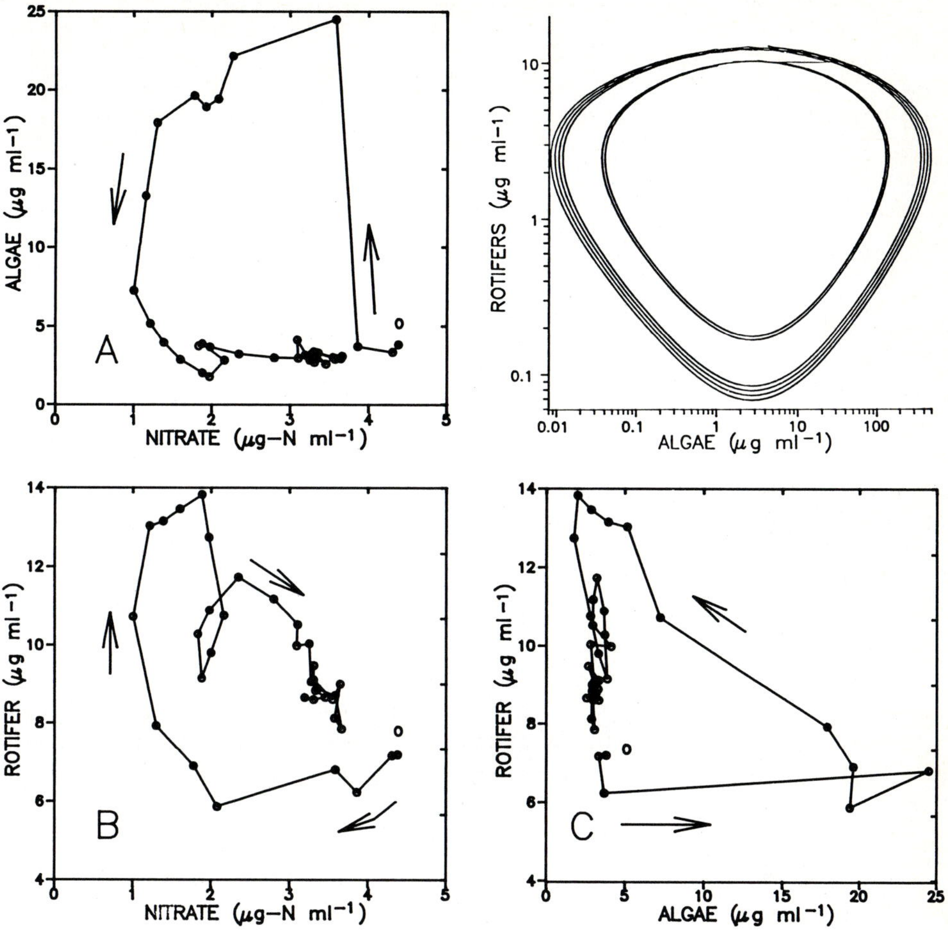

Fig. 3.2.4 Phase planes of the three components. In all two-by-two interactions, the *circle* indicates t_0, and the *arrows* indicate the direction. *A* shows algal biomass vs. nitrate N; *B* shows rotifer biomass vs. nitrate N; and *C* shows rotifer biomass vs. algal biomass. *Top right* a simulation of the Lotka-Volterra model modified to account for saturation of predator growth with increasing algal concentrations as a rectangular hyperbola. Rotifer r_{max} was 1.46 day^{-1}; K_s was 3.35 $\mu g\,ml^{-1}$; algal μ_{max} was 1.8 day^{-1} less the dilution rate of 0.66 day^{-1}; rotifer death was equal to 0.66 day^{-1}. Euler integration with an interval of 0.02 day was used. The inner locus represents initial conditions of 2 $\mu g\,ml^{-1}$ of algae and 10 $\mu g\,ml^{-1}$ of rotifer biomass. After 25 days of simulation, the initial conditions were changed to 20 $\mu g\,ml^{-1}$ of algae and 10 $\mu g\,ml^{-1}$ of rotifers

injected algae were growing at the dilution rate of the chemostat from which they were taken, 0.7 day^{-1}, rather than at their μ_{max} of 1.8 day^{-1}. About 1–2 days after the pulse, a small shoulder in algal concentration became apparent (Fig. 3.2.3A). This may represent a physiological adaptation of these algae to the high-nitrate environment of the mixed-species chemostat. In any case, the algal population declined rapidly thereafter, as the rotifer population increased.

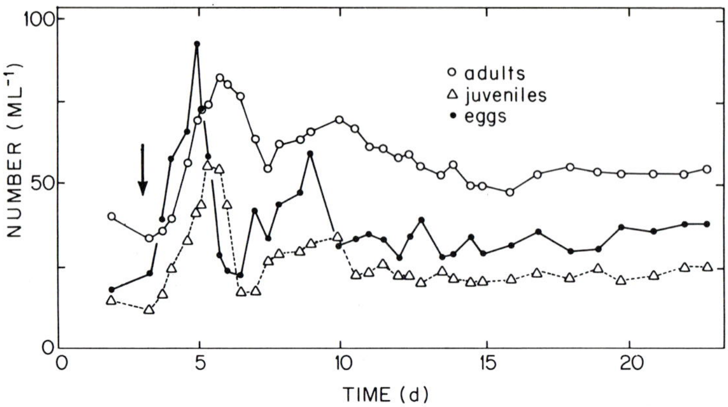

Fig. 3.2.5 Time course of the age structure of the rotifer population following the algal pulse (*arrow*) shown in Fig. 3.2.3

Corresponding to the maximum rotifer level (about 6 days; Fig. 3.2.3b), the algae displayed an undershoot to $0.5\,\mu g\,ml^{-1}$. However, the algal population regained and thereafter maintained its former concentration of $2\,\mu g\,ml^{-1}$ (Fig. 3.2.3A), while both the rotifer (Fig. 3.2.3B) and the nitrate (Fig. 3.2.3C) concentrations continued to fluctuate. In fact, the nitrate time course was virtually the inverse of the rotifer time course, with a lag of about 30 h.

3.2.6 Using a Single-Stage Chemostat to Test the Lotka-Volterra Predator-Prey Model

Since continuous cultures have rigorous constraints on environmental variables, they are particularly suited for testing models of ecological processes. By way of example, the data from this study are presented as a test of the familiar Lotka-Volterra models for predator-prey dynamics. Despite having spawned an immense theoretical literature, the Lotka-Volterra model has seldom been experimentally tested under conditions in which the biological and physical assumptions have been met, or at least approximated to a reasonable degree. As originally formulated, the Lotka-Volterra model makes two major assumptions: (1) prey per capita increase is constant, and (2) predator per capita increase is a linear function of prey density with a constant loss rate (e.g., see Kingsland 1985). No environmental factors (other than the organisms themselves) are explicitly included in the model. An implicit assumption of the model is that the organisms are distributed as if they were chemical concentrations; an assumption that can only be realized if they have very high population densities and are randomly dispersed. Other biologically simplifying assumptions (see Williams 1972) are that the organisms have constant age and size structure.

On close examination, the Lotka-Volterra model is not particularly accurate, nor is it biologically reasonable. For example, May (1975) pointed out that the model has "neutral stability," a condition that is infinitely improbable or "pathological." Neutral stability means that any perturbation will result in new, perpetuated dynamics. May and others have correctly pointed out that virtually any other function(s) for either prey or predator growth, or the inclusion of other environmental factors, will eliminate the condition of neutral stability.

In most experimental systems, the highly restrictive assumptions of the Lotka-Volterra models are difficult to satisfy. However, the assumptions can be reasonably approximated in a single-stage chemostat culture. Specifically, in the data set presented above, algal growth was nutrient saturated. Hence, the algal-specific growth rate, μ, was constant (exponential growth of prey). With a constant mortality rate from system dilution (D), the net specific rate of algal increase in the culture was $\mu - D$. Over much of the range of algal concentrations in this study, the rotifer-specific growth rate, r, was an approximately linear function of algae. The population densities of both species were high, and the culture was well mixed. Finally, the chemostat operation determined the rotifer mortality rate, which was constant and equal to D from dilution.

While quantitative predictions are difficult, it seems reasonable that if the Lotka-Volterra model can ever approximate reality, it should hold qualitatively for a biological system that physically represents the model. At the very least, it should provide qualitative predictions for the stability characteristics of the predator-prey system. The Lotka-Volterra model predicts that the system should be extremely sensitive to perturbations and that it will readily enter into sustained oscillations. If the system is perturbed, then it should not return to its former state, at least not rapidly.

Although both the physical and biological assumptions of the Lotka-Volterra model were satisfied to the extent that they reasonably could be, the outcome of the algal pulse perturbations directly conflicts with two predictions of the Lotka-Volterra model. First, the Lotka-Volterra model predicts that populations will oscillate indefinitely. In this study, when the system was perturbed from a steady state, it returned to the steady state via one or two highly damped oscillations. Secondly, the Lotka-Volterra model predicts that if the predator-prey system is perturbed from one state, it will enter a new state. In this study, the system returned to its former state after being perturbed (or was unaffected in the case of the nitrate pulse). It seems highly unlikely that the return to the former steady state was simply due to chance. The trajectories of the phase plane plots (Fig. 3.2.4) are very direct and unequivocal, with the exception of the rotifer-nitrate phase plane. This plot (Fig. 3.2.4B) shows a small loop, probably due to the lag between the mass transfer of nitrate to the algae to the rotifers. Therefore, the Lotka-Volterra predator-prey model is inconsistent with the observations and must be considered to be falsified, at least with regard to these zooplankton-phytoplankton-nutrient interactions.

The Lotka-Volterra model was simulated by computer to investigate the behavior of the model with measured coefficients. The phase plane patterns resulting from one such simulation are shown in Fig. 3.2.4 (top right). In this particular run, the assumption of a linear growth response of rotifers to algae was replaced with the more realistic assumption of a hyperbolic response. As for the unmodified Lotka-Volterra predator-prey model, the simulations predicted sustained oscillations with increasing amplitude (inner locus of points in Fig. 3.2.4, top right), when the initial conditions were set at a low algal biomass (2 $\mu g\, ml^{-1}$) and the steady-state rotifer biomass (10 $\mu g\, ml^{-1}$). When the initial conditions were reset to a higher algal biomass (20 $\mu g\, ml^{-1}$) with the same rotifer biomass, the amplitude increased. Simulations were qualitatively very similar when the rotifers were assumed to grow as a linear function of the algal biomass.

3.2.7 Stability of the Continuous Culture System not Predicted by the Lotka-Volterra Model

While the Lotka-Volterra model is not very useful for describing this planktonic interaction, its lack of fit has considerable interest. Clearly, the model must be modified or replaced. Adding a saturating growth function for the predator did not substantially improve the fit of the model. Incorporating nitrate as a dependent variable, while essential for describing the system under other environmental conditions, would have little effect for this particular data set, since the algae were nitrate saturated. Evidently, the predator-prey interaction is not accurately described by the Lotka-Volterra model. Specifically, the system is much more resistant to perturbation than predicted by the model.

One likely reason for the stability of this highly interactive system is that the system is sensitive to adverse changes in the rotifer physiological state and demography at low algal levels. These rotifers are not starvation tolerant. Indeed, egg production fell off sharply (Fig. 3.2.5) when the rotifer biomass peaked (Fig. 3.2.3B) and the algal biomass was minimal (Fig. 3.2.3A). In the context of the analysis of the Lotka-Volterra model, the physiological state could profoundly affect both prey capture rate (assumed to be proportional to the product of prey and predator densities) and the birth rate of the predator (assumed to be a simple direct function of prey capture rate). Starving rotifer populations exhibit a reduced ability to capture and grow on algal prey (Boraas 1983; Chap. 5.2). This is revealed as an increase in the estimated value of Ks as *Brachionus* when r declines at low algal food levels. In the Monod-Herbert growth model, K_s is an inverse measure of growth ability (see Chap. 5.2). A high K_s indicates a reduced ability to capture and grow on the limiting food. Changes in the physiological state vs. r are difficult to establish and quantify without steady-state populations, whether in single-state or two-stage culture. However, such physiological variability may be of great importance in the individual species dynamics and maintenance (or its lack) of species diversity in natural systems.

Recently, McCauley and Murdoch (1990) observed that enriched vs. nutrient-poor *Daphnia*-algal interactions were more stable than predicted by models analogous to the Lotka-Volterra predator-prey model. They suggested that this stability may be due either to changes in species content (inedible algae replacing edible algae with enrichment) or "...some universal stabilizing process that is missing from current models." The data presented here indicate that there may be an upper bound to enrichment in two-species predator-prey systems. That is, beyond a certain level of enrichment, no further biological changes will occur, only an increase in nutrient concentration. (In a multi-species system, such "excess" nutrient would doubtless be consumed by other, perhaps subdominant, algal species that may or may not be edible.) Secondly, the biological mechanism for McCauley and Murdoch's "universal stabilizing process" could well be changes in the physiological state of the herbivore. If the feeding and growth potential of the predator declines with the rate of supply of algal biomass per unit predator biomass, then such predator-prey systems should be inherently stable (tending to a steady state), rather than unstable (tending to oscillations). Such a decline in growth potential is discussed in Chap. 5.2 and Boraas (1983) as an increase in the K_s of rotifers with a reduced r in steady-state two-stage chemostats. However, demographic lags, i.e., responses in egg production to changes in algal supply, may add a destabilizing effect.

In summary, above a certain threshold, further enrichment of a well-mixed, spatially homogeneous, nutrient-algal-zooplankton system should have little effects on the overall stability of the system. The primary mechanism of this stability is hypothesized to be alterations in the predator physiological state, expressed as a reduced growth potential, at low predator-specific supply rates of algal food. Increased system stability due to changes in the predator physiological state may overwhelm the potentially destabilizing effects from demographic lags.

This is contribution no. 369 from the Center for Great Lakes Studies.

3.3 Characteristics of Two-Stage Chemostat Cultures of *Brachionus angularis*

NORBERT WALZ

3.3.1 Introduction

Rotifers can grow in a chemostat culture vessel like unicells when provided with a suitable growth substrate, e.g., edible algae. In contrast to bacteria and protozoa with simple cell division, rotifers have a more complicated generation cycle. After a short time of development, rotifers produce eggs carried by the mothers until hatching in most cases. This period of embryonic development (D_e) is often called the "egg development time." After this period, the eggs develop into free-swimming young. After the period of juvenile development, they become adults and lay eggs themselves. The population growth rate is dependent both on the length of these periods and the number of eggs laid per time interval.

According to Paloheimo (1974), the specific birth rate is defined by:

$$b = \frac{\ln[(E/N) + 1]}{d_e} \text{ (day}^{-1}\text{)} \tag{3.3.1}$$

where ln = natural logarithm, E/N = egg ratio = eggs per females, D_e = period of embryonic development (days).

Among monogonetic rotifers, normally all individuals are parthenogenetic, amictic females (Gilbert 1974). Haploid males appear only in special cases depending on environmental conditions (e.g. temperature and unlimited food concentrations; Rothbucher 1988). Males of *Brachionus angularis* do not appear in the food-limited steady-state phases of chemostate cultures. Furthrmore, this strain loses its capability to produce resting eggs with mictic females after a few months of steady-state cultivation.

The birth rate is the potential growth rate of metazoans (derivation in Appendix). It is the rate of production of new self-standing individuals (Fig. 3.3.1). But metazoa, unlike potentially immortal unicells, suffer individual death. Therefore, the increase of density of organisms cannot be calculated by birth rate alone (see below). Increase of the population, in terms of biomass, has to be described by Eq. 3.3.2:

$$\frac{dX}{dt} = r * X, \tag{3.3.2}$$

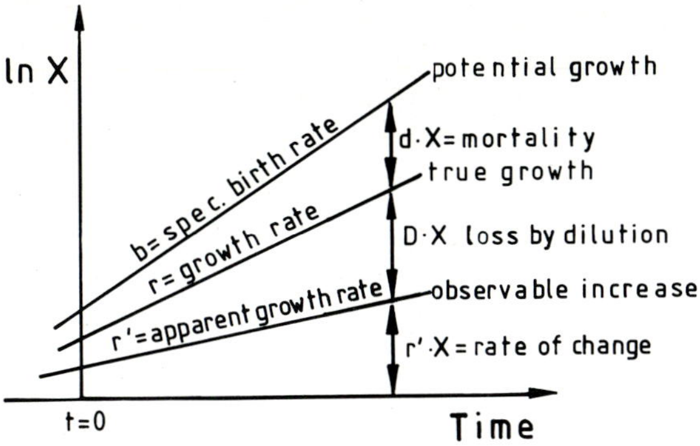

Fig. 3.3.1 Potential and real increase of numbers of metazoan organisms in chemostats

where r = specific growth rate (day^{-1}), X = biomass concentration of rotifers ($\mu g\, C\, ml^{-1}$) (= N∗W, N = number of rotifers, W = mean weight of rotifers).

In a homogeneously mixed, two-stage chemostat, a steady flow of algae suspension is pumped into the rotifer reactor (Fig. 2.2.1). A steady outflow at the same rate leaves the vessel with rotifers and algae of the same concentration as inside the reactor. If Q = hydraulic flow rate ($ml\, day^{-1}$) and V = constant volume content of the reactor (ml), then Q/V = D = dilution rate (day^{-1}). The mean residence time of any organism (rotifers and algae) in the culture vessel = 1/D (days). Therefore, a nongrowing organisms is washed out of the reactor by a negative exponential function:

$$X_t = X_0 * e^{-D \cdot t}$$

and

$$-\frac{dX}{dt} = D * X. \qquad (3.3.3)$$

In a continuous culture vessel, the apparent change of concentration of an orgaism (r′) is described according to Herbert et al. (1956):

$$r' = \frac{dX}{dt} = r * X - D * X \qquad (3.3.4)$$

where r′ = apparent growth rate = observable increase or decrease (day^{-1}) in a flow-through vessel.

For $r > D$, the concentration of organisms increases: $r' > 0$. For $r < D$, the concentration of organisms decreases: $r' < 0$. If $r = D$, no apparent change of concentration of organisms occurs: $r' = 0$. This is the steady state. Therefore, the specific growth rate of organisms may be easily measured by the dilution rate of the system by volume measurements.

Figure 3.3.1 shows the varying increase in numbers according to different growth and birth rates. In continuous cultures, only the apparent growth rate is observable as an increase or decrease of the numbers of organisms. $X*D$ organisms will be washed out of the system.

Monod (1942) stated the following energetic relationship:

$$\frac{dX}{dt} = -Y\frac{dC_2}{dt}, \tag{3.3.5}$$

where Y = yield coefficient, C_2 = food concentration in the reactor.

According to the derivation in the Appendix, in the steady state (Monod 1942)

$$X = Y(C_1 - C_2), \tag{3.3.6}$$

where C_1 = inflow food concentration, substrate concentration in the reservoir, and ingestion is X/Y. Ingestion is higher with higher consumer biomass and with lower yield coefficient, i.e., with more food consumed per biomass developed.

The Monod model functions by a limitation of the substrate. Monod (1942) also found a kinetic relationship between the growth rate of microorganisms and the concentration of the limiting substrate by a saturation function analogous to enzyme-caralyzed chemical reactions formulated by the kinetics of Michaelis and Menten (1913):

$$r = \frac{r_{max} * C_2}{K_s + C_2} \ (\mathrm{day}^{-1}) \tag{3.3.7}$$

where K_s = substrate concentration for $r_{max}/2$ ($\mu g\,C\,ml^{-1}$).

Although the Monod model has no real theoretical basis (in contrast to the Michaelis-Menten equation, due to the nonconstant biomass), it agrees with known observations (Powell 1956); for rotifers, see many examples in this volume. The quality of this relationship is the reason for the stability of the system in the steady state. When a culture is started, food in the rotifer reactor is in overflow. Rotifers grow at their maximum rate, consuming more food until only a particular food concentration is available to balance the outflow and other losses. The growth is a self-regulating process. In many cases, it was possible to describe adequately the growth process of microorganisms by three growth "constants": r_{max}, K_s, and Y (Herbert et al. 1956). This model will be applied to rotifers (*Brachionus angularis*) in a two stage chemostat described in Sect. 2.2.2 with *Stichococcus bacillaris* as food.

3.3.2 The Steady-State Rotifer Density

B. angularis in a steady state doubles between $D = 0.05$ and $0.1\ \mathrm{day}^{-1}$ from ca. 75 to nearly 150 rotifers ml^{-1} (Fig. 3.3.2). There is no significant increase with higher dilution rates, but above $D > 0.3$ (day^{-1}), the density of rotifers drops

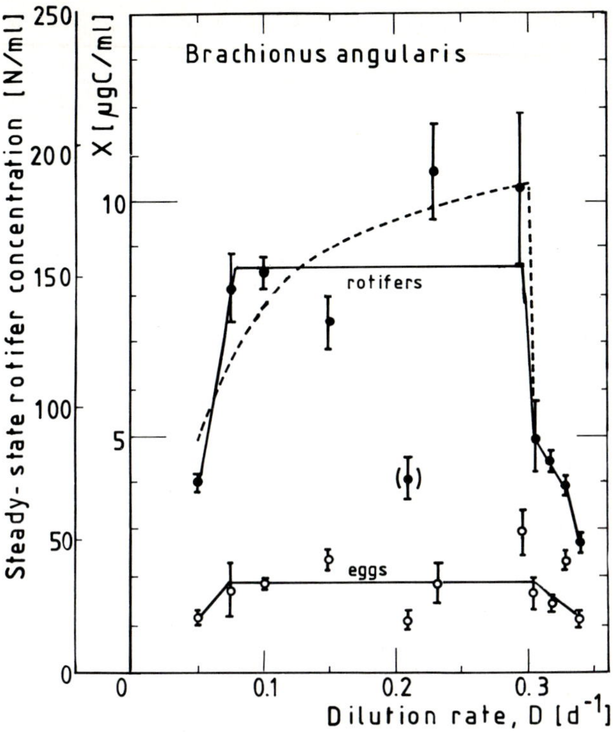

Fig. 3.3.2 Density of rotifers and eggs in the steady state of rotifer chemostats in relation to the dilution rate ($\pm 95\%$ confidence intervals of the means) *solid lines* Acceptance of a constant mean biomass between o = 0.075 and o = 0.3. *Dotted line* Calculated rotifer biomass according to Eqs. (3.3.6), (3.3.8) and (3.3.12); see Sect. 3.3.6

abruptly. These steady states are as stable and constant as the others. At dilution rates >0.34 (day^{-1}) the rotifers are washed out of the reactor. The egg concentration (eggs ml^{-1}) also follows this pattern. The same relationship of rotifer density to dilution rate was found by Boraas (1983) and for some carbon-limited bacteria and fungi by Herbert (1958). In contrast, other bacteria cultures have not shown this density reduction at low dilution rates (Herbert et al. 1956), presumably because they were not C limited (Herbert 1958).

Two parameters can be varied and may control the adjustment of the steady-state rotifer and egg densities: C_1 and D. This shall be studied by a correlation matrix (Table 3.3.1). Keeping one factor constant, partial correlation coefficients result for the rotifer population density as follows:

for C_1 = constant, R = 0.736*.
for D = constant, R = 0.815*.

Therefore, the rotifer density is more dependent on the inflow food concentration than on the dilution rate. Both parameters together produce a multiple linear

Table 3.3.1 Pearson correlation coefficients between steady-state rotifer density (N), egg density (E), egg ratio (E/N), and the independent variables, dilution rate (D) and inflow food concentration (C_1), in chemostat cultures of *Brachionus angularis*

	E	Egg ratio	D	C_1
N	0.46	−0.42	0.72*	0.58
E	–	0.72*	0.44	−0.11
Egg ratio	–	–	−0.03	−0.53*
D	–	–	–	0.10

* $P < 0.05$, significant relationship

correlation coefficient, $R = 0.882$ ($n = 14$, $P < 0.0001$); they determine 78% of the variability of rotifer density.

The following partial correlations were measured for the egg density:

for C_1 = constant, $R = 0.44$,
for D = constant, $R = -0.13$.

Both partial correlationships are not significantly different from zero. The same is true for the multiple correlation with $R = 0.453$. Therefore, the egg density is independent of C_1 and D.

In contrast, the egg ratio (E/N) is inversely dependent on the C_1 inflow concentration and independent of the dilution rate. The following partial correlation coefficients are found:

for C_1 = constant, $R = 0.02$,
for D = constant, $R = -0.60$*.

As the density of individuals increases with C_1 and the egg density remains constant, the egg ratio has to decline with C_1. Further experiments show that the egg ratio (it is proportional to b at constant temperatures, Eq. 3.3.1) increases with the C_2 reactor food concentration (Chap. 4.1).

3.3.3 The Steady-State Algae Density (C_2)

The density of food algae in the rotifer reactor is dependent on the dilution rate. It is low at low D. Its concentration becomes moderately denser until $D > 0.3$, above which very high concentrations of algae are measured (Fig. 3.3.3). The increase when $D < 0.3$ may be described by a linear regression:

$$C_2(\mu g\,C\,ml^{-1}) = 0.61 + 0.35 * D$$

with $R = 0.67$, $n = 14$, $P < 0.01$.

Boraas (1993) found steady-state algae densities in *B. calyciflorus* cultures at low dilution rates to be a little bit higher than at mean D, and opposite to

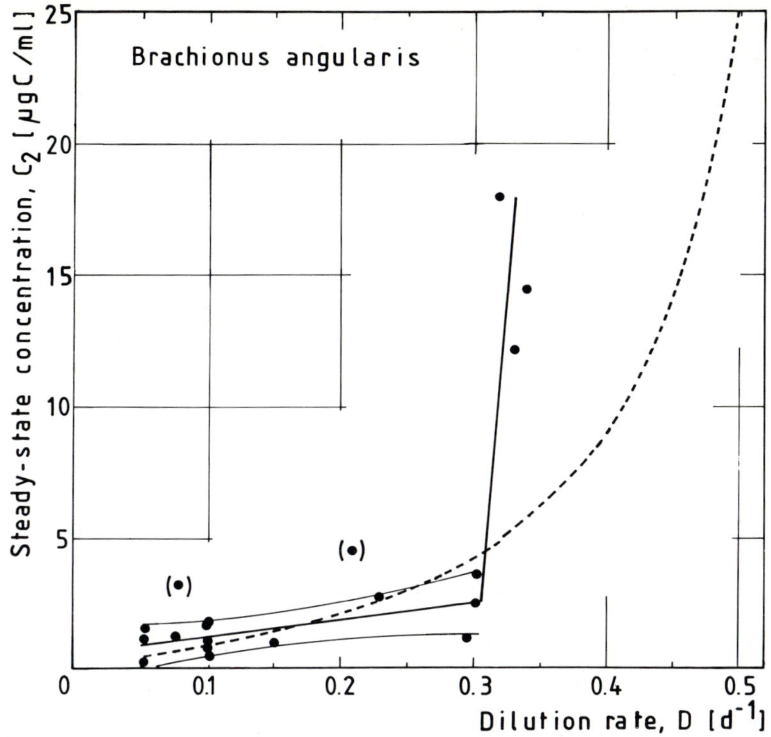

Fig. 3.3.3 Steady-state algae concentration (C_2) in the rotifer chemostat in relation to the dilution rate ($\pm 95\%$ confidence interval of the regression)

rotifer numbers in the reactor. The increase of the reactor concentration at higher dilution rates was also shown for bacteria in axenic cultures, while it was independent of the C_1 infllow concentration (Grady et al. 1972).

With

$$C_2 = \frac{K_s * D}{r_{max} - D} \tag{3.3.8}$$

and $K_s = 4.05$ (μg C ml^{-1}), $r_{max} = 0.58$ (day^{-1}) (Chap. 4.1, steady and transient states), the dashed line in Fig. 3.3.3 was calculated. (Herbert et al. 1956). For $D < 0.3$, the calculated line is inside the 95% confidence limits of the linear regression. For $D > 0.3$, the calculated line follows the data of the transient states, see Chap. 4.1.

C_2 may be clearly accepted as unused algae after grazing, which equals the residual algae concentration:

$$C_2 = C_1 - \text{grazing} = C_1 - \frac{i * X}{D} \tag{3.3.9}$$

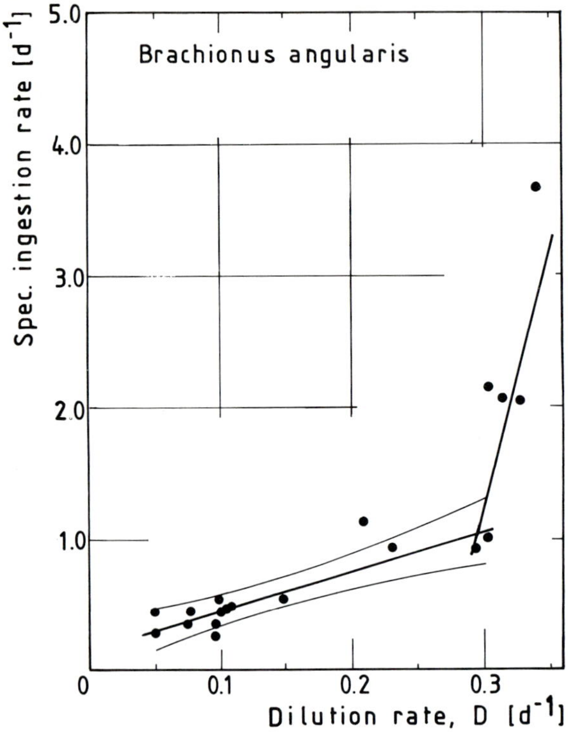

Fig. 3.3.4 Specific ingestion rate of the rotifers in the chemostat in relation to the dilution rate ($\pm 95\%$ confidence interval of the regression)

where i = specific ingestion rate (ingestion $*$ biomass^{-1} $*$ day^{-1}); for the derivation, see Appendix.

This becomes obvious due to the clear correspondence between the course of C_2 to the dilution rate and the course of the specific ingestion rate (Fig. 3.3.4) with this factor. The linear correlation between C_2 and i reveals a highly significant Pearson coefficient of $R = 0.84$ ($n = 19$, $P < 0.0001$).

The ingestion rate first increases slowly with the dilution rate, but above $D > 0.3$, this increase becomes very steep. In contrast, Boraas (1983) observed in chemostats of *B. calyciflorus* a continuous increase in ingestion over all dilution rates. Both curves of Fig. 3.3.4 may also be described by the following linear regressions:

For $D < 0.3(\text{day}^{-1})$, $i = 0.16 + 3.01 * D$
$R = 0.90$, $n = 15$, $P < 0.0001$,

For $D > 0.3(\text{day}^{-1})$, $i = -13.38 + 48.82 * D$
$R = 0.86$, $n = 5$, $P < 0.05$.

3.3.4 Relationship Between Steady-State Rotifer Density and Food

At first, the rotifer population density should depend on the food concentration in the reactor. But in steady states, the following inverse relationship between X and C_2 was observed, where a linear regression gives:

$$X = 7.80 - 0.26 * C_2$$

with $R = -0.51$, $n = 19$, $P < 0.015$.

This relationship is weak and only shows the general trend. The rotifer biomass is better defined by the amount of algae, available, which is the difference between the input food concentration, C_1, and the reactor concentration, C_2 (Fig. 3.3.5).

$$\text{According to Eq. (3.3.9): } C_2 = C_1 - I = C_1 - X/Y \qquad (3.3.10)$$

then it follows from Eq. (3.3.6): $X = Y\ (C_1 - C_2)$. In this context, C_2 is the "unutilizable (residual) food concentration."

Figure 3.3.5 shows the linear relationship between the available algae and the steady-state rotifer biomass:

$$X = -0.16 + 0.22(C_1 - C_2)$$
$$R = 0.64,\ n = 19,\ P < 0.01.$$

When C_2 is small (see Fig. 3.3.3), X is largely determined by C_1, as shown with partial correlation above. The available algae $(C_1 - C_2)$ appear as the "carrying

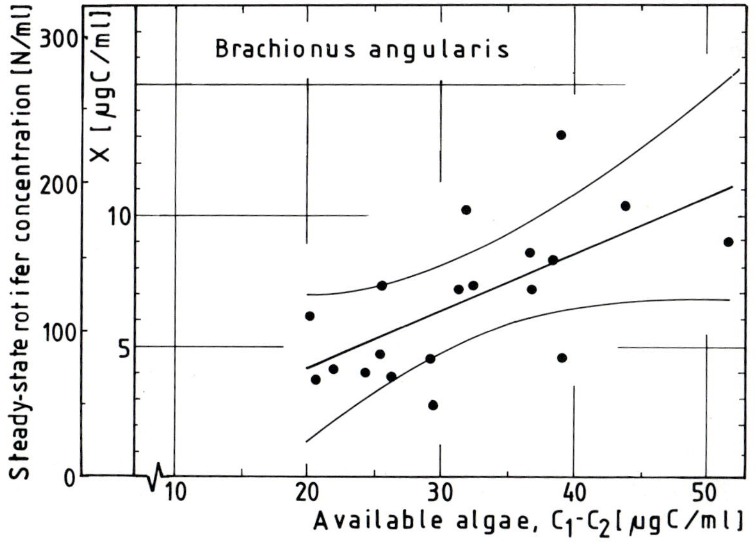

Fig. 3.3.5 The steady-state rotifer density in relation to the available algae (±95% confidence interval of the regression)

capacity" of the system. This factor is nothing else than the ingestion of the population (see Chap. 7.2).

Only 41% of the variation of X is determined by this factor ($P < 0.01$), but all steady states are included. Without the four points outside the 95% confidence limits, one obtains exactly the same slope, but with a correlation coefficient of R = 0.81, explaining 66% of the variability. According to Eq. (3.3.6), the slope is identical to the mean value of Y. If Y were constant, the correlation should be better. However, as we shall see in Sect. 3.3.6, Y is not constant but is dependent on other factors of the population energy balance.

3.3.5 The Utilizable Harvest of a Chemostat

Herbert et al. (1956) introduced two performance criteria of a continuous culture system, the output and the yield. The output rate is the amount of organisms leaving the reactor times the dilution rate ($= D*X$). For many practical purposes, e.g., aquaculture, this value must be maximized. Because of the mortality of metazoa (see Sect. 3.3.7), this output is often lower than the production of the population (for derivation see Appendix production =

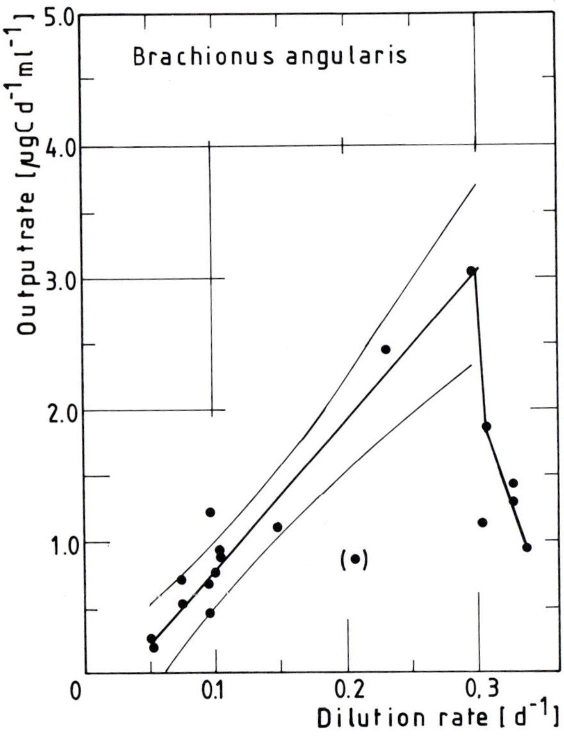

Fig. 3.3.6 The utilizable harvest of rotifers of the chemostat in relation to the dilution rate (± 95% confidence interval of the regression)

output + mortality). Therefore, to point out its connection with the living biomass, this output was termed "utilizable harvest."

In *B. angularis* (Fig. 3.3.6), the output rate increases up to D_m, the maximum dilution rate output for the rotifers. D_m is nearly at $D = 0.3(\text{day}^{-1})$, and the maximum possible output of the rotifers is $3\,\mu\text{g C day}^{-1}\,\text{ml}^{-1}$ at that algal concentration input. Figure 3.3.6 also shows the linear regression calculated for this relationship:

$$\text{Rate of output } (\mu\text{g C day}^{-1}\,\text{ml}^{-1}) = -0.35 + 11.37 * \text{D}$$
$$R = 0.96,\ n = 13,\ P < 0.00001.$$

This line intersects the x-axis at $D = 0.031\,(\text{day}^{-1})$, which is obviously the lowest possible dilution rate assuming no physiological state changes.

This increase and later decrease of the output with dilution rate is similar to chemostat cultures of *B. plicatilis* (Droop 1976) and of bacteria (Tempest et al. 1965). Boraas (1983) found only an increase of the output of *B. calyciflorus* with dilution rate.

3.3.6 The Yield Coefficient of the Energy Metabolism

Y is usually considered to be a constant factor in most chemostat theories (Herbert et al. 1956). It is defined by:

$$Y = \text{yield} = \frac{\text{biomass built up}}{\text{food used}} = \frac{X}{(C_1 - C_2)}. \qquad (3.3.11)$$

In steady states of food-limited rotifers, Y was not constant but increased with the dilution rate as a saturation function up to $D = 0.3(\text{day}^{-1})$. For higher dilution rates, Y values were considerably lower (Fig. 3.3.7).

The saturation function was analogous to Michaelis-Menten enzyme kinetics calculated with lowest squares of deviations (Bliss and James 1966):

$$Y = \frac{Y_{max} * D}{\mu_c + D} \qquad (3.3.12)$$

where

$$Y_{max} = 0.38,$$
$$\mu_c = 0.07(\text{day}^{-1}),$$
$$R = 0.86,\ n = 13,\ P < 0.001.$$

For bacteria, Herbert (1958), too, found that Y decreases at lower dilution rates for a carbon-limiting substrate and postulated a relative increase of metabolic costs for endogenous respiration. Pirt (1965) interpreted the Y–D saturation curve by the following parameters:

$Y_{max} = Y_g$ = "true growth yield", yield for growth (Schulze and Lipe 1964),

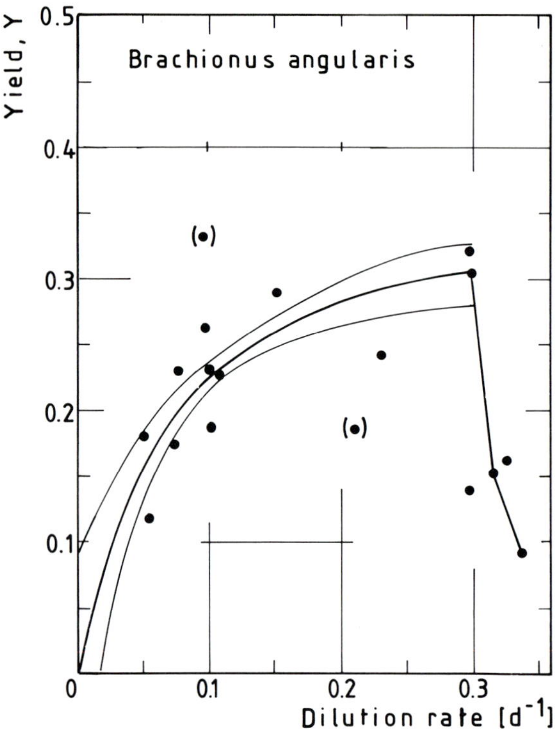

Fig. 3.3.7 The yield coefficient of the growth of rotifers in the chemostat in relation to the dilution rate ($\pm 95\%$ confidence interval of the regression)

μ_c = specific maintenance rate (Marr et al. 1963)
= dilution rate for maintenance.

For derivation of Y_g and μ_c, see Appendix.

Schulze and Lipe (1964) related to μ_c a specific maintenance consumption i_0 as $i_0 = \mu_c/Y_g$. This is the food intake to balance the losses of energetic costs. For *B. angularis*, it was calculated as $i_0 = 0.18$ (day^{-1}), i.e., (μg food-C/μg rotifer-C$*$d).

Hyperbolic relationships as in Fig. 3.3.7 were also found in other rotifers, in *B. calyciflorus* and *B. rubens* by Rothhaupt (Chap. 4.3), in *B. rubens* by Schlüter et al. (1987), and in the marine *Encentrum linnhei* (Scott 1988). For *B. plicatilis*, Droop and Scott (1978) obtained a constant yield over a range of growth rates. Boraas (1983) treated this subject for *B. calyciflorus* rather extensively.

By inserting Eqs. (3.3.8) and (3.3.12) in Eq. (3.3.6), the rotifer density (X) may be calculated with the knowledge of Y_{max}, μ_c, r_{max}, and K_s (dotted line in Fig. 3.3.2). The calculated values (X_y) give a linear regression to the observed values (X_x) of the steady states below the maximum utilizable harvest:

$$X_y = 1.64 + 0.78 * X_x$$

$R = 0.84$, $n = 13$, $P < 0.01$.

The slope at $P = 0.05$ is not signfiicantly different from 1.

3.3.7 Mortality as a Factor of Energy Loss

In contrast to most groups of unicellular organisms, metazoans develop and die. Rotifers reproduce by eggs and have postreproductional senscence, but even younger rotifers may die. Therefore, the "standing crop" of a metazoan population continually loses energy by mortality (Sect. 4.2.1).

Mortality is connected to the birth and growth rates by this basic equation of population dynamics (Lotka 1922):

$$b = r + d \tag{3.3.13}$$

where d = mortality rate (day^{-1}).

The birth rate can be calculated by Eq. (3.3.1). If the mortality were negligible, the birth rate (potential growth rate) would be identical to the actual growth rate. This is not the case for *B. angularis* (Fig. 3.3.8). The difference at lower growth rates becomes greater and is due to mortality. At higher growth rates, the mortality is negligible, and the curve approximates identity. In addition to steady states (r = D), transient states (Chap. 4.1) are considered at higher growth rates in this figure.

The coefficient of the linear regression between *b* and *r* (D) was significantly different from 1:

$$b = 0.25 + 0.57 * r \quad (\text{day}^{-1})$$

$$R = 0.90, \quad n = 25, \quad P < 0.0001.$$

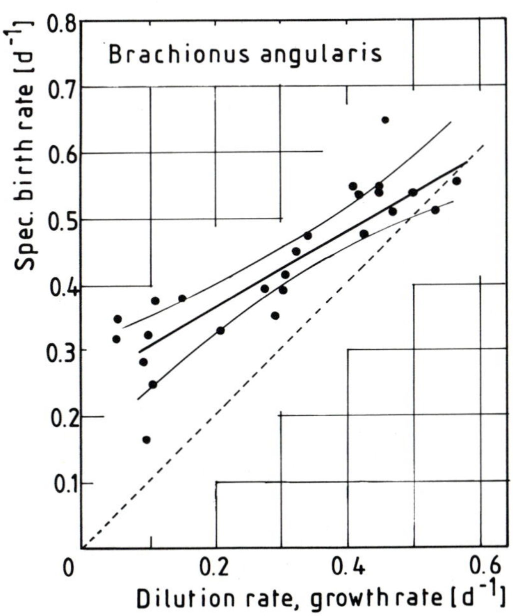

Fig. 3.3.8 Birth (= specific production) rates of rotifers in relation to the dilution or specific growth rate in transient and steady states of rotifer chemostats (± 95% confidence intervals of the regression)

Therefore, in a formulation of a loss rate μ_c (Droop and Scott 1978; Boraas 1983), this term should be involved in the energy balance of a population (Appendix). Beyond this, feces are also contained in this loss (Eq. 4.1.1). Bennett and Boraas (1989c) found no detectable mortality in chemostats of *B. calyciflorus*. They found b and r to be identical, but they worked exclusively on higher growth rates, when mortality is expected to be very low (Fig. 3.3.8).

3.3.8 Limits of the Monod Model for Rotifer Chemostats

According to the chemostat theory of Herbert et al. (1956), the dilution rate of maximum output D_m (see Sect. 3.3.5) is:

$$D_m = r_{max} * \left(1 - \sqrt{\frac{K_s}{K_s + C_1}}\right) \qquad (3.3.14)$$

and D_c, the dilution rate, when the population is washed out:

$$D_c = \frac{r_{max} * C_1}{K_s + C_1}. \qquad (3.3.15)$$

D_m and D_c where calculated with the experimental parameters of r_{max} and K_s (Chap. 4.1) and plotted against a variable C_1 (Fig. 3.3.9). Both lines show first

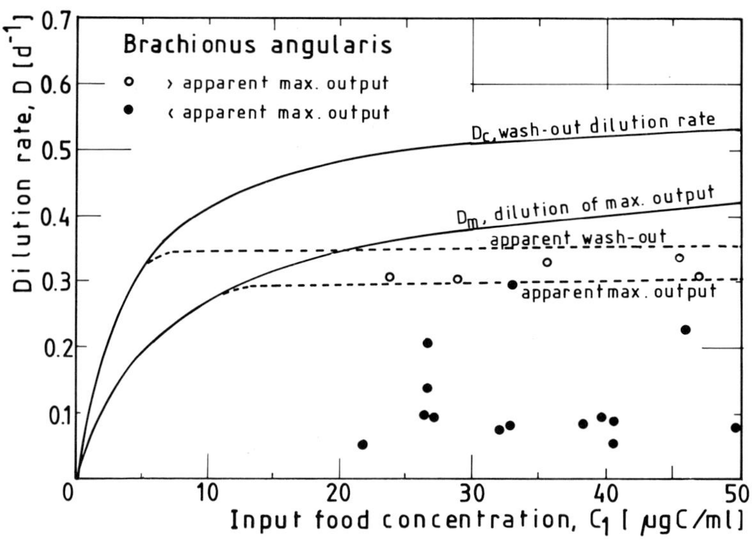

Fig. 3.3.9 Theoretical washout dilution rate (D_c) and theoretical dilution rate with maximum harvest (D_m) in relation to the input food concentration (C_1). Theoretical rates are compared with "apparent" rates

a steep increase. Later on, at higher C_1 values, both curves flatten out. Of course, D_c must be greater than D_m. Recording all experiments, all points are considerably below the theoretical D_m. The apparent D_m is probably (see Fig. 3.3.6) at or near $D=0.3\,(day^{-1})$ (lower dashed line in Fig. 3.3.9). Also, the dilution rate of the washout is lower than the theoretical D_c. The apparent rate seen in Fig. 3.3.2 is nearly at $D=0.34\ (day^{-1})$ (upper dashed line in Fig. 3.3.9). The reason for this disagreement is probably that the Monod constants, r_{max} and K_s, were experimentally determined with transient states of animals growing near r_{max} (Sect. 4.1.4).

Due to the time spent in embryonic and juvenile development, there is a minimum residence time rotifers must be in the reactor to develop their first eggs and to close a generation cycle. A positive net growth of the population requires a balanced mortality of eggs and young on "age of zero growth" (Sect. 6.3.6). It is in the matter of things that such a minimum residence time is only to be found in metazoans.

Derived from the following equations (e.g., in Pianka 1978):

$$R_0 = e^{r \cdot T}$$

where T = generation time and R_0 = net reproductive rate (number of offspring/female during her lifetime), it follows (Droop and Scott 1982):

$$\ln R_0 = r * T, \text{ and } 1/D = T/\ln R_0. \tag{3.3.16}$$

Therefore, $\ln R_0 = 1$, because $r = D$ in the steady state, and the residence time (1/D) must be equal to T. R_0 is the effective multiplication factor before washout, which has to be "e", i.e., 2.72, at any dilution rate (Droop and Scott 1982).

While these theoretical considerations are correct for the case in which the mean age exactly equals the residence time, discrepancies follow due to the broad variances of the mean times of development (both for embryonic and juvenile development) found by Rothbucher (1988) for *B. angularis* and Benker (1992) for *B. rubens* and *B. calyciflorus.* This minimum age is 2.8 days for individual *B. angularis* cultures at 20 °C (Walz 1987a) and relates to a dilution rate of $D = 0.36\ (day^{-1})$; this is exactly the "apparent" dilution rate of washout ($0.34\ day^{-1}$). In chemostat cultures of *B. angularis*, R_0 according to Eq. (3.3.16) and that minimum age is 2.3 days. For *Keratella cochlearis* with an "age of zero growth" of 6.3 days (Walz 1987a), D_c should be $0.16\ day^{-1}$. Nevertheless, a chemostat experiment was operated at $D = 0.15\ (day^{-1})$ with R_0 of 2.57 (Walz 1983a). This R_0 comes near to the above theoretical value.

This minimum age challenges an explanation for the high $r = 1.92\ (day^{-1})$ of *B. calyciflorus* found in the turbidostat bý Bennet and Boraas (1988). From this follows a residence time of 0.52 days. Although the high dilution rate selected drastically for short development periods, a half day seems impossible low. Development times for *B. calyciflorus* given by Benker (1992) suggest that the minimum generation time was not shorter than 1 day. With $T = 0.52$ (days) $r = 1.92\ (day^{-1})$, the fraction not having been washed out at time = T

($e^{-D \cdot T} = e^{-1}$) is 0.36. This fraction sinks with T = 1 (day) to 0.146, while the surviving population requires a R_0 of 6.8 to persist at this dilution rate. As *B. calyciflorus* is able to carry these numbers of eggs, this may compensate for the longer development.

This boundary is a serious limitation for steady-state experiments with metazoans. It intensifies with a longer generation time and with longer embryonic and juvenile development. Cultures of the slower growing *B. angularis* and *K. cochlearis*, therefore, are more difficult to operate than other *Brachionus* species (Chaps. 4.3, 6.1). Disregarding technical obstacles, this renders true steady states in chemostats rather impossible for zooplankton other than rotifers. Perhaps some cladocerans may be able to compensate for their longer minimum age by very high numbers of embryos in their brood chamber. Rothhaupt (Chap. 7.1) mentions further limitations of the chemostat method.

3.4 Turbidostat Culture of *Brachionus calyciflorus*: An Experimental System to Assess Biological Limits on Population Growth

WILLIAM N. BENNETT, MARTIN E. BORAAS, and
DIANNE B. SEALE

3.4.1 Introduction

The turbidostat is a continuous culture device in which the population density of an organism (or its food) is held constant, and the specific growth rate (r) of the population becomes a dependent variable. Turbidostat cultures are powerful tools for assessing biological limits on population growth. In theory, and in practice, if the food level is high enough to alleviate the effects of food limitation, the r in a turbidostat should equal r_{max}, the maximum specific growth rate of the population. Steady states can be readily achieved, in which the growth rate is balanced by losses: dilution, metabolism, and death. Hence, populations can be sustained at r_{max} for extended periods of time to assess key questions concerning biological limits on population growth: What is the maximum growth rate for a species? What are the cost-benefit trade-offs for partitioning energy to achieve maximum growth? How can theory and empirical results be integrated to reveal the nature of the environmental limiting factor? What are practical environmental constraints on population growth?

Bennett and Boraas (1988) were the first to demonstrate that the r of a rotifer population (*Brachionus calyciflorus*) could be controlled in a turbidostat culture and maintained in a steady state indefinitely. In the classical turbidostat, the population density is monitored and regulated by turbidometric means. However, for the rotifer turbidostat, the limiting factor (algal food, as *Chlorella pyrenoidosa* from a separate chemostat) is regulated. The input algal concentration (C_1) is maintained at high concentrations, but rotifers consume algae within the culture vessel, which is held in the dark. The rotifer population is regulated indirectly, by measuring the algal population density optically as turbidity (Boraas and Bennett 1988). A computer is used both to sense turbidity and to regulate the flow rate of the culture. When rotifers cause the algal density to decrease below a set point for residual algal concentrations within the turbidostat (C_2), a volume of algal culture is pumped into the culture, which displaces and washes out the same volume of rotifers and spent media from the vessel. The system rapidly enters a steady state in which r = dilution rate, D (D = media flow rate/culture volume; see Chap. 2.3, this Vol. for details).

We have used chemostat and turbidostat sysems to examine interactions between environmental factors and population growth for rotifers (Boraas 1980, 1983; Boraas and Bennett 1988; Bennett and Boraas 1988, 1989a, b, 1990; Boraas et al. 1990). Our rotifer turbidostat system represents the first turbidostatically controlled metazoan culture. The rotifer turbidostat was used to isolate a population of rotifers with the fastest growth rate recorded for any metazoan (1.85 day^{-1}) after 8 months in the system operated near r_{max} (Bennett and Boraas 1988). The selection process was shown to be genetically determined, with explicit evolutionary trade-offs, not simply a form of physiological adaptation (Bennett and Boraas 1989a, b).

In this chapter, we demonstrate an approach using turbidostat cultures to show how the demographics and physiology of rotifers can change when grown in steady-state turbidostat culture for extended periods. The results will be summarized from rotifer studies both before and after a fast-growing mutant strain appears. Combined turbidostat and chemostat results will be compared with the predictions of a theoretical model to show that these populations are carbon-energy limited. Hence, our data can be used for an empirical analysis of the biological mechanisms for the selection of rapid growth, in terms of energetic cost-benefit trade-offs for achieving maximum growth. We also provide data to show that the selected r_{max} is genetically determined and results in rotifers with a smaller body size, smaller eggs, and a faster egg production time.

The methods and materials follow those described previously in Chap. 2.3 and can be found in more detail in Boraas and Bennett (1988) and Bennett and Boraas (1988). In this chapter, data are presented as dry mass for both rotifers and algae. The steady-state rotifer biomass was computed from measurements of numerical densities of adults, juveniles, and eggs under the microscope. The densities were converted to biovolume and then to biomass. Adult rotifer numbers were multiplied by $1.36 \times 10^6\ \mu m^3$, juvenile numbers by $6.88 \times 10^5\ \mu m^3$, and eggs by $2.86 \times 10^5\ \mu m^3$ to convert microscopic counts (number ml^{-1}) to biovolume (μm^3), after Boraas and Bennett (1988). Rotifer biovolume was converted to dry mass using the constant 0.101 g dry mass cm^{-3} biovolume for *B. calyciflorus* (Boraas 1983; Chap. 3.1, this Vol.). Algal biovolume, measured directly with an electronic particle counter, was converted to biomass as 46 μg dry mass per $10^8\ \mu m^3$.

3.4.2 Selection for Maximum Growth Rate in Rotifers

In theory, since r in the turbidostat is set by the fastest growing individuals, the r_{max} may be subject to selection. As described earlier, rotifers can be maintained in turbidostat culture near r_{max} if they are provided with a high concentration of algal food. This is accomplished easily if the set point for residual algae (C_2) needed to trigger dilution is high. This food concentration must be sufficiently high to saturate feeding and digestion. Selection can occur because slower-

Table 3.4.1. Steady-state values for specific growth rate (r), rotifer density (X), input algae (C1), residual algae (C2), and yield (X/C_1). Values were obtained from two open culture systems: the chemostat ([a]Boraas 1983) and the turbidostat (Boraas and Bennett 1988)

r (day^{-1})	X ($\mu g\,ml^{-1}$)	C_1 ($\mu g\,ml^{-1}$)	C_2 ($\mu g\,ml^{1-1}$)	X/C_1
0.20[a]	10.8	168	18.2	0.064
0.33[a]	15.8	135	20.5	0.117
0.49[a]	22.6	125	13.1	0.181
0.83[a]	17.5	129	11.4	0.136
0.91[a]	22.0	112	18.2	0.196
1.07[a]	23.3	120	20.0	0.194
1.18[a]	35.4	217	13.7	0.163
1.19	6.6	123	39.1	0.054
1.27	7.7	139	48.0	0.055
1.37	3.1	132	101.8	0.024

growing idividuals are unable to match the dilution rate imposed on them by the faster-growing individuals and are removed from the culture by dilution. Differences in the steady-state rotifer population densities and r may be obtained when the system is operated at different values for C_2.

By growing populations of rotifers in turbidostat culture over a period of several months, we were able to document selection for extremely high population growth rates. Initially, measurements of rotifer r_{max} in the turbidostat culture were comparable to previous data in chemostat culture systems. In our first three runs, the rotifers were maintained in turbidostats with three steady-state algal concentrations (C_2; Table 3.4.1). The turbidostatic steady-state r values which resulted (1.19, 1.23, and 1.33 day^{-1}) were very near the chemostatic r_{max} previously reported by Boraas (1983; 1.23 day^{-1}) using the same rotifer species and environmental conditions (food species, temperature, medium, etc.). The earlier value for r was computed from transient data obtained during the approach to a steady state in chemostat cultures and during washout from a chemostat when the dilution rate exceeded r_{max} (see Boraas 1983; Seale et al., Chap. 6.1, this Vol., for calculations and comparable results).

Time courses are plotted for C_2 (residual algal densities) and X (rotifer densities) for cultures at the three different algal concentrations (Fig. 3.4.1). The rotifer populations can clearly maintain a steady state near their r_{max}. Before the steady state, the rotifers essentially grow in batch culture, since $D = 0$. The rotifers consume algae until C_2 reaches the set point, and turbidostat-regulated growth begins. For these particular initial conditions (i.e., high set point and high C_1), C_2 and X attained a steady state soon after the set point was reached (Fig. 3.4.1, particularly C_2).

Corresponding values for r for the three rotifer turbidostats (data in Table 3.4.1, Fig. 3.4.1) were calculated at 10-h intervals using the dilution data

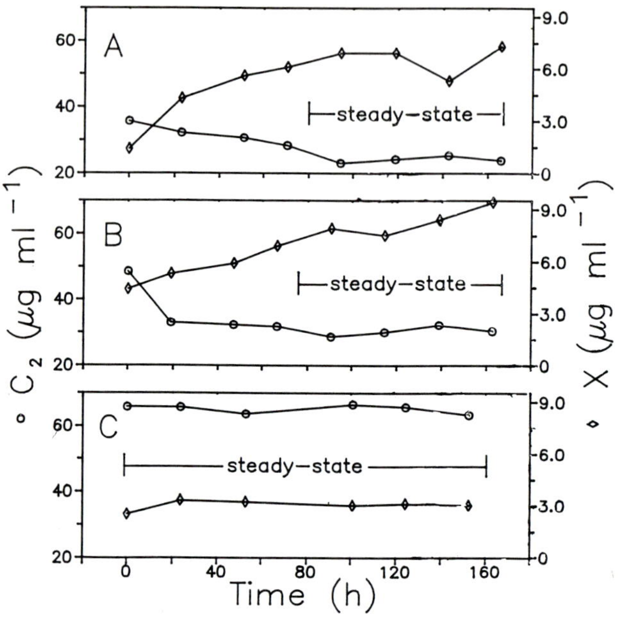

Fig. 3.4.1 A–C Time course of biomass of rotifers (*diamonds*) and residual algae (C_2: *circles*) in three turbidostatic runs. The duration of each steady state is indicated on the figure. Rotifers include juveniles and adults, but not eggs. Average steady-state specific growth rates were 1.19 day^{-1} (**A**), 1.27 day^{-1} (**B**), and 1.37 day^{-1} (**C**); corresponding values for time courses for r are given in Fig. 3.4.2

collected by computer at 15-min intervals (see Sect. 2.3.3). The resulting time courses (Fig. 3.4.2) show that r also tends to reach and maintain a steady state. However, the r of the population grown at the highest C_2 (Fig. 3.4.2C) appeared to vary cyclically in time. The peaks in r for rotifers at the highest C_2, including a rate of 1.7 day^{-1}, were maintained for several generations. These observations are consistent with the hypothesis that the population experienced selection for increased r.

A subsequent experiment showed that sustained growth at or near r_{max} can select for an increased r_{max} (Bennett and Boraas 1988). Rotifers were grown near r_{max} for 32 months. The r_{max} of the population at the start of the experiment was 1.3 ± 0.05 day^{-1} (SD; $n = 13$; doubling time = 0.53 day) (Bennett and Boraas 1988). After about 1 month in culture, the population r_{max} began to increase until values near 1.7 day^{-1} (doubling time = 0.40 day) were observed (Fig. 3.4.3). For practical reasons, the culture was temporarily placed into a chemostat mode at a constant r of 1.23 day^{-1} for 1.7 months. At 5.7 months the culture again was placed into turbidostat mode. The population r_{max} again was observed to increase. However, the rate of increase was faster than earlier observations. At approximately 8 months, the r_{max} reached and maintained a mean value of

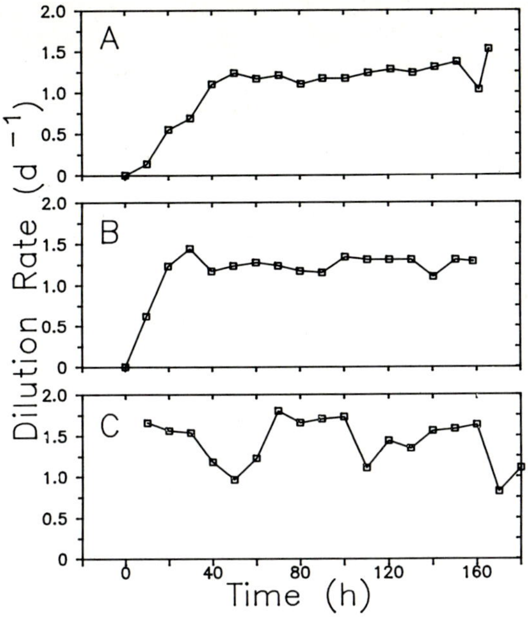

Fig. 3.4.2 A–C Time course for specific growth rate (dependent variable in the turbidostat) averaged over 10-h intervals, corresponding to data on rotifers in Fig. 3.4.1

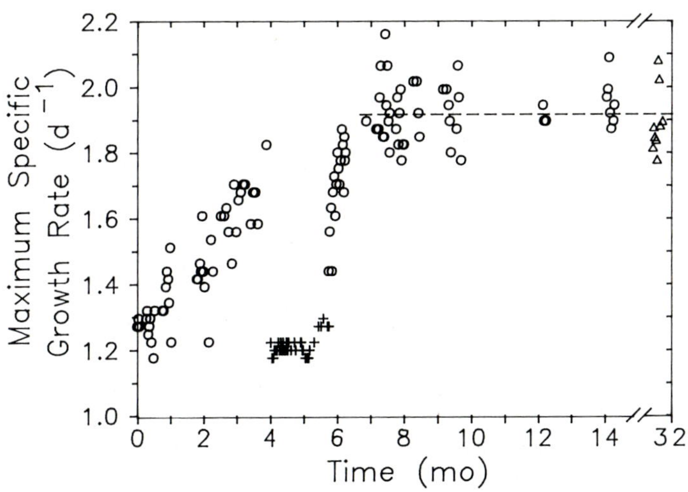

Fig. 3.4.3 Independent measurements (via daily outflow volumes) of the maximum specific growth rate, r_{max}, of the rotifer population in turbidostat culture over a 14-month period (*circles*). The culture was shifted to chemostat mode and maintained at a specific growth rate of 1.23 ± 0.02 day^{-1} (SD; $n = 34$) between 4 and 5.7 months (*crosses*). After reinstating turbidostat mode, the r_{max} reached an apparent plateau of 1.85 ± 0.1 day^{-1} between 8 and 14 months (*dashed line*). After an additional 18 months in turbidostat mode, the r_{max} remained at this value (r_{max} at 32 months: *triangles*)

1.85 ± 0.12 day^{-1} ($n = 8$; doubling time = 0.35 day). This value represented a 51% increase over the initial value and is the fastest growth rate yet reported for a metazoan (Banse 1982b; Bennett and Boraas 1989a, b). After 11.5 months this population was lost, apparently due to a *Vorticella* sp. contaminant.

The culture vessel was then reinoculated with the fast-growing rotifer strain which had been maintained in subculture in a chemostat. A r_{max} of 1.85 day^{-1} again was attained (data shown at 14 months in Fig. 3.4.3). This fast-growing strain had been maintained in a chemostat ($D = 0.22$ day^{-1}) for 24 months when it was reintroduced into turbidostat culture. As in previous cultures, this culture attained the plateau r_{max} of 1.9 ± 0.1 day^{-1} for an additional 31 months (Fig. 3.4.3).

In another test, the fast-growing strain was also subcultured in batch culture, where r_{max} was measured from increases in rotifer biomass with time (data in Fig. 1, Bennett and Boraas 1989a). Two sets of batch cultures were established. One set (2 replicates) received an inoculum from a population which had been maintained at a r of 0.22 day^{-1} for 5 months at that time. The other set received an inoculum from a population growing at 1.85 day^{-1}. There was no significant difference between the calculated r_{max} of these cultures. The calculated r_{max} from batch cultures was that obtained in turbidostat culture within the experimental error.

3.4.3 Integration of Theory and Empirical Results to Evaluate Limiting Factors

The evaluations of biological limits on growth can be improved by using culture systems that allow an independent assessment of the various physiological constraints. Steady-state culture systems, particularly combinations of turbidostat and chemostat culture modes, have both theoretical and practical advantages for examining energetic and other physiological constraints on animal populations. Steady-state growth is a necessary, but seldom realized, assumption for animal bioenergetic models (Seale 1987). Excellent descriptions are available for the basic constructions and operation of a turbidostat (Munson 1970) and a chemostat (e.g., Herbert et al. 1956; Pirt 1975; Boraas 1983). The evaluation of physiological attributes of selected and unselected rotifers can be facilitated by use of the well-developed, mature theory which is available for the mathematical description of continuous culture dynamics (Herbert et al. 1956; Pirt 1975; Boraas 1983). Rates are linear functions of D, allowing a ready computation of growth variables, such as ingestion, production, and respiration rates (Droop and Scott 1978; Boraas 1983; Bennett and Boraas 1989b, 1990; Chapter 4.2, this Vol.). Practical advantages for assessing population attributes (including age structure and fecundity) include (1) volume replenishment and self-cleaning: detritus and potentially inhibitory waste products are removed constantly; and (2) multiple samples may be taken through time from a population that is in a constant physiological state. Therefore, these cultures are powerful tools for studies in which accurate measurements and control of metazoan growth are essential, e.g., assessments of evolutionary change, estimation of population parameters, growth optimization for aquaculture, and growth inhibition in toxicity studies.

Although animal population growth is often assumed to be carbon-energy limited, any trace growth factor (e.g., vitamins or amino acids) can set the upper limit on population growth. In contrast to carbon-energy limitation, in trace growth factor limitation, the limiting factor can be conserved by the organism during periods of slow growth. Carbon cannot be conserved in animal growth; regardless of food levels or growth rates, a proportion of ingested C must be oxidized to CO_2 to maintain basic metabolic needs (Morowitz 1968; Calow 1977b; Seale 1987). Hence, the partitioning of energy for various uses in animals must include a "cost" for basal metabolic loss. Although often not stated, "cost-benefit" analyses for energy partitioning make the implicit assumption that the populations are carbon-energy limited. With steady-state continuous cultures, this assumption can be tested explicitly.

By combining steady-state data from the turbidostat experiments outlined here with comparable data from rotifer chemostat studies (Boraas 1983), we can compare steady-state rotifer and algal populations for a range of specific growth rates (Fig. 3.4.4). Because steady state X, X*, is a function of the input algal density, C_1 (Pirt 1975), which was different for the chemostat vs. turbidostat

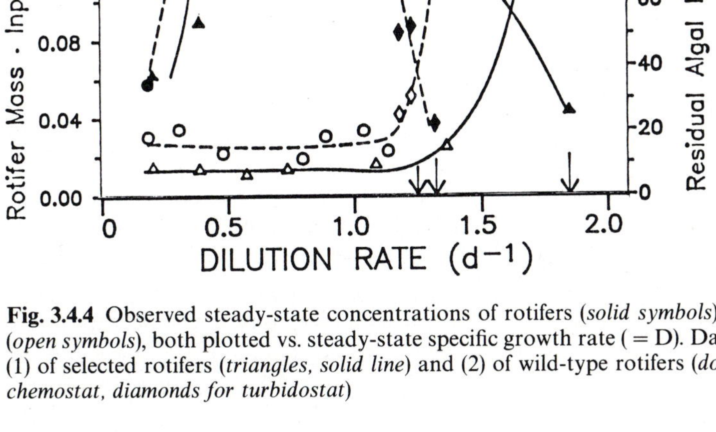

Fig. 3.4.4 Observed steady-state concentrations of rotifers (*solid symbols*) and residual algae (*open symbols*), both plotted vs. steady-state specific growth rate (= D). Data from two studies (1) of selected rotifers (*triangles, solid line*) and (2) of wild-type rotifers (*dotted line: circles for chemostat, diamonds for turbidostat*)

cultures, X^* was normalized to C_1 for all experiments (Boraas 1983; Boraas and Bennett 1988).

Energy vs. trace growth factor limitation can be determined from a straightforward examination of X^* vs. D (Fig. 3.4.4). There are three fundamental patterns for the relationship between steady-state biomass and D in a chemostat (see Boraas 1983). The relationship between rotifer steady-state biomass and D most closely resembles the curve for the condition in which the biomass is reduced by metabolic losses: a carbon-energy limited population. In this case, because of a necessary metabolic cost over all growth rates (maintenance metabolism), the steady-state biomass must decline as D approaches 0 (Fig. 3.4.4). In the case of inorganic nutrient limitation or trace growth factor limitation, no reduction in steady-state biomass is predicted at low D (see Boraas 1983; and Chap 5.2 for further discussion).

3.4.4 Physiological Characterization of the Fast-Growing Rotifer: Apparent Trade-offs

The fast-growing rotifer strain isolated in turbidostat culture was compared with a field population of the same species (Bennett and Boraas 1989a, b). Although fecundity in the fast-growing strain had shifted to younger age classes (Fig. 2 in Bennett and Boraas 1989a), the time to production of the first egg was about the same in the two populations: 22 h. However, the time to production of the second, and subsequent, eggs was about 1 h shorter for the fast-growing population: 24 vs. 25 h (Fig. 5 in Bennett and Boraas 1989a). The shift in fecundity and the shorter egg production times appeared to be the major contributing factors for the increase in r_{max}, based on calculations of r from the life table data (Bennett and Boraas 1989a). In addition, the rotifer life span decreased about 28% for the fast-growing population.

There also were apparent trade-offs between adult and egg sizes and r_{max}. Adult size and egg volume were decreased by 24 and 71%, respectively, for the fast-growing population (Fig. 6 in Bennett and Boraas 1989a). Rotifers of a given species have a constant number of cells (eutely). Therefore, the observed reduction in body and egg size probably was due to a reduction in the volume of cells rather than a decrease in the number of cells in the adults. The reduction in egg size (a single cell) may reflect a decrease in storage reserves for the neonatal rotifer. These observations provide a hypothesis consistent with the increased r_{max} of the fast-growing strain.

In continuous culture, regardless of the growth rate, the storage reserves are less essential than for animals in the field. In any continuous culture, individuals of the population will encounter food at frequent intervals, even if the rate of food input is low (e.g., at low dilution rate in a chemostat). Therefore, in an environment with strong selection for increases in r_{max}, a reduction of energy stores may be a favorable trade-off. In the field, where food levels are much less predictable, production of larger energy stores would, presumably, be favored at the expense of r_{max}. These observations suggest a trade-off between reproduction rate and size. Thus, the apparent mechanism for the increase in r_{max} was a reduction in the time to produce the second, and subsequent, eggs. This increase in egg production rate was probably due to a decline in the mass of storage products per egg.

Another study (Bennett and Boraas 1989b) compared population variables between the fast-growing strain and an unselected strain (Boraas 1983). The results indicated that selection for an elevated r_{max} had little, if any, effect on the physiological state of the rotifers. This information supports the hypothesis that the observed elevation in r_{max} is a function of a relatively small shift in demographic parameters rather than more fundamental changes in their physiology and biochemistry.

The r_{max} of populations can be subject to selection (Bennett and Boraas 1988, and references included). Furthermore, in turbidostat culture, r_{max} is determined by the fastest-growing genetic strain, directed by the following characteristics. First, in turbidostat culture, the population is forced to grow at its maximum rate for the given environmental conditions, a rate that is determined by the population, not the culture device. Secondly, due to inherent variations in r within the population (Doyle and Hunte 1981), individuals with a higher r_{max} will produce more offspring than individuals with a lower r_{max}. Therefore, a rotifer strain with a higher r will proliferate within the population. The faster-growing strains impose a dilution rate on the slower-growing strains that is higher than their r_{max}. Consequently, a slower-growing strains will be "washed-out" of the population, leaving only those strains able to maintain the new imposed r_{max}. Thus, turbidostat culture imposes strongly directional selection for strains with an elevated r_{max}. The upper limit of r_{max} is dependent on the environmental conditions, e.g., temperature, medium composition, and food type (Bennett and Boraas 1989b). The use of turbidostat cultures offers the most straightforward means for determining the optimum environmental conditions for rotifer growth, important considerations in aquaculture.

3.4.5 Evolutionary Changes in Rotifer Populations

The turbidostat can serve as a tool to explore evolutionary changes on rotifer populations in the laboratory. Through several lines of evidence, we have shown that the changes in r_{max} were genetic, not a simple physiological adaptation. The new r_{max} appeared over several generations, which is consistent with the appearance of a mutant that displaced the former strain. Males were lost from these cultures; hence, recombination was not a possible mechanism for change. The batch cultures, together with the rapid reattainment of the elevated r_{max} from the reinoculated rotifers previously grown in slow chemostat culture, confirmed that the observed elevated r_{max} was inherited. The selected r_{max} was retained after many generations of slow growth in the chemostat. When rotifers were reinoculated into the turbidostat, the new r_{max} rapidly became reestablished.

The turbidostat is an unique tool that can be used to explore the physiological and demographic characteristics of rotifer populations in assessing evolutionary trade-offs. Rotifer populations can change genetically when grown in turbidostat culture for extended periods ($>$ 1 month). Comparisons between pre- and postselected strains (Bennett and Boraas 1989b, 1990; Figs. 3.4.3, 3.4.4) can be refined and specialized to reveal significant physiological and ecological differences. Furthermore, experiments can explore different environmental conditions to determine which environmental variables have the greatest influence on r_{max}.

As the most rapidly growing metazoans yet studied, the selected rotifers empirically show cost-benefit trade-offs for achieving maximum growth rates. As suggested by theory (e.g., Reznick 1985; Reznick et al. 1988), both eggs and adults were smaller, indicating that energy is being shunted from individual growth to the production of eggs with a minimum percentage of storage material. Development time was reduced in producing the second egg of a given female. In the turbidostat, the population growth increased the dilution rate, which necessarily resulted in a very high mortality rate and a rapid reproduction rate. Adult survivorship had no adaptive value; the only possible "strategy" was to produce as many offspring as possible before being washed out of the culture. In contrast, in a series of batch culture studies, Bell (1984) found no evidence of a trade-off between adult survivorship and reproduction (also see Reznick et al. 1988). Possibilities not considered by Bell are: (1) the rotifers in these batch cultures were not actually carbon-energy limited, or (2) selective pressures were not sufficient in these studies to cause evolutionary change in the laboratory. In our turbidostat cultures when imposing a high selective pressure for increasing r_{max}, the theoretically predicted trade-offs were obvious.

3.4.6 Potential Use of Turbidostats for Toxicity Assays

Changes in r_{max} of a rotifer population become very important if the turbidostat culture is to be used as a tool in toxicity assessment. Such changes would interfere with measurements of toxicity because in turbidostat culture, r_{max} is the measure of toxicity. Assessment of toxicity using turbidostat culture of algae has been successful (Bennett 1988, 1990); however, no genetic changes in r_{max} were observed. The response of r_{max} to toxins has been shown to be a sensitive measure of toxicity and should be adaptable for toxicity assessment in rotifer populations (Chap. 7.3, this Vol.).

This is contribution no. 370 from the Center for Great Lakes Studies.

4 Metabolism and Energetics in Chemostats

4.1 Carbon Metabolism and Population Dynamics of *Brachionus angularis* and *Keratella cochlearis*

NORBERT WALZ

4.1.1 Introduction

The substrate or food concentration is the primary factor for the energy metabolism and population dynamics of a consumer (Walz 1986b). Its rate of change is highly dependent on the kinetic relationship between food and the functional response of the population.

The substrate consumption mechanism in microorganisms is often considered to be analogous to the enzyme kinetics of Michaelis and Menten (1913) as long as the enzyme concentration is held constant, i.e., if there is no population growth. In bacteria and algae, Monod (1942), Dugdale (1967), Rhee (1980), and others established this mechanism for the intake of soluble limiting organic and inorganic nutrients. This model can also describe the consumption of particulate food, e.g., bacteria or protozoa (Taylor 1978). Similar saturation functions have been applied to feeding and predator-prey interactions of metazoans (for example, the "disc equation" for insects; Holling 1959b; Hassel et al. 1977). This model is also applicable to the filtration or clearance mechanism of zooplankton species (Frost 1975; Mullin et al. 1975) and the predation process of carnivorous copepods (Williamson 1983).

This saturation kinetics is also useful for describing the ingestion of rotifers. Brachionidae are not true "filtrators" (they do not use a "sieve") but "sedimentators" (Naumann 1923). Their food intake may be described by a two-step reaction analogous to the stoichiometry of enzyme kinetics (Segel 1975):

$$(S) + F \underset{k_2}{\overset{k_1}{\rightleftarrows}} (SF) \underset{k_4}{\overset{k_3}{\rightleftarrows}} (S) + (I) \qquad (4.1.1)$$

with the following concentrations and constants:

(S) = free structure of the feeding apparatus
(F) = food
(SF) = structure-food complex
(I) = ingested food
k_1 = retention rate constant

k_2 = rejection rate constant
k_3 = ingestion rate constant
k_4 = irreversible = 0

Enzyme kinetics are formally applicable to ingestion processes (Caperon 1967), under some assumptions like:

- Structure concentration is constant, which means the consumer biomass is constant. This assumption is satisfied in steady-state and tracer experiments.
- Food once filtered is not significantly returned to food suspension, but ingested.

$$-\frac{dF}{dt} = \frac{dI}{dtX} = \frac{k_3(SF + S)*F}{(k_2 + k_3)/k_1 + F} \tag{4.1.2}$$

If the total structure (SF + S) is constant, it may be summarized together with k_3 to I_{max}:

$$I_{max} = k_3*(SF + F). \tag{4.1.3}$$

Reaction rate constants k_1, k_2, and k_3 are summarized to K_m, if there is a steady state between building and decomposing of SF:

$$k_1*(S)*(F) = (k_2 + k_3)*(SF); \quad \text{for} \quad dSF/dt = 0:$$

$$K_m = \frac{k_2 + k_3}{k_1} \quad \text{(Briggs and Haldane 1925).} \tag{4.1.4}$$

From this, and $F = C_2$ = reactor algae concentration, follows:

$$\frac{dI}{dtX} = i = \frac{i_{max}*C_2}{K_m + C_2}. \tag{4.1.5}$$

This two-step ingestion process finds its confirmation by direct microscopical observations (Gilbert and Starkweather 1977). The food particles are moved into the buccal funnel by cilia of the circum apical band. They are partially prevented from entrance into the funnel by pseudotrochus cirri. Funnel cilia (S) catch (k_1) the incoming particles (F). Particles and cilia combine to a reversible food-structure complex (SF) by direct interception (Rubenstein and Koehl 1977). The particles of this intermediate product will be either rejected (k_2) or ingested (k_3).

4.1.2 Kinetics of Ingestion Rate in Tracer Experiments

To study the food consumption of *Brachionus angularis*, two experimental sets with different methods were compared: tracer experiments and continuous cultures.

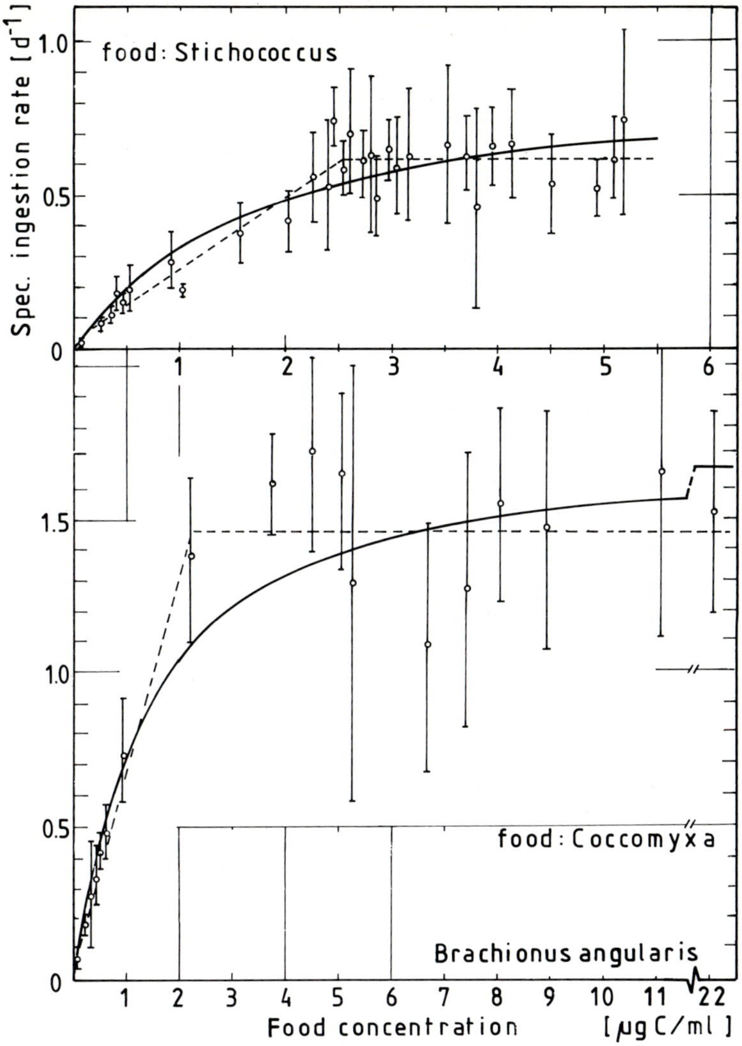

Fig. 4.1.1 Specific ingestion rates of *Brachionus angularis* in the lower range of food concentration by ^{14}C-tracer experiments ($\pm 95\%$ confidence intervals of the means). (After Walz and Gschloessl 1988)

^{14}C-tracer experiments were conducted by Walz and Gschloessl (1988) with *Stichococcus bacillaris* and *Coccomyxa* sp. as food. After a 60-min prefeeding period in an unlabeled food suspension, rotifers were permitted to feed 10 min in radioactive algae. Time-limited experiments showed gut passage times of >20 min (Gschloessl 1985), which was also found by Starkweather and Gilbert (1977a) and Korstad et al. (1989). For details of this technique, see Walz and Gschloessl (1988) and Gschloessl (1985).

Table 4.1.1 Parameters of the Monod equation of *Brachionus angularis* for the ingestion of two food algae measured by ^{14}C-tracer technique. (After Walz and Gschloessl 1988)

Food	*Stichococcus bacillaris*	*Coccomyxa* sp.
i_{max}	0.90 (day^{-1})	1.78 (day^{-1})
K_m	1.67 ($\mu g\,C\,ml^{-1}$)	1.42 ($\mu g\,C\,ml^{-1}$)
k	0.54	1.25
R	0.92, $n = 39$, $P < 0.001$	0.96, $n = 25$, $P < 0.001$

For both algae, the ingestion rate increased with higher food concentration (POC) according to a hyperbolic reaction in analogy to a Michaelis-Menten saturation function (Eq. 4.1.2). Figure 4.1.1 reveals that a plateau is reached above $2\,\mu g\,C\,ml^{-1}$. The height of this plateau is significantly different ($P < 0.01$) for the two algae. Therefore, i_{max} differs considerably (Table 4.1.1). With nonlinear regressions for minimizing the squares of deviations (Bliss and James 1966; see Sect. 2.2.13), the equation was fitted with the following values (Table 4.1.1):

The K_m values do not differ significantly.

At very low substrate concentrations $C_2 \ll K_m$, C_2 may be ignored in the denominator of Eq. (4.1.5), then $i = (i_{max}/K_m) * C_2$ or $k * C_2$, where $k = i_{max}/K_m$, the slope of the function or the first-order rate constant of $i = k * C_2$ (Healey 1980). At concentrations $C_2 \gg K_m$, K_m in the denominator of Eq. (4.1.4) may be ignored, and this equation reduces to the zero-order kinetics $i = i_{max}$ (Segel 1975).

As an alternative to the hyperbolic function, the rectilinear model of Blackman (1905) was often used (e.g., Rigler 1961; Frost 1972; Condrey 1982). This model proposes a linear increase in the lower concentration range until at the "incipient limiting level" a plateau is reached. Indeed, this model is valid for the extreme cases of the more general hyperbolic Michaelis-Menten model and corresponds to the "first order" or the "zero order" range of the reaction.

According to a rectilinear model, the correlations for the linear increase are very high in *B. angularis* with $R = 0.97$ ($n = 20$) for *Stichococcus* and 0.99 ($n = 15$) for *Coccomyxa*. The incipient limiting levels for *Stichococcus* and for *Coccomyxa* are not significantly different (Table 4.1.2), but the plateau values differ at the $P = 0.01$ level. The carbon content per algae cell volume is for *Coccomyxa* 2.7 times higher than for *Stichococcus*. This is nearly the same ratio as the plateau values of the ingestion rates (2.4). From this, one might accept as i_{max} a constant ingested food volume with which different carbon contents are consumed. Comparison of the ingestion rate values alone would act like a food selection with different intake mechanisms. Beyond that, differences in "taste" (DeMott 1986) must also be discussed.

The existence of a plateau and of a "incipient limiting level" for rotifers is dependent on the size of the food particles. Starkweather and Gilbert (1977b) did not find a plateau for the ingestion of *Rhodutorula glutinis*, a small yeast

Table 4.1.2 Characteristics of cell size and ingestion rates of the food algae *Stichococcus bacillaris* and *Coccomyxa* sp. (Walz and Gschloessl 1988)

Parameter	*Stichococcus*	*Coccomyxa*
Mean cell diameter (μm)	3.3	3.6
Mean cell volume (μm^3)	19.3	23.7
Carbon content (pg C/cell)	0.37	1.23
Carbon content (pg C/μm^3)	0.019	0.052
Incipient limiting level (μg C ml^{-1})	2.5	2.2
Plateau ingestion rate (day^{-1})	0.61	1.46

cell (smaller than *Stichococcus* and *Coccomyxa*), but they did for the ingestion of the large *Euglena gracilis*. They supposed rotifers in yeast suspensions would not build a pseudotrochus screen. Therefore, the ingestion rate could increase up to very high food concentrations. In contrast, *B. plicatilis* in the experiments of Schlosser and Anger (1982) showed a plateau with the small yeast *Saccharomyces cervisae*, but only at very high food concentrations. With bacteria (*Aerobacter aerogenes*), *B. calyciflorus* never attained a plateau. On the other hand, for the same species, the ingestion rate fell at very high *Chlorella* concentrations because of a supposed toxin effect of this alga (Halbach and Halbach-Keup 1974). In a literature review by Starkweather (1980), with all larger particles a plateau is induced. Duncan and Gulati (1983) were able to show an inverse relationship between the cell volume and incipient limiting level, when a plateau is reached. They associated small food particles with high critical concentrations for incipient limiting levels. Additionally, these critical concentrations for rotifers were much higher than those for daphnids (Chalk 1981, cited by Duncan and Gulati 1983).

In my own experiments with *B. angularis* the plateau ingestion rate remains constant with climbing food concentration, but above an upper threshold limit of 5 μg C ml^{-1} for *Stichococcus* and 22 μg C ml^{-1} for *Coccomyxa*, the ingestion rates again increase considerably in a secondary step (Fig. 4.1.2). This appears as soon as the regulation of the clearance rate (FR = water volume, theoretically filtered free of particles) comes to a lower constant level (Walz and Gschloessl 1988). Because $i = \text{spec. FR} * \text{POC}$, the ingestion rate has to increase at constant FR with higher food levels.

This secondary step looks very strange at first. However, in many of the above cited experiments, it seems the intermittent plateau could not be found because of too few measuring points. Only Yúfera and Pascual (1985) observed for *B. plicatilis* a secondary increase of the ingestion rate at high food concentrations and simultaneously a constancy of clearance rates. On the other hand, Rothhaupt (1988) did not find such a renewed increase of ingestion rates in the concentration range up to 10 μg C ml^{-1} in a tracer study in *B. calyciflorus* and *B. rubens*.

It is possible that the filtration rate was not lowered at high substrate concentrations because of the coupling of the activity of the "catching" cilia

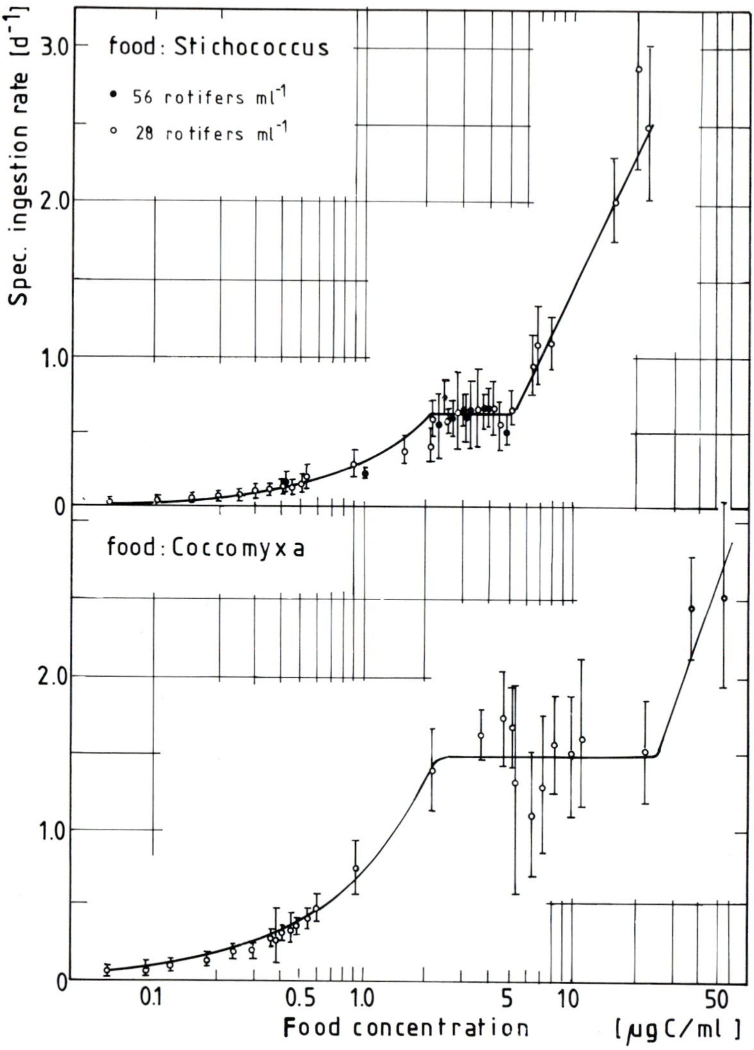

Fig. 4.1.2 Specific ingestion rates of *B. angularis* over the whole range of food concentrations by ^{14}C-tracer experiments ($\pm 95\%$ confidence intervals of the means). (After Walz and Gschloessl 1988)

with locomotion cilia (Erman 1956, cited by Starkweather 1980). Because the pseudotrochus screen could not exclude very small particles efficiently (Gilbert and Starkweather 1977), the only consequence was in accepting "superfluous feeding" (Conover 1966) and in decreasing the assimilation rate. Indeed, Gulati et al. (1987) and Pilarska (1977b) found for *Euchlanis dilatata* and *B. rubens* lower assimilation rates at higher food concentrations. In regulation kinetic terms, this change is consistent with a higher k_3 value in Eq. (4.1.3), which means a higher volume packing of the gut content. As in daphnids, a limitation of ingestion at high food concentrations seems to take place according to food volume (Geller 1975).

In some cases also, a decrease of ingestion rate at a high food concentration was observed. Rothhaupt (1988) found this for food particles greater than the optimum size, e.g., for *B. rubens* feeding on *Chlamydomonas sphaeroides* and *Micractinium pusillum.*

4.1.3 Kinetics of Ingestion Rate in Continuous Cultures

In steady states, the specific ingestion rate may be calculated by (Curds and Cockburn 1971):

$$i = \frac{D*(C_1 - C_2)}{X} \ (\text{day}^{-1}). \tag{4.1.6}$$

For derivation, see Appendix. This equation assumes that no sedimentation or consumption by a competitor takes place. Sedimentation was highly diminished or prevented by the conic form of the culture flasks and by the small cell size of the algae (*Stichococcus*), which could be held in suspension by the turbulence of the bubbling air.

If the ingestion rates are plotted against the reactor food concentration (Fig. 4.1.3), a continuous, nonlinear hyperbola is revealed according to Eq. (4.1.5)

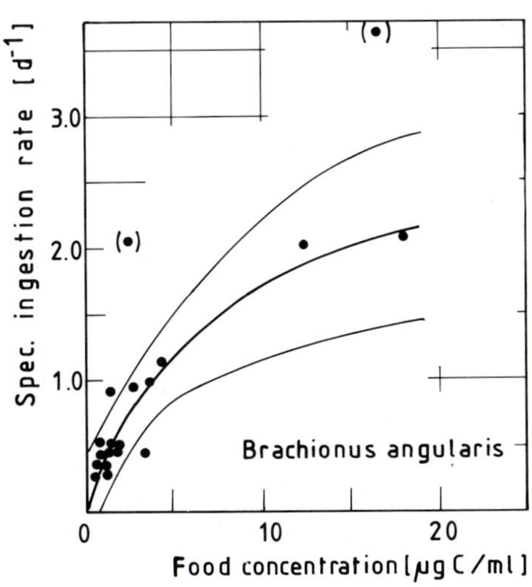

Fig. 4.1.3 Specific ingestion rates of *B. angularis* in steady states of chemostats ($\pm$95% confidence intervals of the regression)

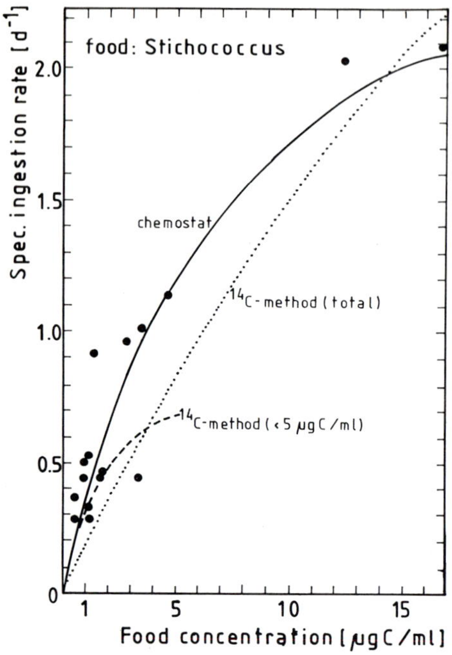

Fig. 4.1.4 Specific ingestion rates of *B. angularis* feeding on *Stichococcus bacillaris*. Comparison of chemostat and ^{14}C-tracer experiments. The points are of the chemostat. (After Walz and Gschloessl 1988)

with the following values of its parameters:

$$i_{max} = 3.01 \ (\mathrm{day}^{-1})$$

$$K_m = 7.50 \ (\mu g \, C \, ml^{-1})$$

$$R = 0.94, \ n = 17, \ P < 0.0001.$$

In comparison with the tracer experiments, the ingestion rate has very high maximum values, until at very high food concentrations the daily consumption is two to three times the body biomass (200–300%). Also, the K_m value is very high. In comparison with Fig. 4.1.2, the chemostat results appear to have "overshot" the plateau because of fewer points. If all tracer ingestion rates are pooled to a general curve with $i_{max} = 7.93$ (day^{-1}) and $K_m = 42.6$ ($\mu g \, C \, ml^{-1}$), they seem, at least below the level of 20 $\mu g \, C \, ml^{-1}$, not significantly different from the chemostat culture values (Fig. 4.1.4). A Kolmogorov-Smirnov test (Sachs 1992) shows no difference ($P > 0.20$) between both curves. This implies that it is admissible to extrapolate 10-min tracer experiments to long-term studies.

In *Keratella cochlearis* cultures with *Stichococcus bacillaris* as food, no steady states, but stable transient states have been achieved (Chap. 2.2). Also in these cases, Eq. 4.1.6 may be applied (Walz 1983b) when the difference $C_1 - C_2$ is large compared with variations of C_2, and X is the geometric mean between 2 days, if these values do not differ too much. The specific ingestion rate of *K. cochlearis* increases linearly with food concentration (Fig. 4.1.5). Below

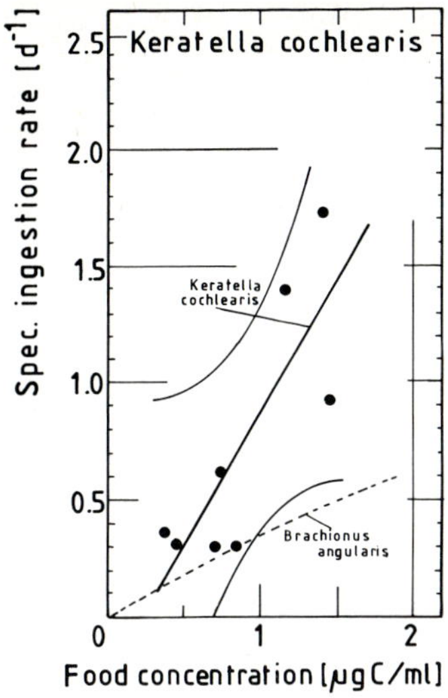

Fig. 4.1.5 Specific ingestion rates of *K. cochlearis* in transient states of chemostats (±95% confidence intervals of the regression). (After Walz 1983b)

$C_2 = 1.5\,\mu g\,C\,ml^{-1}$, according to a linear regression the specific ingestion rate is related to the food concentration by:

$$i = -0.25 + 1.11 * C_2$$

$$R = 0.85,\ n = 8,\ P < 0.01.$$

A plateau and an incipient limiting level have not been found, but the observed range of food concentration is very low. The slope of the increase corresponds to "k" of the above-mensioned first-order reaction $i = k * C_2$. For *K. cochlearis*, this k is much higher than the "k" of *B. angularis* chemostat cultures (1.11:0.40,

Table 4.1.3 Parameters of the Monod functions for growth and birth rates in chemostats of *B. angularis* and *K. cochlearis* and of $k = r_{max}/K_s$ and b_{max}/K_s, the slope of the Monod function at the lowest food concentrations

Parameter	*Brachionus angularis*	*Keratella cochlearis*
r_{max} (day^{-1})	0.58	0.36
K_s ($\mu g\,C\,ml^{-1}$)	4.05	0.66
k	0.14	0.52
R	0.94, $n = 25$, $P < 0.0001$	0.70, $n = 35$, $P < 0.0001$
b_{max} (day^{-1})	0.55	0.37
K_s ($\mu g\,C\,ml^{-1}$)	0.78	0.65
k	0.71	0.56
R	0.84, $n = 27$, $P < 0.0001$	0.81, $n = 39$, $P < 0.0001$
Ingestion rate[a]		
k	0.40	1.11

[a]Chap. 4.1.3.

see Table 4.1.3). This shows that the affinity for food (Button 1986) is much higher for *Keratella*.

4.1.4 Kinetics of the Specific Growth Rate

The above-mentioned relationships are also applicable to growth processes in analogy to enzyme kinetics. Monod (1942) stated the empirical equation:

$$\frac{dX}{dt*X} = \mu = \mu_{max}*\frac{S}{K_s+S} \text{ (time}^{-1}\text{)}. \tag{4.1.7}$$

Growth is considered formally as a two-step process, ingestion and catabolic events into an intermediate pool being the first step. Anabolic synthesis of body tissue determines the velocity of the whole reaction and $k_3 \ll k_1$. Equation (4.1.3) reduces to $k_2/k_1 = K_s$. In the literature of microbial cultures, especially of chemostat cultures, "μ" is used as the organisms' specific growth rate. To demonstrate the full applicability of the Monod function to the population dynamics of metazoans, we want to apply the more usual "r," in population ecology often called the "intrinsic rate of natural increase." Equation (4.1.7), therefore, becomes Eq. (3.3.7). Then S is the C_2 residual algae concentration in steady states. But this relationship is also valid in transient states (Cooney et al. 1981), when C_2 is simply the food concentration in the environment.

Therefore, in studying growth rate kinetics it is possible to combine transient and steady states. In chemostats the steady states become unstable when approaching the washout dilution rate D_c (Sect. 3.3.8). In *Brachionus angularis* transient-state cultures can be held at higher dilution rates $> D_c = 0.34$ (day^{-1}). In continuous cultures, the residence time is 1/D, and the minimal residence time for *B. angularis* in a steady state, therefore, is 2.9 days. This range would be a good case for studying population growth rates in turbidostates (Chap. 3.4). Also in the starting phase of chemostat cultures, populations are maintained in transient states at high food concentrations. In this phase C_2 (= reactor food concentration) is as high as C_1 (= input food concentration).

In transient states, the increase of rotifer biomass was followed for several days, and the apparent increase rate r′ is the exponent in the following exponential function:

$$X_t = X_0 * e^{r' * t}. \tag{4.1.8}$$

After natural logarithm transformation of the biomass, the slope r′ of the linear regression: $\ln X_t = \ln X_0 + r' * t$ was calculated. The true growth rate is $r = r' + D$ (Mateles et al. 1965). In Fig. 4.1.6, transient-state experiments are shown arranged for decreasing r′. C_2 is the geometric mean food concentration between the beginning and the end of the experimental period. At food concentrations $\gg K_s$, the reaction approaches zero-order kinetics ($r = r_{max}$, see Eq. 3.3.7); as a consequence, a difference in C_2 does not carry much weight.

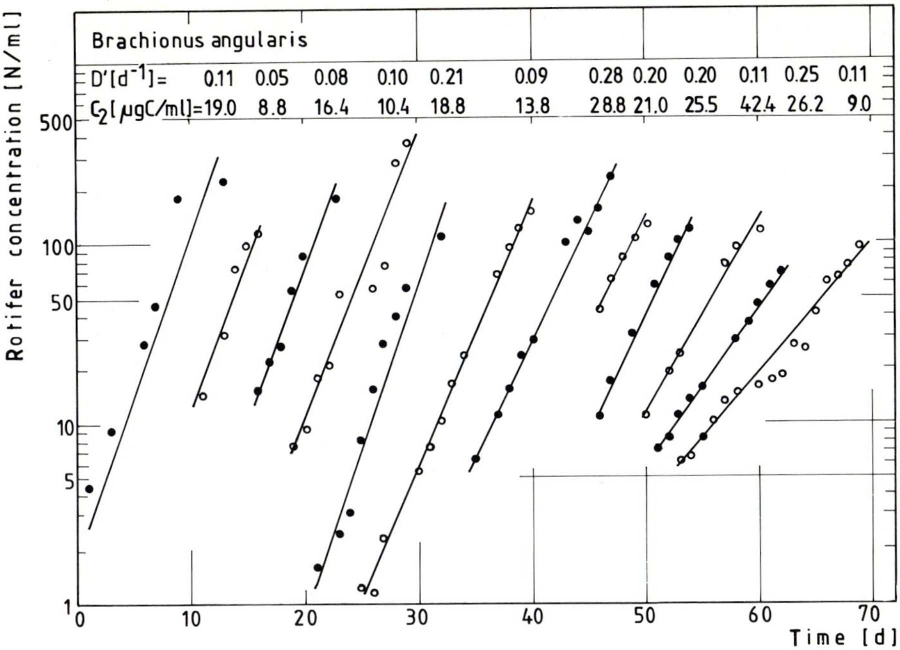

Fig. 4.1.6 Exponential increase (r′) of *B. angularis* density in transient states of chemostats at different food concentrations (C_2) and dilution rates (D)

Transient states and steady states are arranged together in Fig. 4.1.7a. Parameters were calculated for Eq. (3.3.7) by a nonlinear regression according to Bliss and James (1966) with the parameters given in Table 4.1.3. First, it seems doubtful that the simple Monod function is applicable to such transient states. Koga and Humphrey (1967), however, showed in model simulation a good correspondence to a modified Monod model (with variable Y).

Figure 4.1.7a shows that the transient states of *B. angularis* are near the maximum growth rates. There are three steady states at high food concentrations, but they do not reach the saturation growth rates of the transient states. In these cultures, the required residence time is too short to close a generation cycle (Sect. 3.3.8, Chap. 6.3).

K. cochlearis cultures were in transient states only. The reason for this is given in Sect. 2.2.12. Growth rates were calculated from $r = r' + D$ and r′ from:

$$r' = \frac{\ln X_t - \ln X_0}{dt} \text{ (day}^{-1}\text{)}. \tag{4.1.9}$$

C_2 was measured regularly in the range of first-order kinetics only, when r is mostly dependent on this factor. In the plateau range, when the kinetics approaches zero-order, only a few C_2 values were measured directly; others

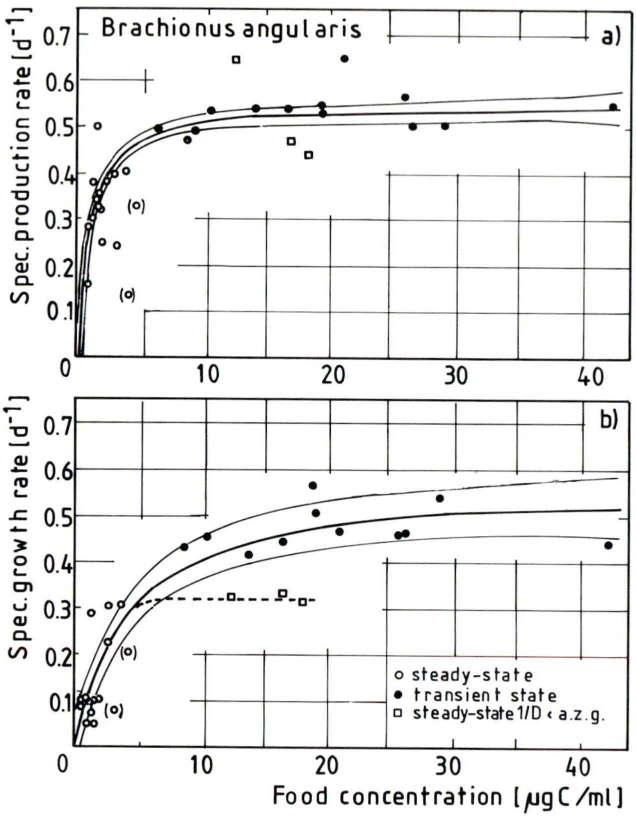

Fig. 4.1.7 a, b Specific production (**a**) and growth (**b**) rates of *B. angularis* in steady and transient states of chemostats ($\pm 95\%$ confidence intervals of the regressions). *a.z.g.* Age of zero growth; see Sects. 3.3.8, 6.3.6

were interpolated or replaced by C_1 (see above). Also in *K. cochlearis* (Fig. 4.1.8a) the relationship between specific growth rate and food concentration follows the Monod function (Eq. 3.3.7). The parameters were calculated by a nonlinear, least-squares regression (Sect. 2.2.14) and listed in Table 4.1.3.

As for the growth of microorganisms, the saturation kinetics of Monod (1942) are valid for rotifers. The regressions were calculated without assuming a threshold concentration. The points in Figs. 4.1.7 and 4.1.8 show more or less clearly the existence of a threshold. In order to consider this fact, the regressions were alternatively calculated in the following modified form:

$$r = y_0 + \frac{r_{max} * C_2}{K_s + C_2} \ (\text{day}^{-1}), \tag{4.1.10}$$

where y_0 = section of r with ordinate, where $C_2 = 0$.

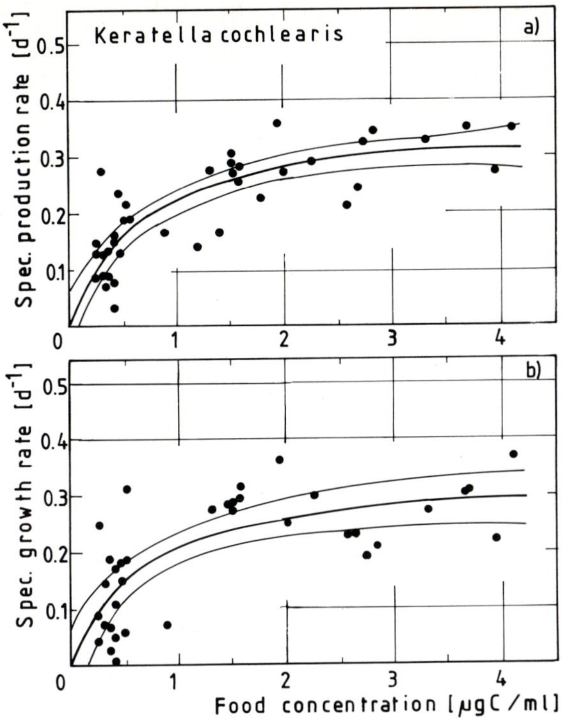

Fig. 4.1.8 a, b Specific production (**a**) and growth (**b**) rates of *K. cochlearis* in transient states of chemostats (±95% confidence intervals of the regressions)

But as the Y values were never significantly different from zero, the saturation function was finally applied in its original form.

Continuous cultures are very unsuitable to study threshold concentrations. At low growth rates, the residence times become very long. For the lowest dilution rates in *B. angularis* of 0.05 (day^{-1}), rotifers and algae have to spend a mean time of 20 days in the reactor. The biochemical nutrient content of the algae after such long dark periods is unknown. Additionally, because the cultures cannot be maintained at dilution rates = 0, the regression has to be extrapolated.

Energetic considerations suggest that a metazoan population has to use substrate for energy losses of feces, metabolism, and mortality in addition to energy to be used for growth (for derivation see Appendix). Figure 4.2.3 shows these theoretical relationships for the specific growth (r) and production rate (b) in relation to the food ration. The same theoretical relationships hold true in relation to food concentrations. As $b = r + d$ (d = mortality), the threshold for growth must be higher by the amount of substrate necessary for maintaining mortality. Hartmann (1987) determined a growth rate threshold for 15 °C cultures of *B. angularis* of 0.08 (µg C ml^{-1}). These experiments were carried out in "rotating cultures," where the sedimentation of the food suspension in small 20-ml vessels was prevented by rotation on a wheel (rpm = 1 min^{-1}). Such a

Monod function with a threshold (Sects. 4.2.8, 6.3.4) was likewise found for the growth of many other rotifer populations (Stemberger and Gilbert 1984, 1985).

An inhibition of growth rate at high food concentrations appeared repeatedly. The growth rate of *K. cochlearis* was inhibited by dense suspensions of *Rhodomonas*, but not of *Cryptomonas* (Stemberger and Gilbert 1985). In this case, a soluble toxin of the algae was excluded by filtration of the nutrient solution. In contrast, filterable *Chlorella* toxins inhibited the growth and fertility of *B. calyciflorus* (Halbach and Halbach-Keup 1974). On the other hand, no continuous culture study showed an inhibition of the rotifer population growth rate at high food concentrations (Figs. 3.3.1, 6.1.3).

4.1.5 Kinetics of the Specific Production Rate

The specific production rate is the secondary production of the population per unit biomass, which is, according to the derivation in the Appendix, congruent with the population birth rate. This derivation is strictly for steady states, but Romanovsky and Polishchuk (1982) give a general derivation for rotifers. Rotifers have after birth a small body size and reach their adult size rapidly. The specific production rate, therefore, is approximately the population birth rate (b). In Fig. 4.1.7a, the combined steady and transient states of *B. angularis* chemostat cultures with *Stichococcus* are fitted by the method of least squares to a saturation curve in analogy to the Monod function (Sect. 2.2.14):

$$b = \frac{b_{max} * C_2}{K_s + C_2} \ (\mathrm{day}^{-1}). \tag{4.1.11}$$

In Fig. 4.1.8b, the same procedure is shown for transient chemostat cultures of *K. cochlearis* in suspensions of the same alga. Again, the course of the curve follows a hyperbola. In the literature, there are only a few examples showing the relationship of birth rate to food. In suspensions of *Monoraphidium*, the egg ratio of *Brachionus rubens* climbs linearly (Rothhaupt 1985) or hyperbolically according a saturation function (Vuković 1981). At constant temperatures, the egg ratio proceeds parallel with the birth rate. *Chlorella* in high concentrations inhibits the production of *B. plicatilis* (Hirayama et al. 1973) and of *B. rubens* (Pilarska 1977b). The fitted parameters of the hyperbola are arranged together with correspondent values of *B. angularis* in Table 4.1.3.

In chemostat cultures with *Stichococcus* as food, *B. angularis* shows higher specific growth and birth rates than *K. cochlearis*. *B. angularis* obviously has an advantage at high food concentrations. On the other hand, *B. angularis* has higher K_s values for both rates. This means that the slope of increase (k) in the lowest range of food concentrations is higher for *K. cochlearis*, which is true for the growth rate (Table 4.1.3). This determines the better success at low food concentrations, as the higher "k" for the ingestion rates indicates the higher affinity to food and the higher grazing pressure to food concentration in the surrounding environment. The higher food affinity of the *B. angularis* birth rate

Table 4.1.4 C_2 steady state food concentrations (calculated)

D (day^{-1})	*B. angularis* ($\mu g C ml^{-1}$)	*K. cochlearis* ($\mu g C ml^{-1}$)	*Brachionus/Keratella*
0.05	0.38	0.11	3.4
0.10	0.84	0.27	3.1
0.15	1.41	0.51	2.7
0.20	2.13	0.91	2.3
0.25	3.07	1.74	1.8
0.30	4.35	4.42	1.0
0.35	6.18	$=C_1$	–

(Table 4.1.3) is not used for growth because of the higher mortality of this population (see next paragraph).

In steady-state cultures, C_2 is the result of Eq. 3.3.8 [$C_2 = K_s * D/(r_{max} - D)$]. This is the food concentration at which consumer growth equals the respective dilution rate. Low K_s values and, simultaneously, relative high growth rates (high k values) produce low C_2 steady-state concentrations. With the experimentally ascertained constants (Table 4.1.3), C_2 concentrations were calculated (Table 4.1.4).

K. cochlearis is superior to *B. angularis* at low food concentrations and at low dilution rates. Solving Eqs. (3.3.7) and (3.3.8) for C_2 or D, it can be shown that *K. cochlearis* is the better competitor below $C_2 = 4.32$ ($\mu g C ml^{-1}$) and $D = 0.3$ (day^{-1}). Above this dilution rate and at higher food concentrations, *B. angularis* wins. In comparison, both species are oriented to different life strategies (Walz 1987a). At high D and food concentrations, a high r_{max} is of advantage (*B. angularis*, r_{max} strategist). At low dilution rates and low food concentrations, K_s strategists (e.g., *K. cochlearis*) may better use the limited food resources. Seale et al. (Chap. 6.1) and Rothhaupt (Chap. 4.3) studied other rotifer life strategies in continuous cultures.

Obviously, all experiments show a physiological relationship between K_s and r_{max}, which is not accidental. Also, studies with bacteria suggest a positive correlation with both parameters. In chemostats, the organisms were characterized by high r_{max} and K_s values and vice versa. In many culture competition experiments, one or the other competitor wins as it is dependent on the dilution rate (Veldkamp and Kuenen 1973; Veldkamp 1976). There was only one exception to this rule, where a small r_{max} was related to a high K_s (Veldkamp and Jannasch 1972).

4.1.6 Kinetics of the Specific Mortality Rate

The mortality of the population, as $d = b - r$, is dependent on the food concentration. At higher food concentrations, the populations have better feeding conditions. For this reason, in *B. angularis* the natural mortality, without the intervention of a predator, declines with increasing food concentration and

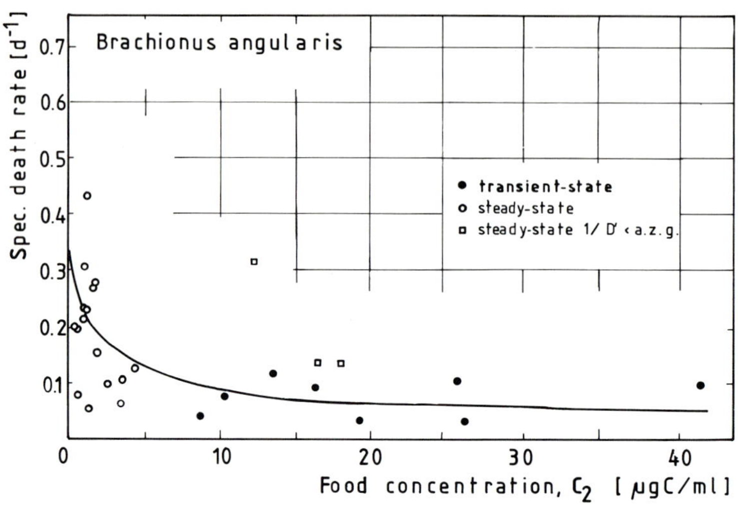

Fig. 4.1.9 Specific mortality rates of *B. angularis* in steady states of chemostats

reaches a plateau at a low motality level (Fig. 4.1.9). The curve may be approximated to a hyperbola by a method of Bliss and James (1966) with the parameters to be taken from Tale 4.1.5.

$$d = d_{min} + \frac{c}{c' + C_2} \quad (\text{day}^{-1}). \tag{4.1.12}$$

For lim $C_2 \rightarrow 0$, $d_{max} = 0.304$ (day^{-1}) is extrapolated, which for $b = 0$ replaces growth rate as "decreasing rate" (Fig. 4.2.3). It is self-evident that a population cannot survive for a long period under these conditions.

In *K. cochlearis*, a relationship between the mortality rate and food concentration was not found. As with r_{max} and b_{max}, both K_s values are hardly different, and the mortality rates scatter around zero. This scatter is higher at

Table 4.1.5 Parameters of the inverse hyperbola (Eq. 4.1.12) relating mortality rate (d), production efficiency (PE), and yield (Y) to food concentration in steady states of *B. angularis* chemostats [$y_{min} = d_{min}$, Y_{min}, PE_{min}, c and c′, see Eq. (4.1.13)]

Parameter	Mortality (d)	Yield (Y)	Production efficiency (PE)
y_{min}	0.037	0.136	0.031
c	0.69	0.241	2.363
c′	2.507	1.157	2.686
R	−0.68	−0.61	−0.82
n	21	19	17
P	< 0.01	< 0.01	< 0.001

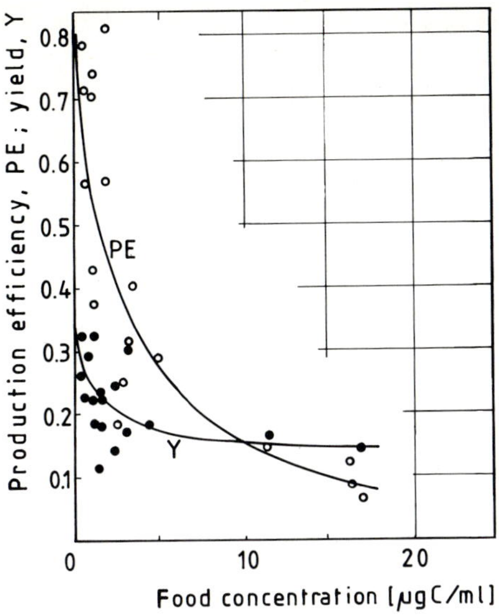

Fig. 4.1.10 Production efficiency (PE) and yield (Y) of *B. angularis* in steady states of chemostats

lower food concentrations. At higher concentrations ($> 1.0\,\mu g\,C\,ml^{-1}$), a mean value stabilizes at $d = 0.007$ (day^{-1}), with a 95% confidence interval of the mean of ± 0.019 ($n = 17$).

4.1.7 Kinetics of Metabolic Efficiencies

Although in the lower concentration range, the mortality drastically increases, the efficiencies of metabolism in production and growth (Sect. 4.2.6, derivation in the Appendix) are higher (Fig. 4.1.10). The utilization of food for the production of new biomass in the form of young (PE) and for growth of the population (Y) is especially high. At low concentrations, more than 70% of the ingested food is used for production, the rest is eliminated as feces and respiration. But due to mortality, the yield for turnover in the biomass is only 30%, decreasing for higher food concentrations. Both curves were fitted by an inverse hyperbola according to Bliss and James (1966):

$$Y = Y_{min} + \frac{c}{c' + C_2} \quad \text{and} \quad PE = PE_{min} + \frac{c}{c + C_2}. \tag{4.1.13}$$

The calculated values of the parameters are given in Table 4.1.5.

For $\lim C_2 \to 0$, $Y_{max} = 0.341$ and $PE_{max} = 0.908$ were extrapolated. In reality, this is not possible, and in Fig. 4.1.10 the lowest part of the hyperbola has to decline sharply to $Y = 0$ and $PE = 0$. By definition, the crossing point must be exactly at the threshold food ration (i_0, where $r = 0$, see Fig. 4.2.3).

4.2 Elements of Energy Balance of *Brachionus angularis*

NORBERT WALZ

4.2.1 Introduction

The ecological success of an organism is largely dependent on the characteristics of its metabolism. All metabolic activities have to adapt to the conditions a species experiences in its environment, or the species cannot survive. For energy metabolism, this is especially valid for the food conditions as an environmental factor. From studies of the energy balance, therefore, it is possible to deduce the food demands an organism makes on its habitat. Different species in the same habitat will have different requirements (qualitative and/or quantitative), thereby avoiding competition and using all available resources. Many publications deal with the qualitative aspects of this competition, e.g., the use of different sized particles as food in zooplankton filtrators (e.g., Gliwicz 1969; Hall et al. 1976; Bogdan and Gilbert 1984). However, also the quantity of food is characteristic of a niche. For this reason, the reaction of the energy balance to different food inputs should be studied. Energy is best represented by organic carbon as the basic element of energy-rich molecules.

The balance of energy on the basis of populations may be formulated by:

$$\textit{ingestion} - \text{feces} = \textit{production} + \text{respiration}$$
$$\textit{production} = \textit{growth} + \text{mortality}$$

(The parameters printed in italics were measured in this study.)

Ingestion of particulate matter is the only method (Fig. 4.2.1) whereby higher heterotrophic organisms obtain energy. But feces are not incorporated in the body. That is the reason why assimilation is the input channel for energy and matter to be incorporated into the metabolism of animals (assimilation = ingestion − feces). The sum of the output quantities, all other parameters standing on the right-hand side in the equation above, has to be equal to the inputs. Growth is understood as population growth, integrating somatic growth and reproduction. In rotifers, the somatic growth from young to adults is typically small.

It was not possible to record all parameters, and the respiration and feces have not been measured. In contrast to the energy balance of individuals, on

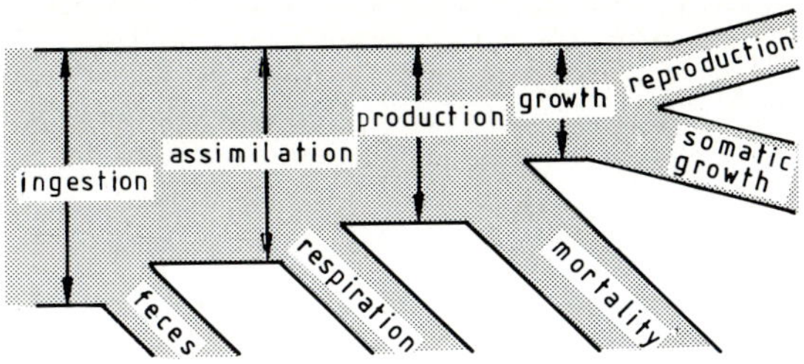

Fig. 4.2.1 Model of the energy balance of a population

the population level mortality has to be added, because dead rotifers had synthesized body tissues in the period before. This is the case for all metazoans, which finish their life by death, in contrast to most unicellular microorganisms (bacteria, fungi, and protozoans), which end their individual life by cell division (under nonlethal conditions).

To establish an energy balance, the energy output must be related to the energy input. For that purpose, the output parameters of this balance are plotted against food ration, representing the energy intake. Calculating specific rates (μg C/ μg C body weight * time), both axes are comparable with the same dimensions (day^{-1}). The relationships are comparable to curves like the saturation functions in relation to food concentration. This may be reasonable since ingestion as well as all energy outputs behave as Monod functions against food concentration (Chap. 4.1). Nevertheless, this interrelation between both sides of the balance is not self-evident.

4.2.2 Growth and Food Consumption

The more food that is consumed, the more energy is invested in the growth of the population up to some limit (Fig. 4.2.2). But the population can grow only if a minimum amount of food is ingested to compensate for losses in feces, metabolism, and mortality (see balance equation). This threshold is the food ration for "zero growth" ($r = 0$). What here is represented in relation to food ration is analogously true for food concentration, too (Chap. 4.1). From this relationship, the following modified Monod function was calculated according to a "three-parameter" solution of a hyperbolic regression (Sect. 2.2.13):

$$r = y_0 + \frac{r_{max} * i}{K_s + i} \ (\text{day}^{-1}), \qquad (4.2.1)$$

where

y_0 = intersection with y-axis = -0.197 (day^{-1}), at $P < 0.01$ significantly different from zero

r_{max} = 0.626 (day^{-1}), effective r_{max} = 0.426 (day^{-1})

i = specifiic ingestion rate = food ration (day^{-1})

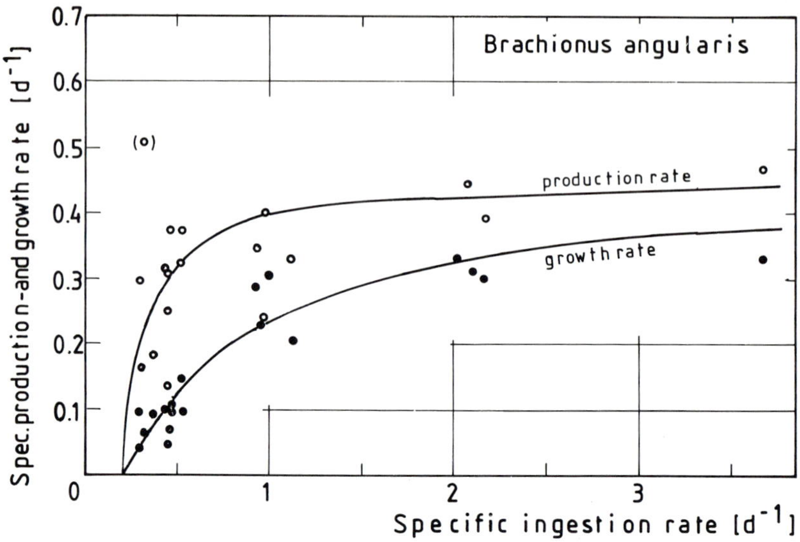

Fig. 4.2.2 Specific production and birth rates in steady states of chemostats in relation to the food ration

$K_s = 0.451$ (day^{-1})
$R = 0.946$, $n = 19$, $P < 0.001$.

This procedure extrapolates to an effective r_{max} (as difference to y_0) of 0.426 day^{-1} (Fig. 4.2.2) somewhat lower than the r_{max} value of the regression to food concentration (Table 4.1.3). This is why the higher transient-state growth rates could not be used for ingestion rates. The threshold for growth is represented by the intersection with the x-axis. This was determined by iterations as $i_0 = 0.207$ (day^{-1}; Sect. 4.2.5). This is an extrapolation of this curve, as at $D = 0$ continuous cultures turn into batch cultures.

Such relationships of growth to food rations are found for other metazoans, for instance, for the individual growth rate of fishes (Brett et al. 1969; Elliot 1976) and for mussels (Thompson and Bayne 1974; Walz 1978a).

4.2.3 Production and Food Consumption

For theoretical reasons i'_0, the threshold consumption to achieve a positive production ($b > 0$, b cannot be negative), must be lower than i_0, the respective threshold for growth (Fig. 4.2.3, for derivation b = spec. production rate, see Appendix). As it seems difficult to obtain significantly different values for i_0 and i'_0 from extrapolations, for practical reasons the same value of i_0 was used for production. The calculation of the production rate in analogy to Eq. 4.2.1 was not successful, as the y-intersection was not significantly different from zero.

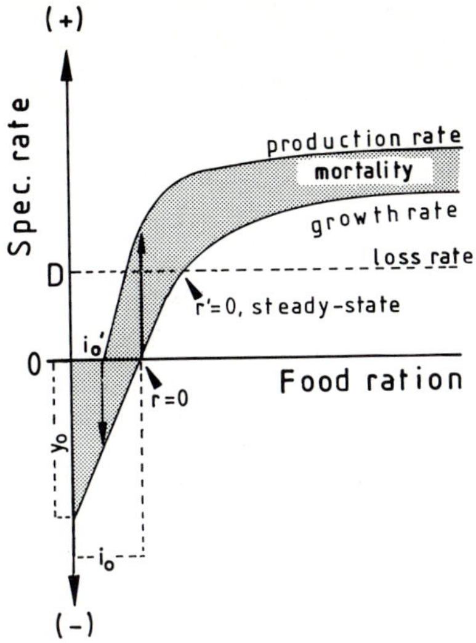

Fig. 4.2.3 Theoretical relationship between food ration (or concentration) and production and growth rates with food thresholds. i_0, i'_0, threshold food rations; *D*, dilution or loss rate; *grey area*, mortality rate. If food concentration is the independent variable: $i_0 = C_0$, $i'_0 = C'_0$

The following variant of the Monod function was applied, which has been used for phytoplankton growth on silicate (e.g., from Paasche 1973):

$$b = \frac{b_{max} * (i - i_0)}{K_{s'} + (i - i_0)} \text{ (day}^{-1}). \tag{4.2.2}$$

The regression with $i_0 = 0.207$ (day^{-1}) is shown in Fig. 4.2.2. With the fitting methods by Bliss and James (1966) the parameters are:

$b_{max} = 0.452$ (day^{-1})
$K'_s = 0.12$ (day^{-1})
$K_s = K'_s + i_0 = 0.319$ (day^{-1})
$R = 0.607$, $n = 19$, $P < 0.01$.

On the production level much energy has to be used to compensate for the population losses by mortality.

4.2.4 Mortality and Food Consumption

The mortality rate (d) is the difference between b and r (Fig. 4.2.4). This means $d = -r$ below the threshold food ration i'_0 for $b = 0$, and $d = b$ at the other threshold ration i_0 for $r = 0$. At low food rations, therefore, the mortality rate reaches very high values and diminishes with higher food consumption until it falls to a constant, food-independent level (Fig. 4.2.4). This may be described by an inverse hyperbola of this type:

$$d = d_{min} + \frac{c}{c' + i} \text{ (day}^{-1}). \tag{4.2.3}$$

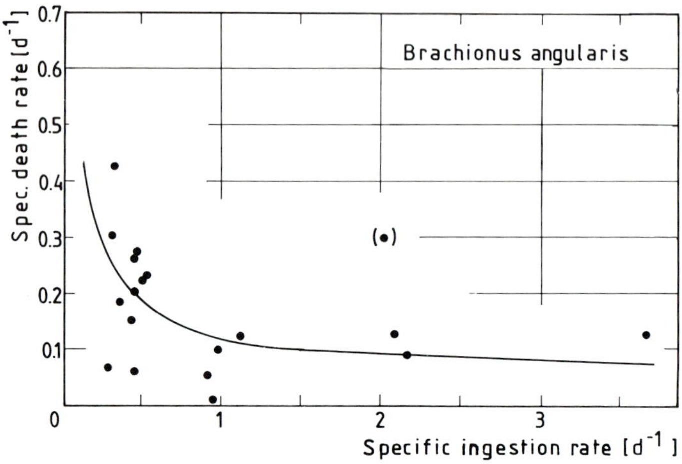

Fig. 4.2.4 Specific mortality rate in steady states of chemostats in relation to the food ration

The following parameters were calculated for least squares deviations according to Bliss and James (1966):

$d_{min} = 0.058$ (day^{-1})
$c = 0.069$ (day^{-1})
$c' = 0.067$ (day^{-1})

The asymptote d_{min} is attained at high food rations, whereas the maximum mortality is extrapolated for $i = 0$ and is $d_{max} = 0.310$ (day^{-1}). As steady states are not extant under these conditions, the mortality rate may even be higher. These rates relate to natural mortality only; no predators are included under the experimental conditions.

4.2.5 Food Consumption for Population Maintenance

The food ration for zero growth ($r = 0$) represents the demand of the population to maintain all other life processes. This is $i_0 = 0.207$ (day^{-1}) for a population of *Brachionus angularis* (see above). In other words, a population must ingest daily a food weight in carbon of 20% of its body weight for zero growth. It is important that all experiments with continuous cultures deal with large populations ($> 10^3$ individuals). Because of this, all energy demands of a population are involved with all losses that have not been converted to growth. In Chap. 3.3, another approach to determine the food intake for maintenance was presented. According to this and to the derivation in the Appendix, the specific maintenance consumption is: $v = \mu_c/Y_g$ (Schulze and Lipe 1964). Pirt

(1965) showed $Y_g = Y_{max}$, the maximum yield coefficient and μ_c = specific maintenance rate for balancing all losses. This calculation leads to $i_0 = 0.179$ (day^{-1}; Chap. 3.3). These values are very close to each other.

According to the balance equation, there are a number of different threshold levels. They depend on the levels of the output processes for which a population has to compensate (Fig. 4.2.3). The threshold for r = 0 is ecologically the most important (Lampert and Schober 1980). At this point, metazoans balance the energy demands and losses of feces, of metabolism (respiration), and of mortality. This is the maintenance rate "μ_c." In rotifers this threshold seems to increase with body size (Stemberger and Gilbert 1987a; Sect. 4.2.8), although it must also depend on the food type quality.

The second threshold is at b = 0. At this point, assimilation equalizes the losses by feces and of the energy demand of the metabolism (aasimilation = respiration). No effforts are made for growth and mortality. Therefore, the threshold is lower. This threshold can be measured in individuals (Lampert 1977) and is the maintenance ration normally given for higher animals, e.g. fishes (Brett 1976) and molluscs (Walz 1978a).

The third threshold is where assimilation = 0, when consumption is only converted to feces. At this food consumption level, individuals cannot live long, so this threshold is trivial.

On the contrary, in both chemostats and nature a fourth threshold is very important. This is the projection of r′ = 0 for food consumption, when r = D (Fig. 4.2.3). This equilibrium is only to be achieved in steady states. In chemostats and some flowed through natural biotopes, D is represented by hydraulic dilution. In other systems, D may be a symbol for the predation rate (see Chap. 7.2). Both processes correspond to losses from the population for which energy has to be expended if a steady state is to be maintained. This threshold is equivalent to R* of Tilman (1982), the resource requirement at equilibrium, corresponding to his ZNGI, the zero net growth isocline.

4.2.6 Efficiency of Energy Turnover

The efficiencies of production and growth are dependent on the characteristics of both input and output parameters. The question is which feeding conditions lead to the optimum energy utilization? The production efficiency (PE) is the quotient between production and ingestion (b/i). According to Fig. 4.2.5 it increases steeply, having passed the maintenance requirement between a food ration of 20–30% per day, up to maximum values of 0.8. This means 80% of the ingested energy is used for production. The actual food ration at this point is about 50% C day^{-1}. Above this maximum, PE decreases with increasing food intake, first rapidly, then slowly until a minimum level of about 15% efficiency is reached at high food rations. This means efficiency is much higher when the rotifers are food limited than when they feed on excess. This is very important for optimizing culture conditions for rotifer aquacultures.

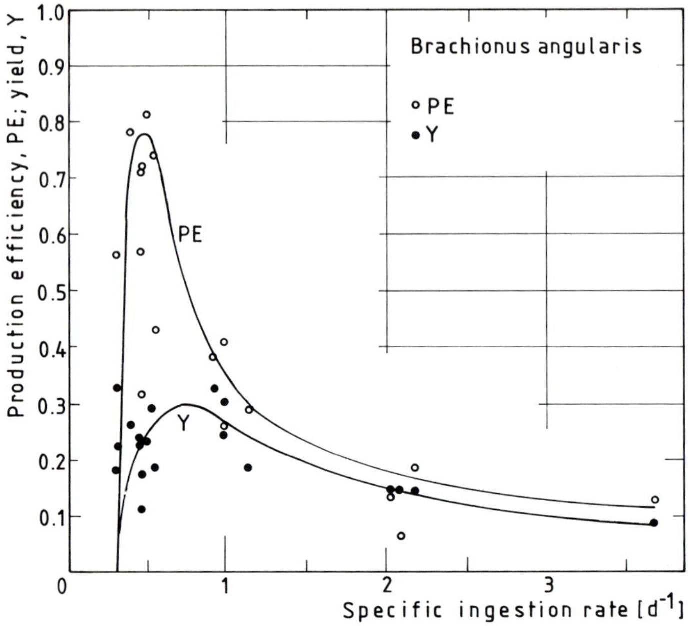

Fig. 4.2.5 Production efficiency (PE) and yield (Y) in steady states of chemostats in relation to the food ration

The efficiency with which food is converted to biomass (Y = yield = r/i), is identical with gross growth efficiency (K_1, see derivation in Appendix) and is much lower. At the maintenance requirement of 20–30% C per day, when r = 0, by definition, Y is zero. The maximum Y is only 30% at a food ration of about 70% C day^{-1}. At this food ration the mortality is already considerably lower than at maximum PE. Above its maximum, the yield decreases to a minimum of about 10%. Aside from food quantity, Y depends on food quality as well (Porter 1977). The value of Y determines whether or not a reaction is energetically possible. Additionally, kinetic factors constrain this reaction.

4.2.7 Energy Balance and Body Size

As rotifers grow up very quickly in their juvenile phase, they reach their maximum size within a very short time relative to most metazoans. However, they do not stop somatic growth definitely as adults, contrary to popular belief. The volume of an adult *B. angularis* increases by 30%, at least until the laying of the third egg (Walz and Rothbucher 1991). Nevertheless, the intraspecific differences are difficult to measure when all stages are present, e.g., in continuous cultures. This is the reason why the interspecific variation is of interest in this chapter only.

All bioenergetic rates change with body size. Stemberger and Gilbert's (1987a) data for the relationship between filtration rate and body mass are available for only a few species. The regression between the maximum filtration rate and body mass is:

$$FR = 31.288\, W^{0.417} \tag{4.2.4}$$

where

FR = filtration rate ($\mu l * rotifer^{-1} * h^{-1}$)
W = body mass (μg dry weight/rotifer)
$n = 6$.

This gives a mass-specific exponent of − 0.58. Stemberger and Gilbert (1987) give reasons for the assumption why the exponent of maximum filtration rate should be similar to the exponent of ingestion rate.

For metabolic rates (respiration) Starkweather (1987) compiled a list of rotifers from which Stemberger and Gilbert (1987a) calculated the following regression:

$$Resp = 7.551\, W^{0.664}, \tag{4.2.5}$$

where

Resp = respiration rate (nl $O_2 * rotifer^{-1} * h^{-1}$)
W = body mass (μg dry weight/rotifer)
$n = 15$.

For this example, the numerical value of slope should be treated with caution as most points are in the center of body weights. It should be wise to hold to Hemmingsen's (1960) exponent: $W^{0.75}$, as the slope in Eq. 4.2.5 is not significantly different from 0.75. At least, it should be noted that the metabolic rate increases with body weight. Because the exponent is significantly less than 1, the weight-specific rate decreases with body size (by $W^{-0.25}$, if Hemmingson's exponent is correct).

The intrinsic rate of natural increase (r) is a body mass-specific rate: $dX/dt * X$, see Eq. (3.3.3). The relationship of this specific growth rate to body size is surprising. Table 4.2.1 shows the body dimensions of *Keratella cochlearis* and *B. angularis* in comparison to maximum specific growth and production rates. This shows that the specific growth rate of rotifers cannot decrease with the higher body weight of rotifers as concluded from the well-known interspecific "bacteria-mouse-elephant" curves in the literature (Fenchel 1974; Blueweiss et al. 1978; Banse and Mosher 1980; Peters 1983). This assumption concerning one taxonomic unity is strongly confirmed by the figures of Stemberger and Gilbert (1985; Fig. 4.2.6) and for rotifers and crustaceans in Lake Constance (Fig. 6.3.6). Romanovsky (1984) found a positive trade-off between body size and specific birth rate in cladocerans similar to Fig. 6.3.6.

Table 4.2.1 Parameters of body size and weight-specific maximum growth and birth rates of *Brachionus angularis* and *Keratella cochlearis* (±95% confidence interval of the means; after Walz 1983b)

Parameter	*Brachionus angularis*		*Keratella cochlearis*	
Body length (μm)	151.3 ± 1.6	$n = 124$	88.1 ± 0.8	$n = 97$
Body width (μm)	136.2 ± 1.8	$n = 124$	61.1 ± 0.7	$n = 97$
Body volume (μm^3)	738 000 ± 24 000	$n = 124$	88 000 ± 200	$n = 97$
Egg volume (μm^3)	125 000 ± 6 600	$n = 124$	57 300 ± 1 300	$n = 97$
Carbon content/rotifer (μg C)	0.056 ± 0.008	$n = 14$	0.019 ± 0.004	$n = 10$
r_{max} (day^{-1})[a]	0.58		0.36	
b_{max} (day^{-1})[a]	0.55		0.37	

[a] Sect. 4.1.5.

For the relationship between r_{max} to body size of rotifers the following regression holds (Stemberger and Gilbert 1985):

$$r_{max} = 1.019\, W^{0.381}, \qquad (4.2.6)$$

where

r = specific growth rate (day^{-1})
W = body mass (μg dry weight/rotifer)
$n = 9$.

This "strange" relationship (Sect. 6.3.6.3) seems to stand in opposition to the above cited weight specific decrease of food intake and of respiration rate. This has to be understood in context with the balance equation:

ingestion − feces = assimilation = respiration + growth + mortality

Weight specific production rate (growth + mortality) intensifies with body size, too (data of Fig. 6.3.8). The same data, compared to growth rate, also show an increase of relative mortality. For this reason, the right hand side of the equation as a total doesn't vary with body size very much. With the present state-of-the-art

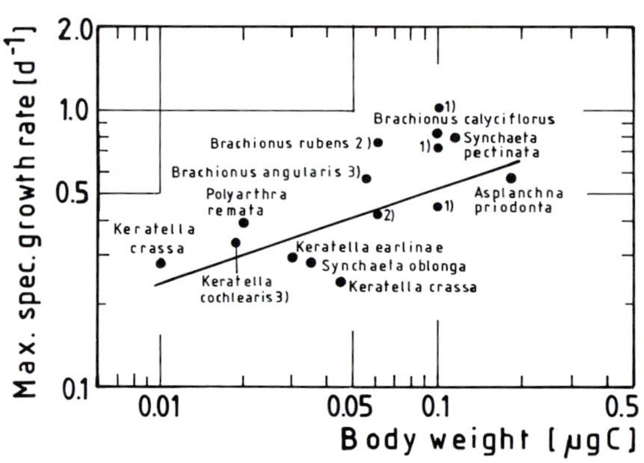

Fig. 4.2.6 Maximum specific growth rates of rotifers in relation to body mass: *1*, *B. calyciflorus* according to Rothhaupt (1988), different algae; *2*, *B. rubens* according to Rothhaupt (1988), different algae; *3*, *B. angularis* and *K. cochlearis* according to Walz (Chap. 4.1). (After Stemberger and Gilbert 1985)

the only solution for the equation above is a strong reduction of feces and a small enlargement or no change of the assimilation rate with body size.

Christiansen and Fenchel (1977) emphasize the well-known relationship of $W^{-0.25}$ of the "bacteria-mouse-elephant" curve was established for cultivated species only. The more detailed analysis, therefore, shows that the exponent of the weight-specific growth rate of closely related species groups within small ranges may be opposite to the general allometric relationship. This was not observed for protozoans, where the maximum specific growth rate does decrease as $W^{-0.25}$ (Fenchel 1974; Banse 1982a).

4.2.8 Energy Balance and Costs of Swimming

The energetic costs of swimming are a size-dependent challenge for metabolism. Zooplankton species swim at lower to middle Reynold numbers ($Re = 10^{-3}–10^{0}$; Wieser 1986):

$$Re = \frac{V * L}{v} \tag{4.2.7}$$

where

V = swimming speed ($mm\,s^{-1}$)
L = length of the organism (mm)
v = kinematic viscosity of water ($= 1 \times 10^{-6}\,m^2\,s^{-1}$, 20 °C)

In the transition range, up to $Re = 10^1$, drag and viscosity forces are prominent (Nachtigall 1977). With increasing Reynolds numbers, the drag forces become more important, and viscosity forces become increasingly negligible. This relationship is significant as the Reynold numbers increase with body size. At the small body size of protozoans, the Reynolds numbers are small, and the inertial effects are insignificant (Wu 1977). The organisms must overcome the viscosity forces of water only. The demands on metabolism to overcome the mechanical work of swimming are only 1% of the aerobic metabolism for protozoans below 100 μm (Purcell 1977; Fenchel and Finlay 1983). Protozoans and rotifers are comparable in this respect, having the same swimming mechanisms (cilia) and about the same size (at least the small rotifers).

Aerobic metabolism increases with body size according to (Hemmingsen 1960):

$$Resp = a * W^{0.75} \tag{4.2.8}$$

(see Eq. 4.2.5). According to Fenchel and Finlay (1983), the coefficient "a" is the same for protozoans and for metazoans in this range of body sizes.

Apparently, the costs for swimming increase more steeply with body size than can be provided by metabolism. For *B. calyciflorus* males, the energy costs for swimming amount to 62% of the aerobic metabolism (Epp and Lewis 1984).

The absolute swimming velocity (mm s^{-1}) seems to be constant for all organisms moving by flagella (Wu 1977) and cilia (protozoa, rotifers, marine larvae) at about 1 mm s^{-1}, not dependent on body size (Sleigh and Blake 1977). Epp and Lewis (1984) found a broad range of absolute swimming speeds between 600 and 800 μm s^{-1} with no clear relationship to body size for different clones of *Brachionus*. However, they demonstrated very clearly mass-specific speed (mm s^{-1} μg^{-1}) decreases with body size. A very similar relationship was also found by Stemberger and Gilbert (1987a) for different rotifer species (Fig. 4.2.7) with the regression:

$$V = 0.589\, W^{-0.971} \qquad (4.2.9)$$

where

V = weight-specific swimming speed (mm s^{-1} μg^{-1})
W = body dry weight (μg)
$n = 15$.

This decrease of relative swimming speed is parallel to the decreasing hydrodynamic efficiency with body size (Sleigh and Blake 1977).

Raising the costs of mechanical work with a larger body size requires a higher rate of food consumption and, as a consequence, a higher specific food supply rate which, in turn, requires a higher food concentration. For that reason, the threshold food concentration to maintain the population (r = 0) should increase with body size. Indeed, this was found for rotifers by Stembeger and Gilbert (1985; Fig. 6.3.2). Additionally, the K_s value, the food concentration at

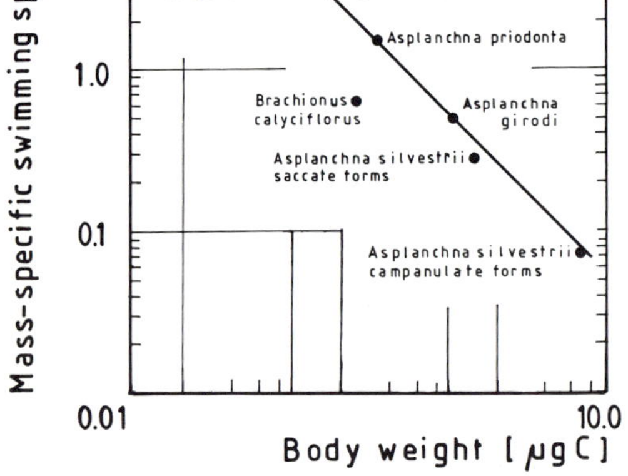

Fig. 4.2.7 Specific swimming speed of rotifers in relation to body mass (after Stemberger and Gilbert 1987a)

half the maximum growth rates, increases with body mass (Stemberger and Gilbert 1985; Fig. 6.3.1). These authors argue (1987a) that smaller rotifers with a relative high swimming speed have disproportionately higher encounter rates with food at diluted suspensions. At high food concentrations, this difference no longer plays a role. Then larger-sized rotifers filter enough food when swimming slowly.

These general relationships may also hold for cladocerans at higher swimming speeds. They swim at higher Reynold numbers where cilia are inefficient. Therefore, in the evolutionary sense they had to overcome these difficulties by the invention of a new swimming technique by locomotion with extremities. Canoeing in the middle range of Reynold numbers is far more energy saving and efficient. The costs for swimming increase with body size again, but cladocerans reach a much higher body size (Wieser 1986). This may be the reason why lower K_s and threshold concentrations of cladocerans compared with rotifers have been energetically established. This is contrary to a foraging model of Gerritsen and Kou (1985) which predicts a decrease of minimum food densities corresponding to body size when the full range of body sizes of the zooplankton community is considered.

4.2.9 Energy Balance and Population Size

The steady-state biomass should be directly constrained by the food consumption of the population. In Fig. 4.2.8, the density of *B. angularis* is related to the specific ingestion rate. The highest population densities are not found at the highest food rations but in a middle range. For a first approximation, the population density seems to be related to the production efficiency (PE; Fig. 4.2.5), but these parameters are not correlated with each other (Pearson correlation $R = 0.139$, $n = 19$, $P > 0.30$).

On the other hand, there is a good relationship with the growth efficiency (Y). If the following equation is calculated (Fig. 4.2.9):

$$X = I * Y^a \tag{4.2.10}$$

where

X = population density as biomass (μg C)
Y = growth efficiency
I = ingestion of the population (μg C)
(see Eqs. 3.3.6, 3.3.10)
a = weight specific exponent

one obtains the following regression:

$$X = 30.92 * Y^{1.012},$$

It can be shown that the exponent "a" is not significantly different from 1,

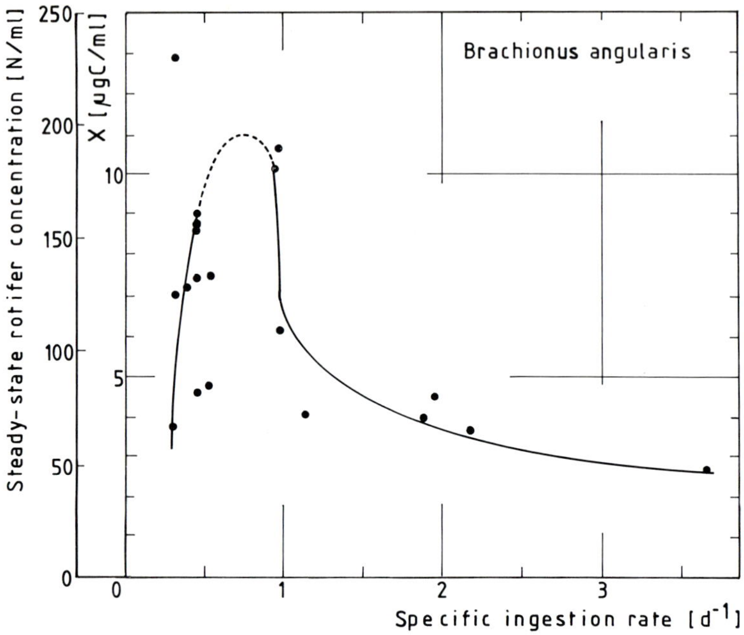

Fig. 4.2.8 Steady-state rotifer density in relation to food ration

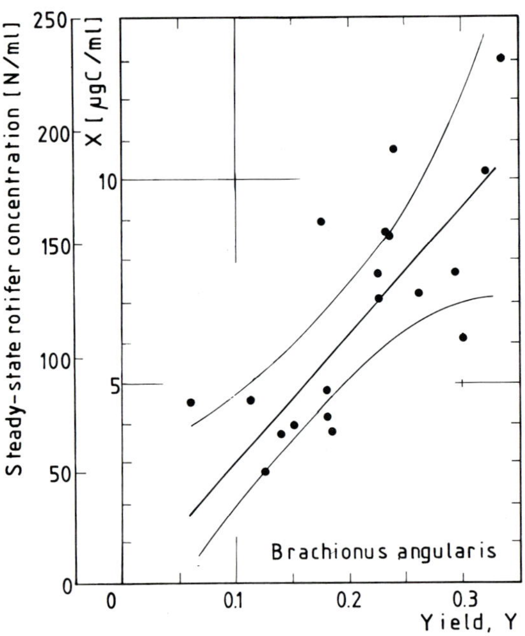

Fig. 4.2.9 Steady-state rotifer density in relation to the yield (±95% confidence interval of the regression)

indicating a linear relationship. The Pearson correlation is $R = 0.803$, $n = 19$, $P < 0.0001$, or 64% of the variation is explained by Y.

Equation 4.2.10 derivates from Eq. (3.3.6), as $I = C_1 - C_2$. Substitution of C_2 by Eq. (3.3.8) gives:

$$X = Y\left(C_1 - \frac{K_s * D}{r_{max} - D}\right). \tag{4.2.11}$$

Y by itself is dependent on D and may be substituted by Eq. (3.3.12). Temperature may also influence Y (Sect. 4.2.10). If the species-specific parameters are known, the steady-state population densities may be calculated. The only explicit environmental parameters that control the population density are C_1, the input food concentration = photosynthetic activity in the field, and D, the turnover of the community (see Chap. 7.2). To obtain the density in terms of individual numbers, X has to be simply divided by W, the individual weight of the organism.

4.2.10 Energy Balance and Temperature

Temperature highly influences all parameters of the energy balance, but in different ways. Walz et al. (1989) studied the ingestion rate of *B. angularis* on labeled *Stichococcus bacillaris.* At a food concentration of $5\,\mu g\,C\,ml^{-1}$, the specific ingestion rate reaches a maximum at 15 °C and decreases at higher temperatures (Fig. 4.2.10). At $2.5\,\mu g\,C\,ml^{-1}$, the maximum is only achieved at 20 °C. From another study (Walz and Gschloessl 1988), it is known that all ingestion rates between 2 and $5\,\mu g\,C\,ml^{-1}$ reach a plateau at 20 °C. Therefore, the 2.5 and $5.0\,\mu g\,C\,ml^{-1}$ ingestion rates are not significantly different at 20 °C. If the temperature is further increased to 25 °C, the $5\,\mu g\,C\,ml^{-1}$ ingestion rate remains unchanged. The decrease of the $2.5\,\mu g\,C\,ml^{-1}$ ingestion rate at 25 °C is not significant, while the 25 °C ingestion rates at both temperatures differ significantly ($P < 0.05$). In contrast to *B. angularis*, the ingestion rate of the warm stenothermal *B. calyciflorus* increases continuously with temperature (Galkovskaya 1987). The filtration rate of *B. plicatilis* also reaches a plateau at higher temperatures (Hirayama and Ogawa 1972).

The effect of temperature on the respiration rate may depend on the optimum temperature range for which a species is adapted, particularly if a continuous increase of metabolic rate or a plateau in oxygen consumption occurs (Miracle and Serra 1989). In *Rhinoglena frontalis* (Pourriot 1973b) and in *B. calyciflorus* (Galkovskaya 1987), an increase of the metabolic rate with temperature is found. In contrast, Epp and Lewis (1980) ascertained an optimum temperature range over which the oxygen consumption reached maximum values for a clone of *B. plicatilis.* This seems to be an exception to the rule, as the metabolic rate of the majority of zooplankton species hitherto studied

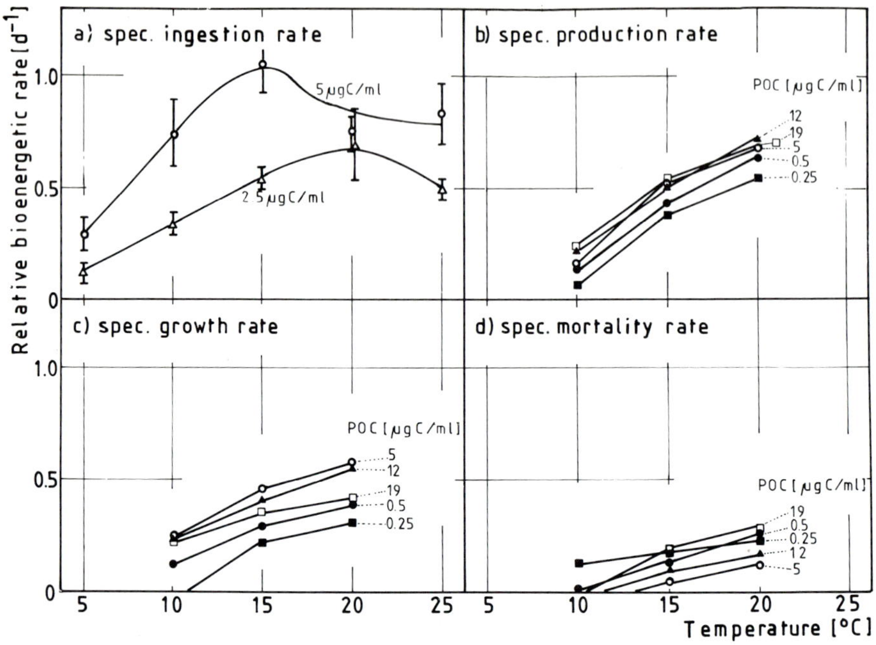

Fig. 4.2.10 a–d Temperature relationship of bioenergetic parameters at different food concentrations: *solid squares*; 0.25 μg C mol^{-1}; *solid circles*, 0.5 μg C ml^{-1}; *open triangles*, 2.5 μg C ml^{-1}; *open circles*, 5.0 μg C ml^{-1}; *solid triangles*, 12.0 μg C ml^{-1}; *open squares*, 19.0 μg C ml^{-1} (± 95% confidence intervals of the means; after Walz et al. 1989)

apparently increases continuously with temperature up to an upper lethal limit (Lampert 1984).

A continuous increase with temperature was found for the specific production rate of *B. angularis* (Walz et al. 1989). These studies were made in "rotating cultures" (Hartmann 1987), a special kind of fed-batch culture (Chap. 2.1). The rotifers were held at constant food concentrations in 20-ml vessels, which rotated on a wheel at 1 rpm to prevent sedimentation. The rotifers were transferred daily into new food suspensions. Figure 4.2.10 shows the temperature increase in the specific production rate for different food concentrations. As for *B. angularis*, the production rate also increased continuously for *Keratella cochlearis* (Walz 1983a). In this case, the birth rate is identical to the production rate (Romanovsky and Polishchuck 1982). In contrast, the egg ratio, the basis for the calculation of the specific production rate (Paloheimo 1974), has a maximum at intermediate temperatures and decreases thereafter in both species. This is also the case for *B. calyciflorus* (Halbach 1970a). The decreasing egg ratio of *B. angularis* and *K. cochlearis* at higher temperatures is more than compensated for by the increased rate of embryonic development (Walz 1983a, 1987a). In contrast to these species, the egg ratio of *B. rubens* declines with climbing temperature (Vucković 1981).

The specific growth rate of *B. angularis* (Fig. 4.2.10) increases with temperature in the observed range up to 20 °C (Walz et al. 1989). According to the proportion of production going into mortality or growth (see the balance equation), the growth rate increases continuously if the mortality rises constantly. This is the case in *B. angularis* (Fig. 4.2.10). Also in *B. calyciflorus* (Halbach 1970) and *B. dimidiatus* (Pourriot and Rougier 1975), the growth rates increase up to the highest measured temperatures. Alternatively, in cases in which the mortality rate increases progressively with temperature, the growth rate declines after reaching a maximum (*Keratella cochlearis*, Walz 1983a; *B. plicatilis*, Hirayama and Kusano 1972). *B. rubens* seems to follow a more complicated pattern. At high food levels, the growth rate decreases with temperature and increases under low food conditions (Vucković 1981).

The efficiencies of the energy turnover of *B. angularis* at 5 $\mu g\,C\,ml^{-1}$ increase with temperature, climbing from 10 to 15 °C (Table 4.2.2; Walz et al. 1989). With a temperature increase to 20 °C, at lower ingestion but higher growth and production rates, the efficiencies decrease again. The efficiencies for food use at 2.5 $\mu g\,C\,ml^{-1}$ are higher than at 5 $\mu g\,C\,ml^{-1}$ and are a little less than 0.5 at 10 and 20 °C. At 15 °C, no experiments were made at this food concentration.

The development of the energy balance with increasing temperature is best shown by Q_{10} values. Q_{10} is the quotient between bioenergetic rates at $T_1 = x\,°C$ and $T_2 = x\,°C + 10\,°C$. For *B. angularis*, such a comparison was made for 10 and 20 °C (Table 4.2.3; Walz et al. 1989).

The Q_{10} values are different for each individual bioenergetic parameter. Each parameter climbs at a different rate with increasing temperature. The production rate shows the greatest increase, with a mean Q_{10} of 4.0. It is interesting to see that some life parameters with a narrow correspondence to production have similar relationships to temperature: These are parameters describing the length of the individual life period. When treated as reciprocals to time, the Q_{10} of the life span is 3.3, of embryonic development is 4.0, of juvenile development is 4.9, and of egg laying interval is 3.1 (Walz 1987b).

Table 4.2.2 Efficiencies of energy turnover of *Brachionus angularis* at different temperatures (after Walz et al. 1989)

Food concentration	Efficiency	Temperature (°C)		
		10	15	20[a]
2.5 μg C/ml	PE	0.49	–	0.42
	Y	0.48	–	0.48
5.0 μg C/ml	PE	0.22	0.49	0.28
	Y	0.30	0.44	0.18

[a] From chemostat experiments, others from "rotating cultures".
PE, production efficiency (b/i); Y, growth efficiency (r/i).

Table 4.2.3 The Q_{10} values of bioenergetic parameters of *Brachionus angularis* between 10 and 20 °C (after Walz et al. 1989)

Parameter	Food concentration ($\mu g C ml^{-1}$)							
(day^{-1})	0.5	2.5	5.0	8.5	12.0	15.5	19.0	Mean $Q_{10} \pm SE$
Ingestion rate	–	2.6	2.5[a]	–	–	–	–	2.5 ± 0.2
Production rate	4.7	4.4	4.2	3.8	3.5	4.1	3.4	4.0 ± 0.2
Growth rate	3.1	4.1	2.5	2.7	2.4	1.9	2.0	2.7 ± 0.3
Mortality rate	3.6	–	4.8	–	2.6	–	2.2	3.0 ± 0.6

[a] Between 10 and 15 °C, because of maximum at 15 °C.

Population growth rate and mortality rate have lower Q_{10s} of 2.7 and 3.0, respectively. The only available Q_{10} of respiration rate (for *B. calyciflorus*) is also 2.7 (Galkovskaya 1987). In contrast to the energy output rates, the Q_{10} of the ingestion rate is still lower ($Q_{10} = 2.3$). This means that the energy balance grows more unfavorable at higher temperatures. At 25 °C, the *B. angularis* population approached its thermal tolerance limits. At this point, it was difficult to keep the culture going, and no growth experiments were carried out. *B. calyciflorus*, an even more pronounced warm eurythermal rotifer, could be cultured up to 40 °C (Galkovskaya 1987). In contrast, the net reproductive rate (R_0) of the cold stenothermal *Notholca caudata* falls dramatically above 10 °C (Laxhuber and Hartmann 1988). Also, *N. squamula* could be maintained in culture only under 10 °C (May 1987).

4.3 Steady-State Growth and Carbon Metabolism of *Brachionus rubens* and *B. calyciflorus*

KARL O. ROTHHAUPT

4.3.1 Introduction

Among the planktonic rotifers, ecophysiological traits markedly depend on the species' body size. Large species usually have higher threshold food requirements but, in turn, reach higher maximal growth rates when the food is not limiting (Stemberger and Gilbert 1985). Part of this phenomenon is ascribed to the fact that rotifers appear to be on the upper size limit for efficient ciliary motion, the costs of locomotion increasing, and the relative speed and thus encounter rate with food items decreasing with body size (Stemberger and Gilbert 1987a; Chap. 4.2). In the genus *Brachionus*, food size preferences are related to body size (Rothhaupt 1990a); smaller species prefer smaller, and larger species larger food items. *B. rubens* (adult body size 120–180 μm) most efficiently ingests particles about 5 μm in diameter, whereas *B. calyciflorus* (adult body size 220–280 μm) prefers particles about 10 μm in diameter.

This study compares the steady-state dynamics of both *Brachionus* species over a wide range of growth rates obtained in chemostats with *Monoraphidium minutum* as a food alga (equivalent spherical diameter = 3.5 μm, longest linear dimension = 5 μm). The objective was to compare the experimental results with predictions derived from the above-mentioned body size trends. The smaller species, *B. rubens*, was expected to have lower threshold food requirements. This, along with the fact that food algae are more efficiently grazed by *B. rubens*, should be reflected by lower reactor algal concentrations (C_2) of this species, especially at low dilution rates (D). *B. calyciflorus*, in turn, was expected to be superior with respect to maximal growth rates and to establish steady states at higher dilution rates than the smaller species. Moreover, it was of interest to determine whether the higher metabolic costs of the larger species can be deduced from parameters of the carbon balance.

4.3.2 Special Methods

B. rubens (strain F in previous work, Rothhaupt 1990a) originated from a culture at the University of Frankfurt kept for several years. *B. calyciflorus* was hatched from resting eggs that had been supplied by Dr. M. Schlüter (Kernforschungsanlage Jülich, FRG). The food alga was the coccale green alga *Monoraphidium minutum* (Culture Collection, University of Göttingen; strain 243–1). The medium was a modified CHU-12 (Müller 1972).

The two-stage chemostat system was as described by Rothhaupt (1985). The temperature of both stages was held at 20 ± 1 °C. The first algal stage was run at a dilution rate of 0.75 day^{-1}. Flow rates to the second (rotifer) stage were checked with measuring pipettes installed at the connecting hoses (Lampert 1976). Samples were usually taken daily. After blocking the outlet tubes, air pressure forced the medium out of a special sampling tube. Rotifers were counted alive under a dissecting microscope from an aliquot of at least 200 animals.

Table 4.3.1 Steady-state values for *Brachionus rubens*: dilution rate (D) in day^{-1}; input algae (C_1), reactor algae (C_2), and rotifer biomass (X) in μg C ml^{-1}; female density (N) in ml^{-1}; E/N is the average number of eggs per female

D	C_1	C_2	X	N	E/N
0.039	51.2	0.86	3.19	92.3	0.069
0.084	46.5	0.83	8.84	203.6	0.119
0.114	53.2	1.91	11.45	208.5	0.155
0.138	43.4	1.08	6.16	141.4	0.169
0.140	46.5	1.17	12.49	255.2	0.128
0.144	48.6	1.87	14.03	269.0	0.155
0.156	49.5	1.61	10.75	219.0	0.183
0.174	44.7	0.77	11.48	222.0	0.192
0.187	42.1	1.30	12.93	285.3	0.221
0.187	42.1	1.02	7.87	196.0	0.200
0.206	38.4	1.81	8.78	186.3	0.238
0.218	44.7	1.15	13.35	255.2	0.217
0.238	44.7	1.10	14.57	230.6	0.322
0.243	43.1	1.23	14.28	271.6	0.240
0.255	20.4	2.25	7.00	122.8	0.227
0.262	42.3	1.66	13.91	302.2	0.221
0.318	47.0	1.59	16.06	276.0	0.349
0.336	33.4	1.55	11.85	225.7	0.340
0.390	33.4	1.18	10.10	208.6	0.346
0.415	44.7	1.70	18.39	265.0	0.465
0.517	31.1	1.70	13.28	202.0	0.460
0.520	33.4	1.92	10.19	170.6	0.576
0.629	33.4	1.71	15.60	202.0	0.661
0.700	30.7	2.40	11.10	144.1	0.654
0.725	31.1	2.91	11.88	123.0	0.775
0.770	–	–	–	–	–

Algal densities were routinely estimated by light extinction (800 nm) using previously established calibration curves.

A steady state was defined when the population density and extinction did not change statistically by an coefficient of variation (CV) ≤ 0.15 over 5 consecutive sampling days. For each steady state, on at least two occasions, the following parameters were detemined: carbon content of the input algae (C_1), carbon content of the rotifer chemostat, and carbon content of the reactor algae (C_2) by removing the animals with 52-μm screens. Rotifer biomass in terms of carbon (X) was taken as the difference between the total carbon content of the rotifer chemostat and the fraction $< 52\,\mu$m (C_2). Carbon samples were filtered on Whatman GF/C glass fiber filters and analysed by combustion and measurement of infrared absorption (Krambeck et al. 1981).

Input algal concentrations (C_1) differed for the various steady states (Tables 4.3.1, 4.3.2). Steady-state rotifer biomasss (X) and rotifer densities (N) were therefore standardized to an average algal input concentration of $40\,\mu g\,C\,ml^{-1}$:

$$X_{40} = X\frac{40 - C_2}{C_1 - C_2} \quad \text{and} \quad N_{40} = N\frac{40 - C_2}{C_1 - C_2}. \tag{4.3.1}$$

Ingestion and production rates were calculated according to the Appendix. Additionally, a loss rate (L) was calculated:

$$L = \frac{D(C_1 - C_2 - X)}{N}. \tag{4.3.2}$$

Table 4.3.2 Steady-state values for *Brachionus calyciflorus*. Symbols and dimensions as in Table 4.3.1; additionally, the male density (males ml^{-1}) is given

D	C_1	C_2	X	N	E/N	Males
0.056	39.3	1.25	2.41	31.7	0.176	0.72
0.098	46.9	2.52	5.91	41.7	0.251	0.77
0.109	46.9	2.96	4.00	43.8	0.210	0.66
0.157	41.4	1.32	5.62	54.8	0.227	1.04
0.169	40.1	1.62	9.22	103.9	0.185	1.20
0.172	18.2	0.91	1.77	30.1	0.233	0.56
0.206	39.7	1.71	9.13	89.1	0.241	4.90
0.238	38.8	1.27	8.45	88.2	0.264	4.06
0.278	37.7	2.41	10.09	78.2	0.290	7.05
0.333	38.9	2.27	8.96	84.1	0.296	9.25
0.363	35.4	3.31	6.63	62.9	0.272	8.25
0.403	39.7	2.02	8.98	95.5	0.274	7.76
0.500	39.7	2.97	12.66	111.4	0.319	8.75
0.592	41.2	2.47	13.41	144.4	0.352	5.80
0.630	43.2	3.75	12.18	138.3	0.551	8.66
0.675	41.2	8.08	11.68	102.3	0.411	12.00
0.722	–	–	–	–	–	–

This loss rate is not precisely defined. It corresponds to the proportion of ingested carbon that is not transferred into rotifer biomass, and it comprises respiration as well as excretion and egestion losses. According to the energy balance formulated in Sect. 4.2.1, L = feces + respiration. If, in the denominator, the population density (N) is substituted by the rotifer biomass (X), one obtains the specific loss rate (spec. L).

A rectangular hyperbola with a threshold for zero population growth was fitted to the rate data. Curves were fitted by nonlinear iterative regression (STSC Inc. 1985). The significance of the difference of parameter estimates between both species was tested using a nonlinear analogue of ANCOVA (Ratkowsky 1983).

4.3.3 State Variables

A total of 25 steady states for *B. rubens* and 16 steady states for *B. calyciflorus* were obtained at various dilution rates (Tables 4.3.1, 4.3.2). The larger species did not reach higher steady-state growth rates. The steady state with the highest growth rate was at $D = 0.724\,\text{day}^{-1}$ for *B. rubens*; the animals were washed out at $D = 0.77\,\text{day}^{-1}$. *B. calyciflorus* reached its highest steady-state growth rate at $D = 0.675\,\text{day}^{-1}$, and the population was washed out at $D = 0.722\,\text{day}^{-1}$.

While *B. rubens* never showed any signs of mixis, mictic females and males (Fig. 4.3.1) were usually present in chemostats with *B. calyciflorus*. Since mictic females can only be distinguished by their eggs (male or resting eggs), the proportion of females with mictic eggs among all ovigerous females should be

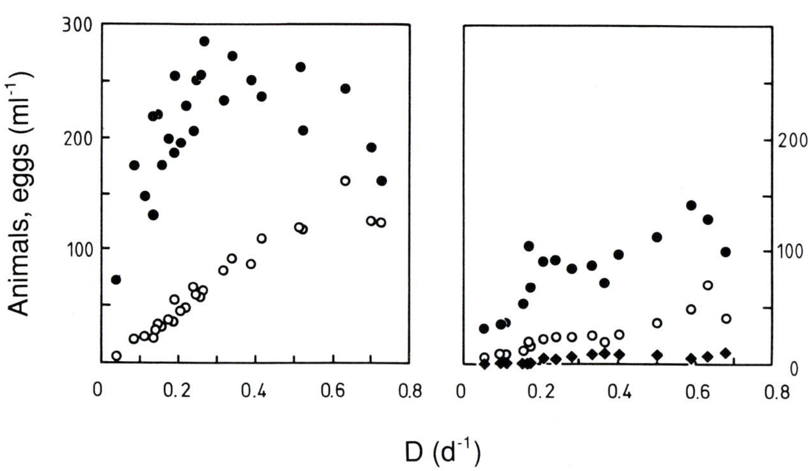

Fig. 4.3.1 Steady-state female densities (N; *dots*), egg densities (E; *circles*), and male densities (*diamonds*) standardized to an input algal concentration of $40\,\mu g\,C\,ml^{-1}$. *Left panel, Brachionus rubens; right panel, B. calyciflorus*

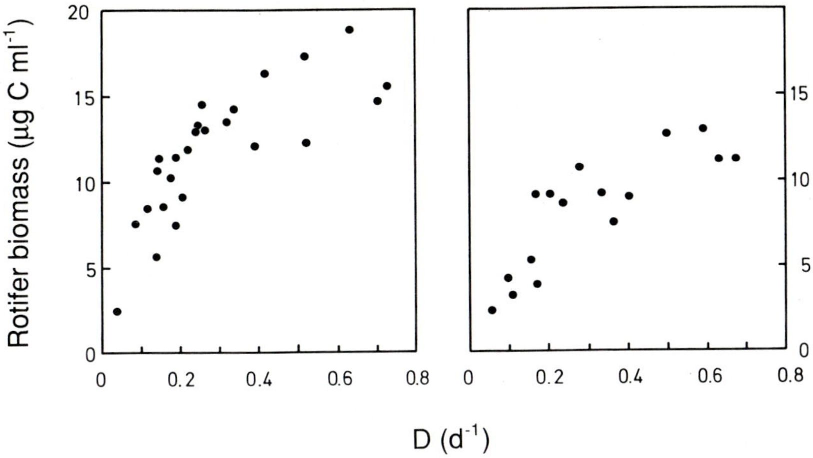

Fig. 4.3.2 Steady-state rotifer biomass (X) standardized to an input algal concentration of 40 $\mu g\,C\,ml^{-1}$. *Left panel, Brachionus rubens; right panel, B. calyciflorus*

an estimate of the proportion of mictic females in the whole population. Mixis appeared to increase with the steady-state dilution rate; however, a linear regression was only marginally significant ($P = 0.068$). The average proportion of mictic animals among all ovigerous females was $14.9 \pm 5.7\%$ (mean $\pm$ 1 SD).

The steady-state population densities (N) showed a convex pattern with increasing D (Fig. 4.3.1). This is caused by the influence of metabolic losses (Herbert 1958), which was predicted by a metabolic model (Rothhaupt 1985), and was similarly found by other authors (e.g., Boraas 1983; Schlüter et al. 1987; Walz 1987b). *B. rubens* had the highest population densities at intermediate dilution rates of about 0.3–0.4 day^{-1}. Generally, *B. calyciflorus* population densities were lower; the maximal population densities of this species were reached at higher D (around 0.6 day^{-1}). Egg densities (E; Fig. 4.3.1) and egg ratios (E/N; Tables 4.3.1, 4.3.2) increased with D and were clearly related to the increasing growth rates. Egg ratios increased less strongly with D for *B. calyciflorus*, which is due to the shorter egg development times of this species (Rothhaupt 1990b).

Rotifer biomass (X; Fig. 4.3.2) increased for both species with D. For *B. rubens*, the individual weight (W) in terms of carbon was related to the steady-state growth rate (Fig. 4.3.3). A linear regression was significant ($P < 0.001$). Body carbon (including eggs) increased from about 40 ng C per individual (Ind) at low growth rates to about 80 $ng\,C\,Ind^{-1}$ at high growth rates. There was no significant relationship between individual body carbon and growth rate for *B. calyciflorus*; the individual body mass fluctuated around 100 $ng\,C\,Ind^{-1}$ (Fig. 4.3.3).

Both the proportion of juvenile animals and the egg ratio generally increase with population growth rate (cf. Edmondson 1965). High proportions of juvenile

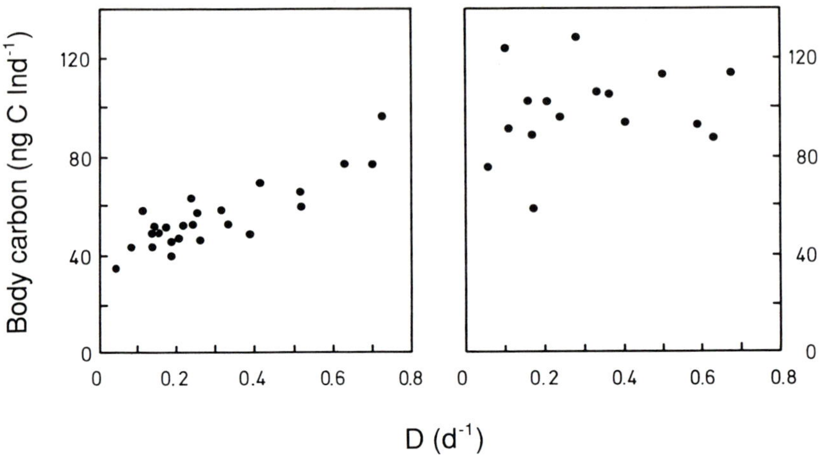

Fig. 4.3.3 Rotifer individual weight (W) in terms of carbon. *Left panel, Brachionus rubens; right panel, B. calyciflorus*

animals tend to lower, whereas high numbers of eggs carried tend to raise the average individual body mass. Due to the shorter egg development times, the egg ratios increase less strongly for *B. calyciflorus*. Furthermore, *B. calyciflorus* juveniles are relatively small (unpublished data); see also Boraas (1983) for the size distributions of bivolumes. A high proportion of relatively small juveniles thus tends to decrease the average body weight under high steady-state growth rates and appears to compensate for the effect of higher egg numbers in this species. The relative juvenile weights are higher and egg development times longer for *B. rubens*. Therefore, the effect of increasing the egg numbers with growth rate appears to be dominating in this species and hence leads to increasing average body weights.

Reactor algal concentrations (C_2) increased with D (Fig. 4.3.4; $P < 0.05$ for both species). In spite of the fact that there is much variance in the data, the food levels were significantly higher for *B. calyciflorus*, thus confirming the general expectations. This was expressed by a significantly higher ordinate intercept (= food threshold for zero population growth, C_0) of this species (1.82 μg C ml^{-1} vs. 1.28 μg C ml^{-1} for *B. rubens*; F = 24.64, 2,37 df, $P < 0.001$). There was no significant difference in the slope parameter, however. Qualitatively similar findings (higher food requirements of the larger species) are reported from semicontinuous growth rate determinations involving the same rotifer strains (Rothhaupt 1990b).

Quantitatively, however, the values obtained with both methods differ considerably. Thresholds for zero population growth (C_0), obtained with the "rotating culture" technique (Walz et al. 1989; Rothhaupt 1990a; Chap. 7.1) and the same algal food (*Monoraphidium*), are 0.07 μg C ml^{-1} for *B. rubens* and 0.24 μg C ml^{-1} for *B. calyciflorus*. The thresholds obtained from the chemostat

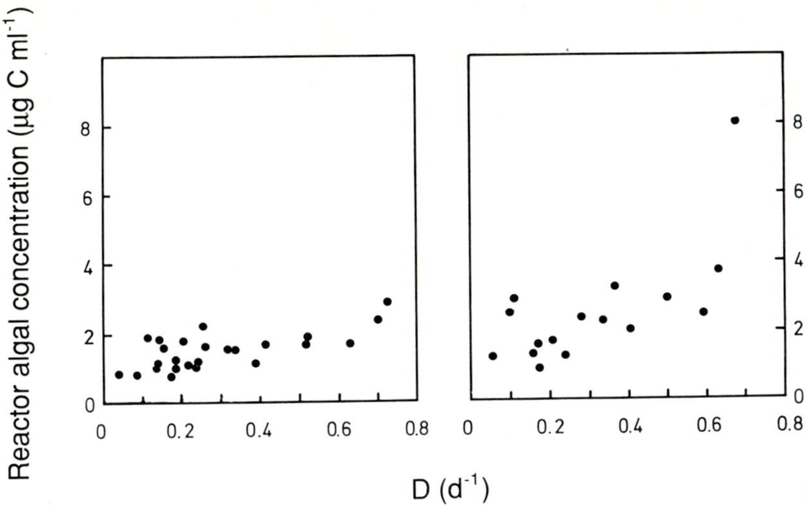

Fig. 4.3.4 Reactor algal concentrations (C_2) in terms of carbon. *Left panel, Brachionus rubens; right panel, B. calyciflorus*

experiments are in a range where population growth would have been close to maximal in the "rotating culture" experiments. While the "rotating culture" experiments were designed in a way to keep the quality of the food as close as possible to the original algal cultures, the same is impossible for the second stage of a rotifer chemostat. In the second stage, the nutritional quality of the particulate carbon in the food size range (C_2) appears not to be identical to the original algal culture. This point and its significance for the interpretation of continuous culture experiments with rotifers will be discussed in more detail in Chap. 7.1.

4.3.4 Rates and Efficiencies of Carbon Metabolism

Per individuum ingestion and loss rates (Fig. 4.3.5), as well as the difference between both, the production rate (see Eq. 4.3.2 and the energy balance in Sect. 4.2.1; production = ingetion − feces − respiration), increased with D (Fig. 4.3.6). The increase of these rates was curvilinear for *B.* rubens, whereas it appeared to be linear for *B. calyciflorus.* Carbon-specific ingestion (i) and loss rates (L) increased linearly for both species (Fig. 4.3.7). The curvilinear increase of the per individual rates for *B. rubens* probably arises from the fact that the individual body weight in this species changes with the growth rate, whereas it does not do so in *B. calyciflorus* (cf. Fig. 4.3.3). For a small D, the ingestion and loss rates converge towards a common ordinate intercept for both species

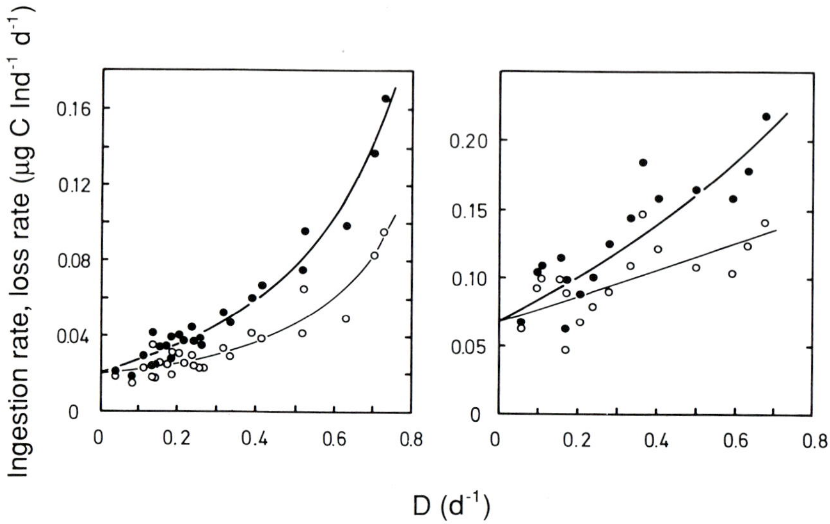

Fig. 4.3.5 Per individuum ingestion rates (IR; *dots*) and loss rates (L∗X; *circles*). *Left panel, Brachionus rubens*; *right panel, B. calyciflorus*

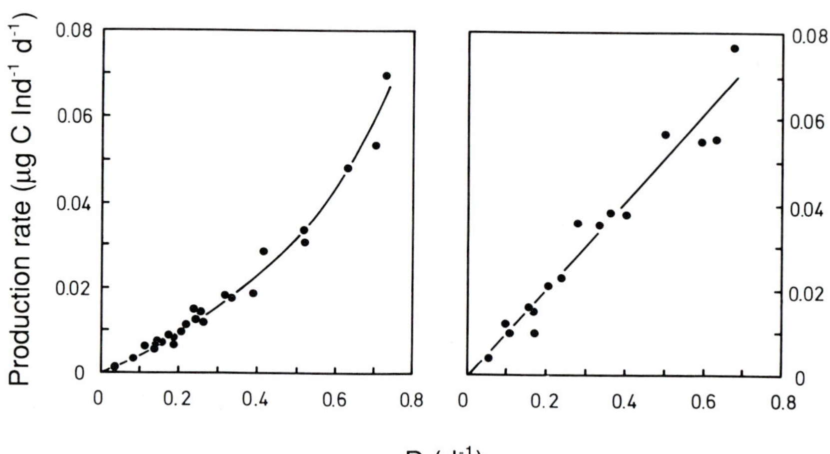

Fig. 4.3.6 Production rates (PR). *Left panel, Brachionus rubens*; *right panel, B. calyciflorus*

(Fig. 4.3.5; in both cases the intercepts of the ingestion and loss rates are not significantly different). This intercept can be taken as an estimate of the minimum ingestion rate for zero population growth (threshold food ration, i_0). This ingestion threshold amounts to $0.021 \pm 0.003\,\mu g$ C Ind^{-1} day^{-1} (0.875 ng C Ind^{-1} h^{-1}) for *B. rubens* and to $0.068 \pm 0.018\,\mu g$ C Ind^{-1} day^{-1} (2.8 ng C Ind^{-1} h^{-1}) for *B. calyciflorus*.

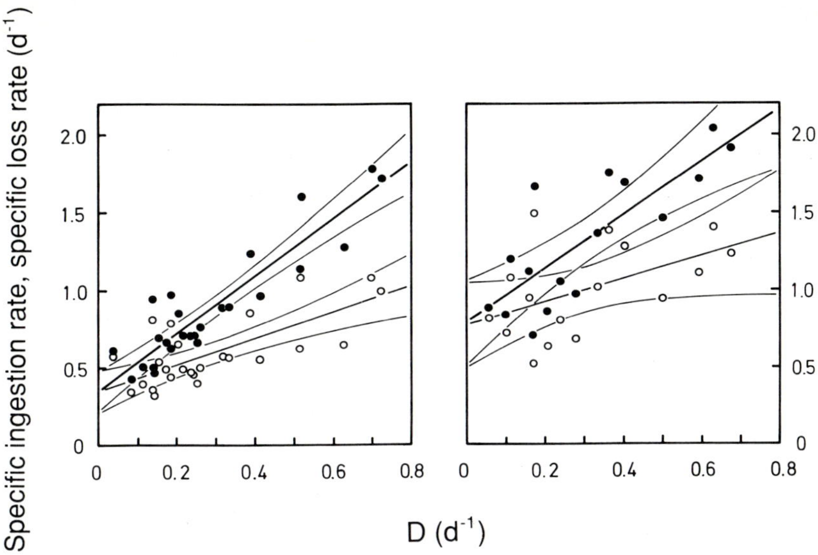

Fig. 4.3.7 Specific ingestion rates (spec. IR; *dots*) and specific loss rates (spec. L; *circles*). *Left panel, Brachionus rubens; right panel, B. calyciflorus*

These values reflect the higher threshold food requirements of the larger species (Stemberger and Gilbert 1985) and correspond closely to estimates for the same species that were obtained with different methods (0.77 ng C Ind^{-1} h^{-1} and 2.4 ng C Ind^{-1} h^{-1}, respectively; Rothhaupt 1990b). At this threshold, the ingested food covers the metabolic losses and permits a birth rate that compensates population losses due to senescence, thus leading to zero net production. Consequently, the curves of the production rates intersect the origin (Fig. 4.3.6; both intercepts not significantly different from 0).

The growth yield (Y) as the ratio of rotifer biomass developed to food consumed, increased hyperbolically with D for both species (Fig. 4.3.8), similarly as described by previous authors (Boraas 1983; Schlüter et al. 1987; Walz 1987b). The higher specific loss rates (L) of *B. calyciflorus* obviously resulted in lower growth yields of this species. Maximal gross growth efficiencies in terms of carbon were about 0.35 for *B. calyciflorus* and 0.45 for *B. rubens.* In a review of the literature, Walz (1987b) found K_1 values for rotifers between 0.11 and 0.38 (see Table 6.3.1). Schlüter et al. (1987) report maximal carbon-turnover efficiencies of 0.43 for *B. rubens.* Of the parameters of the Pirt model (Pirt 1965) fitted to both data sets, the maximum growth yield (Y_{max}) did not differ significantly for both species, whereas *B. calyciflorus* had a significantly higher specific maintenance rate (μ_c) than *B. rubens* ($F = 38.8$, 1,38 df, $P < 0.001$). This again supports the view that both species differ in their basic metabolic costs and conforms with the general body size trend in rotifers (Stemberger and Gilbert 19887a).

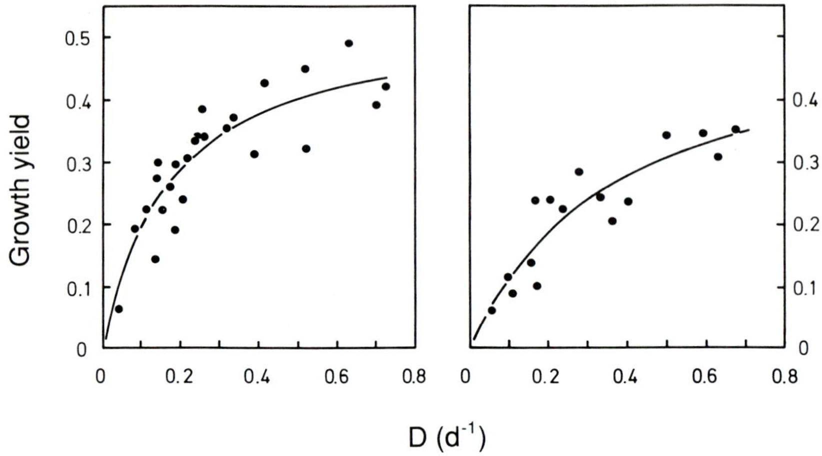

Fig. 4.3.8 Growth yields (Y, K_1). Estimated parameters of the Pirt equation (± 1 SE) are $Y_{max} = 0.552 \pm 0.051$, $\mu_c = 0.328 \pm 0.053$ for *Brachionus rubens* (*left panel*); and $Y_{max} = 0.524 \pm 0.092$, $\mu_c = 0.676 \pm 0.129$ for *B. calyciflorus* (*right panel*)

4.3.5 Rates and Rotifer Body Size

In conclusion, one basic prediction was confirmed by the results: The larger species, *B. calyciflorus*, had higher food requirements than the smaller species, *B. rubens*, and these were apparently caused by higher metabolic losses. A second body size trend, however, was not confirmed. The larger species did not reach greater maximal growth rates in the present experiments. This may have been due to the nonoptimal quality of the algal food. In a comparison of the numerical responses of both rotifer species to different food algae, *B. calyciflorus* reached significantly higher maximal growth rates than the smaller species only when one food alga that proved to be the highest quality food (*Cyclotella meneghiniana*) was offered (Rothhaupt 1990b).

5 Growth Models of Rotifers

5.1 Regulation Models in Rotifer Chemostats

NORBERT WALZ

5.1.1 The Monod Function as a Basis for Models in Chemostats

The classical basis for modeling growth in chemostats is the Monod model (Eqs. 3.3.5 and 3.3.7; Monod 1942). It seems to be the best starting point compared with other models which come into question (e.g., Teissier 1936; Moser 1957; Contois 1959) to describe the growth of microorganisms. The Monod model has been tested in many variations in chemostat studies of bacteria vs. organic substrates (e.g., Bungay and Bungay 1968), of algae vs. inorganic nutrients (e.g., Benndorf 1980; Burmaster 1979; Rhee 1980), of protozoa vs. bacteria (e.g., Hamilton and Preslan 1970; Curds and Cockburn 1971), and of protozoa vs. algae (Gold 1973). For phytoplankton development, a Monod model is widely accepted (Dugdale 1967) as long as the nutrients are not stored. In the latter case, intracellular nutrient pools become important (Droop 1974, 1983). The Monod model adequately describes the growth of many unicellular organisms that have simple cell division.

Rotifers, however, have a more complicated propagation mode. Eggs are produced, although the complications of sexuality are simplified by parthenogenesis. Heterogenesis did not appear in continuous cultures of *B. angularis* and *K. cochlearis* (Chap. 4.1). The strain of *B. angularis* lost its ability to develop resting eggs after several months of culture. Sometimes males were found, but not in steady states of chemostats. They apparently require higher food concentrations than amictic females (Snell et al. 1983; Rothbucher 1988). Nevertheless, the Monod model was successfully applied to rotifer cultures by Boraas (1983) and Walz (Chap. 4.1). Boraas (1983) found a good correspondence of the model in steady-state, but less so in transient-state cultures. Walz (Chap. 4.1) successfully applied the Monod model for transient states, as well.

5.1.2 Regulation Principles in Chemostats

The second stage of a two-step chemostat is, like the first stage itself, a self-regulating system. In the steady state, the rotifer density as well as the concentration of algae are regulated at constant values. This is managed by a

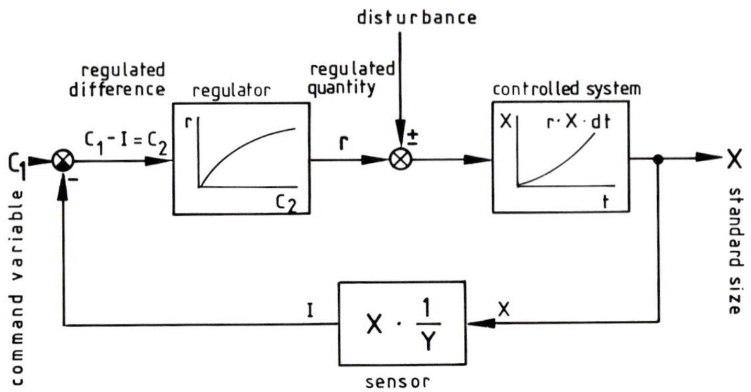

Fig. 5.1.1 The control loop in the second stage of a rotifer chemostat

simple control loop (Merz and Jaschek 1990). In this loop, a "standard size" (the rotifer density) is held constant by a "regulator" (Fig. 5.1.1).

This correcting element for the rotifer density is determined by the metabolism of the rotifers, i.e., by the relationship of the rotifer growth rate to the C_2 algae concentration (see Chap. 3.3, Fig. 3.3.3). The individual shapes of the Monod curves, therefore, their r_{max}, K_s, and C_0 parameters (see below), determine the special characteristics of the regulator for any organisms. The regulator may positively or negatively influence the growth rate as a "regulated quantity". In the "controlled system," the growth rate is integrated over a small time interval (dt) to rotifer biomass (X), the "standard size." This process may be influenced by other factors which together may be characterized as "disturbance variables."

In a "feedback loop," the rotifer biomass has a negative effect on the food concentration. This is accomplished by the grazing activity of the biomass, which is the "sensor" for the algae density. Ingestion (I) of food is subtracted from the food input (C_1). The residual algae concentration (C_2) is the result of this diffference (see Eqs. 3.3.9, 3.3.10).

In ecological control loops, it is often difficult to relate an individual element of the loop to a physical mechanisms (Siebeck 1986). By no means are these elements lacking, nor is a sensor; but this is often more an abstract quantity (Wilbert 1962). The regulation loops are open outwards, but with constant exchange rates. Except for the input of algae (C_1), there is an output of rotifers and algae.

Disturbances of the C_2 algae density in the chemostat, for example, by an increased C_1 input, are balanced out by the regulation loop. If rotifers grow faster, the population density rises, ad the grazing pressure on the algae intensifies. The C_2 concentration declines, slowing the rotifer growth rate, with the rotifer density remaining at a higher level. The C_2 returns to nearly its original value. The C_1 input of algae is the "command variable" of the system. It governs the height of the resulting rotifer biomass. The C_2 algal concentration in the reactor follows as a difference from the grazing rate (Eqs. 3.3.9, 3.3.10). This is the "regulated difference" and the entrance variable to the regulator. Therefore, the system has the ability to adapt to variable C_1 inputs and to regulate different disturbances (Röhler 1973).

5.1.3 Presentation of a Regulation Model

The following cybernetic model was derived from the Monod model and the regulating principles in the chemostat. The rate of change of the rotifer (X) and algae (C_2) concentrations is described by two differential equations:

$$\frac{dX}{dt} = r*X - D*X \tag{3.3.4}$$

$$\frac{dC_2}{dt} = C_1*D - C_2*D - i*X. \tag{5.1.1}$$

The Monod function relates the kinetics of the growth to the energetics of the food concentration (Eq. 3.3.7, correcting element for rotifer growth). Equation 3.3.4, therefore, becomes nonlinear. The population increases exponentially with the actual growth rate, and the population density is stepwise integrated analytically by:

$$X_t = X_0 * e^{r' * dt} \tag{5.1.2}$$

(with $r' = r - D$).

The ingestion rate of rotifers is dependent on the algal density, as determined by the saturation kinetics of Eq. 4.1.5 [$i = i_{max} * C_2/(K_m + C_2)$]. This is the correcting element for algae density. The parameters of both Monod functions for *B. angularis* are given in Sect. 4.1.3 and in Table 4.1.3. The algae density is integrated according to $C_2 = C_1 - X/Y$ (Eq. 3.3.10, according to Herbert 1956) after having substituted Y by r/i (Appendix). In practice, numerical integration is better done by a Runge-Kutta or an Euler-Cauchy version of:

$$C_2 = C_2 + (C_1 - C_2)*D*dt - i*X*dt \tag{5.1.3}$$

The equations are connected in a flow chart (Fig. 5.1.2) of a simple model simulation. After the input of the constants and the starting conditions, the equations are calculated in 50 iterations per day (dt = 1/50 which is a compromise between small step size and rounding error) with a daily graphic and numerical output of X, C_2, and Y (Walz 1992).

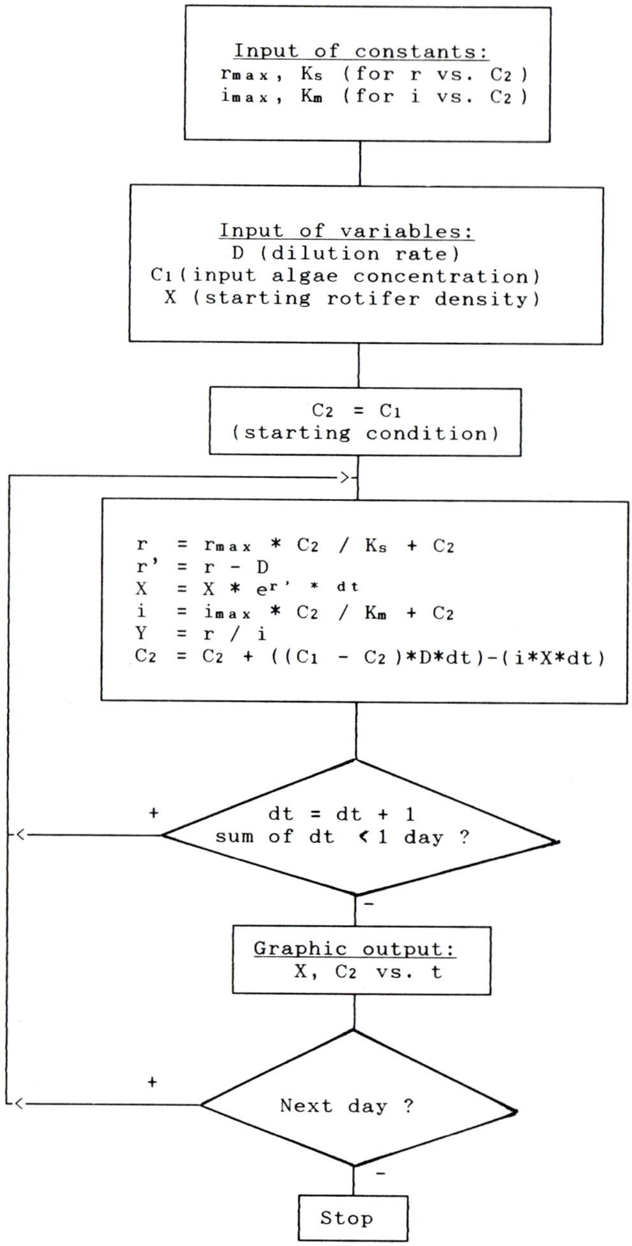

Fig. 5.1.2 Flow chart of the computer simulation model (after Walz 1992)

5.1.4 Model Validation

5.1.4.1 Model Validation of Steady States

The stable outcome of the calculated rotifer biomass (X_{-calc}) is compared in a linear regression with the observed rotifer biomass (X_{-obs}; see Chap. 3.3):

$$X_{-calc} = 3.616 + 0.956 * X_{-obs} \tag{5.1.4}$$

$$R = 0.746,\ n = 14,\ P < 0.01.$$

At $P = 0.05$ the slope is not significantly different from 1, and the intercept is not different from 0. This means there is a 1:1 correspondence between model and observations, and by proof with this method the model cannot be refused.

5.1.4.2 Model Validation of Transient States with Constant C_1 Algae Input and Dilution Rates

The correspondence of simulations with real chemostat runs was exclusively tested with data sets that had not been used for the determination of the constants. Figure 5.1.3a,b shows two simulations (N_{-calc}, C_{2-calc}) and two respective experimental runs (N_{-obs}, C_{2-obs}) in transient states of *B. angularis* with *Stichococcus bacillaris* as food that attained steady states (Walz 1992). In these cases, the C_1 algae input and dilution rates were constant. The simulated rotifer densities follow a similar development to the experimental data and attain similar steady-state plateaus. The calculated C_2 food concentrations are always higher than the experimental ones at the beginning of the experiments, but soon diminish to the observed algae concentrations. The model, as in the culture, reaches stable steady states. This may be approached by an overshoot or a damped oscillation (Fig. 5.1.3b).

5.1.4.3 Model Validation of Transient States with Variable C_1 Input Concentrations

The algae input concentration into the system does not remain constant under natural conditions. $D * C_1$ is equivalent to the rate of primary production (Chap. 7.2). Therefore, a simulation with a variable C_1 was performed (Walz 1992). C_1 as the command variable was altered daily. The simulation model and the real run correspond well (Fig. 5.1.3c); however, from the 18th day when C_1 begins to increase, the observed and calculated time series of the rotifer densities diverge. The model follows the increase of C_1. The observed rotifer densities, on the other hand, decrease to a greater extent. Although from the 20th day on the C_2 food concentration in the reactor increases as well, the population density of the rotifers does not recover. It may be possible that the input concentration had attained a lower threshold, so that the rotifer population entered a poor physiological condition and could not recover immediately. A

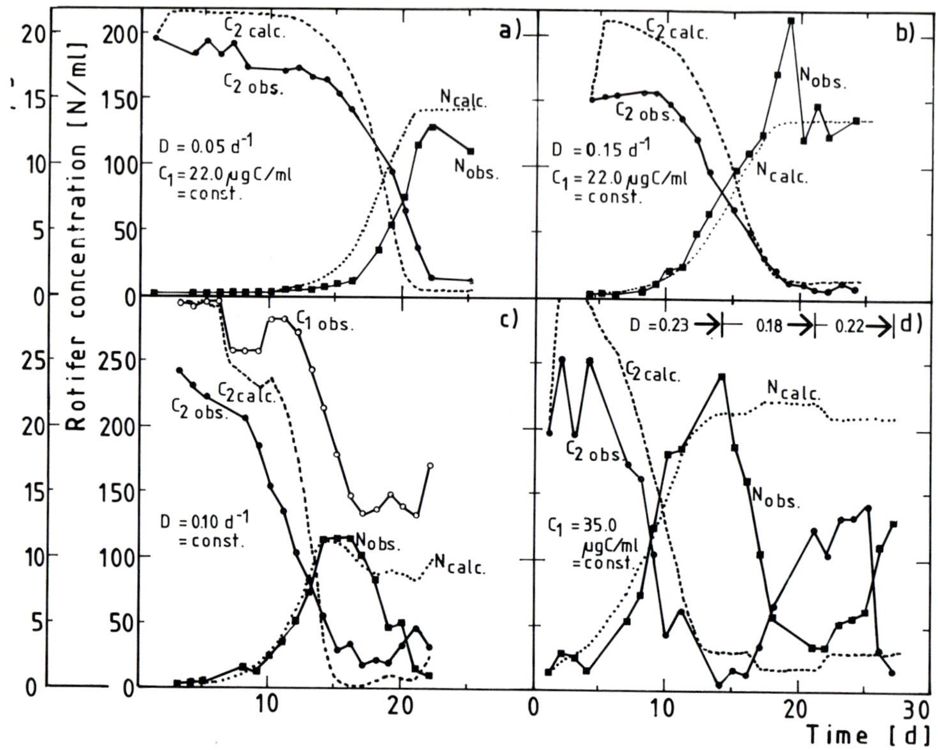

Fig. 5.1.3 a–d Model validation. Simulation and comparison with transient states of a *Brachionus angularis* chemostat: **a**, **b** C_1 and D are constat; **c** D = constant, C_1 = variable; **d** D = variable, C_1 = constant. N_{obs}, C_{2-obs}, observed rotifer and algae density in the reactor; N_{calc}, C_{2-calc}, calculated rotifer and algae density. (After Walz 1992)

sudden incease of substrate may require the induction of further enzymes or macromolecular synthesis if the perturbation is high. These processes necessarily have associated time lags (Barford et al. 1982).

5.1.4.4 *Model Validation of Transient States with Variable Dilution Rate*

The same phenomenon of a "nonrecovery" appears when the dilution rate is changed (Fig. 5.1.3d) (Walz 1992). Such a change, in this case a diminution from D = 0.23 to 0.18 (day^{-1}), has little effect in the simulation model (see also Bergter 1983). The calculated rotifer density (N_{-calc}) even increases somewhat and the C_{2-calc} food concentration slightly decreases. The observed rotifer density (N_{-obs}), however, dramatically decreases after the change of the dilution rate and recovered only when the previous dilution rate was restored. During the low dilution rate, the rotifer density should have recovered since the C_{2-obs} food concentration meanwhile had reached high values.

A sudden decrease of the dilution rate (shift down) seems to be especially effective as now abruptly for fewer algae enter the reactor, and the rotifers are used to a higher food supply. Although now the rotifers have a lower food supply rate, they also have a lower dilution loss rate. The simulation responds by a small increase in the rotifer density. Overall, the energy metabolism must have entered a starvation condition with elevated mortality; also, a longer holding time (1/D) means increased biomass loss per individual by respiration (Boraas 1983). The rotifers need some time to switch back into the growth metabolism. Such time delays may be integrated into the model (Halbach and Burckhardt 1972; MacDonald 1976). They are elements of many regulation processes (Röhler 1973), see below.

5.1.5 Model Simulations for Monoxenic Conditions

5.1.5.1 Stability of Steady States

The stability of steady states is studied in model simulations of *B. angularis* with *Stichococcus bacillaris* as food as functions of the residence time (1/D) and C_1 food input concentration. The apparent maximum dilution rate is introduced from Sect. 3.3.8. Without a time-delay the approach to the steady state values of rotifer density and of food concentration is monotonic, i.e., without oscillations and overshoot. All steady states inside the field, which is not washed out, are stable (Fig. 5.1.4). No overshoots, damped or cyclic oscillations occur in these simulations (simple case, without time delays). This, however, is not always supported by the observations in the chemostat (Fig. 5.1.3b), where overshoots and oscillations are found frequently.

In contrast, single-stage models tend to become unstable at lower dilution rates. In a Monod-based model of a single-stage chemostat with glucose,

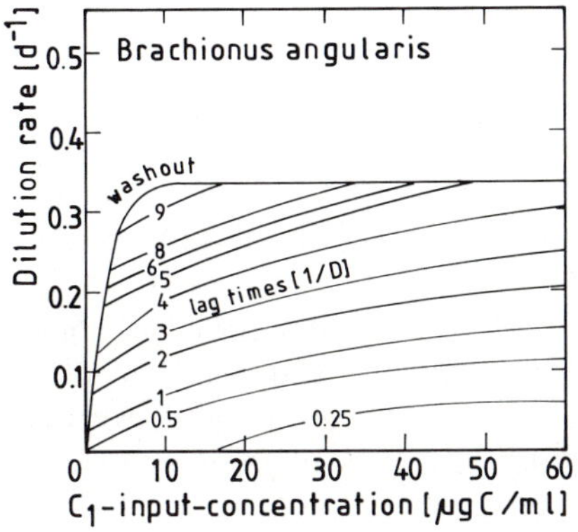

Fig. 5.1.4 Characteristics of steady states of *B. angularis* in simulation models as dependent on C_1 input concentration, dilution rate, and lag time (in residence times, 1/D). Lag times lower than indicated allow damped oscillations resulting in stable steady states., (After Walz 1992)

Escherichia coli, and the amoeba *Dictyostelium discoideum*, different stability regions were found depending on the input concentration and the residence time. In one of these regions, the densities approach monotonically to steady states. At higher residence times, on the other hand, populations show damped and periodic oscillations (Tsuchiya et al. 1972). In a "multiple saturation" model (Jost et al. 1973b), the oscillations disappeared when D decreased in accordance with the experimental findings.

5.1.5.2 Stability with Time Lags

A stand-by reaction of rotifers under changing food conditions is not very probable. A time lag between the ingestion of food and growth is more realistic (Caperon 1969; Sect. 7.1.2). Time lags have often been used in Lotka-Volterra models by substituting N by $N_{(t-delay)}$ in Eq. (5.1.7) (May 1973; Barclay and Van den Driessche 1975). MacDonald (1982) discussed time lags in the logistic and the Monod equation.

Time lag is introduced into the regulation model by the following modification of the Monod equation:

$$r = \frac{r_{max} * C_{2(t-delay)}}{K_s + C_{2(t-delay)}}. \tag{5.1.5}$$

The growth rate is based on the substrate concentration at the time (t – delay). The ingestion rate, in contrast, follows directly the nondelayed food concentration $[C_2(t)]$ as before.

At low time lags, rotifer densities and C_2 food concentrations show damped oscillations and a monotonic approach to steady states. Both densities ultimately reach the same values as in simulations without time lags. At higher lags than these indicated in Fig. 5.1.4, populations enter into stable limits cycles with periodic oscillations and do not ever reach stable steady states. At even higher time lags, not indicated in Fig. 5.1.4, chaotic conditions occur in the model with the final extinction of the rotifers. The extent of a possible time lag to allow damped oscillations depends on the dilution rate and C_1 input food concentration. Time lags (in residence times; Fig. 5.1.4) may be longer at greater dilution rates and higher input concentrations.

This operating diagram does not fit the expectations of May (1981; see also Chap. 7.1.2), according to which populations should become unstable, exceeding the critical values of the product $r_{max} * T$ (T = time delay or generation time), either at great dilution rates or at long time lags. Perhaps these differences are explained by the various underlying models. May (1981) derived this expression from the time-delayed logistic equation.

Long time lags derived from Eq. (5.1.5) depend on the previous algae concentrations. Such long time lags become more and more independent of the actual population development and function rather as a changing command variable. For that reason, it seems to be understandable why the system at higher growth rates is rather stable at longer time lags.

Other kinds of time lags integrate previous C_2 concentrations. In the Monod growth equation, the actual C_2 was substituted by a progressing mean of C_2 between t = 0 and t − delay:

$$C_2 = \frac{\sum_{t=0}^{t-\text{delay}} C_2(t)}{\text{delay}}. \tag{5.1.6}$$

Operation diagrams and time lags are very similar to Fig. 5.1.4 for time lags allowing damped oscillations. In contradistinction to the method above, limit cycles do not appear, but populations get into chaotic conditions exceeding the indicated critical time lags. Cunningham and Nisbet (1983) demonstrated how the stability of transients depends on the applied time-lag function.

5.1.5.3 Model Applications to Transient States

The Monod model, the basis for the regulation model, was originally formulated from batch cultures (Monod 1942). The development of the regulation model assumed steady states, in which $dX/dt = 0$ and $dC_2/dt = 0$. This is not the case in transient states and in nature. However, in the regulation model, change of dt is minimized by frequent iterations, and by these means, deviations of dX/dt from zero are small. The transferability to transient states has to be verified by comparisons of simulations with experimental runs in chemostats or, as a further step, in nature. The results in chemostats with *B. angularis* (Fig. 5.1.3) are in first approximation in concordance with the model. Later, more realistic models with time delay have been incorporated, which could modify the temporal behavior of the model.

As a response to changed parameters (input food concentration, dilution rate, temperature), first the age structure of the population has to adjust. From this, new distributions of the age-specific l_x and m_x (Sect. 6.3.6) follow, which result in new birth and mortality rates. Time delays are well known in systems theory, and it depends on the kind of the regulator how quickly and how exactly the regulated quantities reappear.

In the literature on bacterial and protozoan chemostats, the transferability of models to transient states is accepted to a large degree and supported by model simulations and experiments. Experimental studies on bacteria show differences in the adjustment of the new steady states after disturbances, which depend on the kind and size of the disturbance. After small disturbances of the dilution rate, the systems proceed to a new steady state without a perceptible change of the limiting substrate in the reactor (Mateles et al. 1965). Only in this way are steady states obtained in practice, as small disturbances always have to be "buffered". After greater disturbances of the dilution rate, on the other hand, the reactor substrate concentration changes, too. The Monod model predicts in these cases often a more rapid adjustment than actually occurs (Cooney and Wang 1976; Bergter 1983).

Also, if the C_1 substrate is changed, the time of return to a steady state depends on the dilution rate. Fast adjustments arise in fast-growing cultures and vice versa (Höfle 1984). In the long term, only the cell concentration changes; the growth rate returns to its former value (Regan and Roper 1971). Often damped oscillations around the new steady state follow the adjustment (Gilley and Bungay 1967; Koga and Humphrey 1967).

Hitherto, computer simulations with bacteria and protozoans were only operated with constant Y values in contrast to the models presented here. Such simulations showed no immediate adjustments of a new growth rate of the bacteria after a rise of the input substrate concentration. On the contrary, the growth rate diminished instantly when the substrate concentration was decreased (Curds and Bazin 1977). In a Monod model for protozoans, the reactions was independent of the starting protozoan density, which is the case in experiments, whereas in a Lotka-Volterra model, the development was governed by the starting conditions (Tsuchiya et al. 1972). Jost et al. (1973b) replaced a Monod model with a "multiple saturation model" which gave a better fit to the observed data. For this case, Bonomi and Fredrickson (1976) could still employ the Monod model with a good fit to the experiments by allowing wall growth, which gives the prey some protection. Pickett (1982) had the best success with a structured Monod model to adapt step and continuous perturbations of the input limiting substrate. Curds (1974) supposed that the size of the energetic variables (Y, K_s) has a greater influence on the population development than the specific form of the kinetic function. He recommended basing the equations directly on ecological realities.

The kinetic-energetic regulation model of Sect. 5.1.3 and the model of Rothhaupt (1985), which is based on similar assumptions, meet these demands the best. Rothhaupt calculates the growth and filtration rates of rotifer populations with the explicit input of several metabolic parameters, whereas in the regulation model, both rates are derived from the Monod functions depending on the food concentration. Rothhaupt (1985) observed a faster approach to the steady state in his experimental chemostat cultures of *B. rubens* than in his model. In contrast, the regulation model approach to a steady state is a little bit faster than in *B. angularis* chemostats (Fig. 5.1.3). In experimental runs, chemostat cultures of both authors reached the steady state often only after an overshoot.

5.1.6 Model Simulation with Competing Rotifer Species

5.1.6.1 Steady-State Solutions

The maximum number of populations in a system depends on the individual Y_i (see below). In the steady state, the population which has the lowest Y wins. This drives the C_2 concentration to its lowest values (see Eq. 3.3.10), with which this population is able to survive (Sect. 6.3.4; Armstrong and McGehee 1980).

For this reason, Y is parallel to K_s. The second condition is $r_{max} \geq D$. The following model shows this relationship. If there are two populations competing for the same substrate (which is the case in the pelagial for many consumers; Sect. 6.3.3), they should have different r_{max} and K_s characteristics (Sect. 6.3.7). If $r_{max2} > r_{max1}$, $K_{s2} > K_{s1}$, and $D > r_{max1}$, then population 2 wins (r_{max} strategy, high K_s value), and the residual algae density (C_2) remains high. If $D < r_{max1}$, population 1 should be successful (K_s strategist, low K_s value). This is confirmed by the rigorous mathematical analysis of Hsu et al. (1977). The residual algae density should be lower than in the first case. The outcome of competitive situations depends on the dilution rate. In the aquatic ecosystem, this dilution rate is often represented by the pressure of predators (Chap. 7.2; not to mention a lake outflow). The knowledge of the life strategy of the consumers gives important informations on the characteristics of the correcting elements in the regulation loops. These are identified by the metabolism of the organisms.

The above-described relationships are shown in simulation models with two competitors, *B. calciflorus* and *B. rubens*, feeding on one common food, *Cyclotella meneghiniana* or *Monoraphidium minutum*, respectively. The constants of growth and feeding were taken from Rothhaupt (1988):

1. With *Cyclotella* as food (r_{max} # K_s # C_0) for *B. rubens* = 0.838 # 0.138 # 0.107, for *B. calyciflorus* = 1.020 # 0.434 # 0.193.
2. For *Monoraphidium* as food (r_{max} # K_s # C_0) for *B. rubens* = 0.794 # 0.175#0.071, for *B. calyciflorus*: 0.785 # 0.669 # 0.236.

C_0 is the threshold food concentration, where $r = 0$, to be subtracted from C_2 in the Monod equation (in analogy to Eq. 4.2.2):

$$r = \frac{r_{max} * (C_2 - C_0)}{K_s + (C_2 - C_0)}. \qquad (5.1.7)$$

C_2 is integrated into the competition model by Eq. (5.1.9), see below.

At the beginning of any simulation, *B. calyciflorus* (r_{max} strategist, Sect. 6.3.7) starts to grow more intensively in *Cyclotella* cultures than *B. rubens* (K_s strategist), but soon *B. calyciflorus* disappears except at high dilution rates, where it replaces *B. rubens* with stable steady states (Fig. 5.1.5). At even higher dilution rates and at very low input concentrations, both species are washed out. A coexistence of both species appears only at $D = 0.761$ (day^{-1}) over the entire range of input food concentrations with steady-state rotifer biomasses ($\mu g\,C\,ml^{-1}$) increasing from 0.4 *B. rubens* and 0.3 *B. calyciflorus* at $C_1 = 5\,\mu g\,C\,ml^{-1}$ up to 1.2 *B. rubens* and 9.3 *B. calyciflorus* at $60\,\mu g\,C\,ml^{-1}$. The yield coefficient (Y) of *B. rubens* decreases from 0.13 to 0.02, whereas the Y of *B. calyciflorus* increases from 0.10 to 0.16. In practice, it should be extremely improbable that one can maintain that dilution rate exactly.

In all other cases, that population wins whose "Michaelis-Menten constant is smallest in comparison with its intrinsic rate of increase" (Hsu et al. 1977). As the section of both Monod curves is at $C_2 = 1.47$ ($\mu g\,C\,ml^{-1}$), the residual

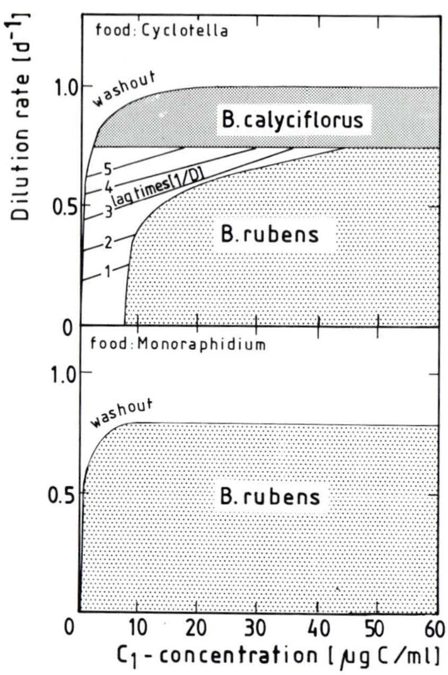

Fig. 5.1.5 Characteristics of existence and coexistence of two rotifer species in simulation models as dependent on C_1 input food concentration, dilution rate, and lag time (in residence times, 1/D). *Dashed areas*, stable steady states of only one species. Lag times lower than those indicated allow the coexistence of both species in periodic oscillations. *Above*, *B. rubens* and *B. calyciflorus* feeding on *Cyclotella meneghiniana*. *Below*, *B. rubens* and *B. calyciflorus* feeding on *Monoraphidium minutum*

algae concentrations are lower when *B. rubens* wins. With *B. calyciflorus*, the residual concentrations are higher than this value.

In contrast, with *Monoraphidium* as food, we find another pattern (Fig. 5.1.5). There is not even a theoretical steady-state coexistence between both rotifer species, and *B. calyciflorus* always disappears after the initial growth phases. *B. rubens* wins at all times. The maximum growth rates are too close together, and both Monod curves do not intersect. Therefore, the lower K_s of *B. rubens* is the deciding term in any case.

Many theoretical studies (Hardin 1960; MacArthur and Levins 1964) derived the competition exclusive principle mathematically for continuous cultures. Taylor and Williams (1975) and Stewart and Levin (1973) found that stable mixed populations of competing species could not live in coexistence unless the number of growth-limiting substrates is equal to or greater than the number of species. This is also predicted and confirmed experimentally by the theory of Tilman (1982) and its application to zooplankton (Rothhaupt 1988). The outcome may be predicted in advance, in contrary to the Lotka-Volterra theory, "from measurements made on each species grown by itself on the resources" (Hsu et al. 1981).

5.1.6.2 Competition Under Fluctuating Conditions

If two populations compete on two complementary growth-limiting substrates, simulations with monotone increasing kinetics (to which also Monod kinetics belong) give stable equilibria under specific conditions. The outcome, however, may be dependent on the initial conditions (Butler and Wolkowicz 1985). When at least one competitor is inhibited at high substrate concentrations, periodic oscillations result (Butler and Wolkowicz 1987). In the simulation studies of Powell and Richerson (1985) and Taylor and Williams (1975), the equilibria adjusted very slowly (> more than 10 generation times) on the basis of Monod kinetics. Increased loss rates led to damped or even long-term oscillations and limit cycles with simultaneous (but counter-coursing) existence (to avoid the term "coexistence") of both competitors for a long time. Such a "dynamic coexistence" maintained by sustained oscillations or seasonal variations is found in both Lotka-Volterra (Armstrong and McGehee 1980) and Monod models (Stewart and Levin 1973). These fluctuations contributed more to the coexistence of phytoplankton species in nutrient-pulsed chemostats than would be possible under steady-state conditions, where the competitive exclusion principle is valid (Sommer 1984).

Fluctuations in the model are driven by time lags (see above). Both calculation methods have been applied. Coexistence is possible in the model with time lags defined by Eq. 5.1.5, but only in a small band which is increasing with higher dilution rates and lower input concentrations. At high dilution rates, higher time lags are possible which allow coexistence (Fig. 5.1.5, for *Cyclotella* as food). Above $D = 0.761$ (day^{-1}), no coexistence was found. Both populations coexist, if the boundary conditions are in the given range, in limit cycles relieving one another. On the contrary, with *Monoraphidium* as food (Fig. 5.1.5), no coexistence was found with any time lag and *B. rubens* always wins. In that case, *B. calyciflorus* is defeated by all characteristic parameters of the Monod equation (see above). For coexistence, both curves must have a crossing point, which these do not. Fluctuations may only enlarge the range of possible coexistence.

Similar operation diagrams for model simulations of competing species were found by Baltzis and Fredrickson (1984) for two competitors in a state of sustained oscillations. At the respective minimum of the density in the cycle, there is a probability for all populations to become extinct. For cases like this, the authors recommend a stochastic formulation of the problem. The stability of competition was also studied by Aris and Humphrey (1977).

Models with negative feedback and with a long time lag compared with the natural time scale of the system tend to become unstable (May 1973). Unlike predictions put forth by May (1973) and May et al. (1974), instability in the presented regulation model appears more often at lower dilution rates.

At low amplitudes of a periodically varying input concentration of one nutrient model, simulations show that the K_s strategist of two competitors wins. As the amplitude is increased, a threshold is exceeded at which both competitors coexist in periodic cycles. If the amplitude is further increased, a second threshold is reached beyond which the r_{max} strategist wins (Smith 1981). The necessary condition seems to be, as above, that both Monod growth curves intersect.

Another way to achieve coexistence is the introduction of a predator (Smouse 1981). Addition of the ciliate *Tetrahymena* as a predator of the bacteria *Escherichia coli* and *Azotobacter vinelandii* stabilized the community, especially *Azotobacter*, which otherwise would be outcompeted by *E. coli* (Jost et al. 1973a). Fredrickson and Stephanopoulos (1981) give further conditions making a coexistence between two microbial competitors possible. Sinking of organisms and diffusion of resources may allow coexistence in situations in which the dominance of one organism is predicted (Powell and Richerson 1985).

5.1.7 Regulation, Logistic, and Lotka-Volterra Models

The study of population dynamics is extensively based on the logistic model (Verhulst 1838; Pearl and Reed 1920). This model presupposes that populations grow first exponentially in a rarefied environment (i.e., few organisms = few intraspecific competitors for food or space) with the growth rate $dN/dt * N = r_{max}$. Regulating factors in this phase are density independent (e.g., abiotic factors). If the population grows further, the growth rate diminishes linearly with the population density (N) and finally reaches the carrying capacity (K) of the system, above which no furthr growth occurs. Therefore, it has to considered a direct density-dependent term, and the logistic equation is:

$$dN/dt = r_{max} * N * (K - N)/K. \qquad (5.1.8)$$

As N approaches K, r declines. At $N = K$, $r = 0$. Then, the population is regulated by density-dependent factors. Logistic models have also been used for predator-prey and competition interactions (Wangersky 1978) and applied to the population dynamics of rotifers (e.g., Halbach 1979) as well.

In favor of the regulation model, the logistic equation has not been used here because:

1. There is no linear relationship between N and r (Smith 1952; Jacobs 1984). In steady states of rotifers, no linear relationship was found between N and D, the dilution rate, which is equivalent to r in a steady state (Chap. 3.3, Fig. 3.32).

2. The logistic model is too inflexible. In its enlarged form for competition between two species (Lotka 1925; Volterra 1926), it defines competition by linear coefficients with which each population affects the other. Boraas et al. (1990) claim to have falsified the Lotka-Volterra competition model. Linear coefficients between species changed significantly with changes induced in the species composition of the remainder of the community (Neill 1974). Attempts to introduce nonlinear relationships into the logistic equation require an extremely complex mathematical treatment very quickly (Goel et al. 1971). Beyond this, the Lotka-Volterra models are only descriptive and give no insight into the underlying mechanism of the processes.

3. The logistic model tries to describe the limitation of the population density with a regulation term embedded in the equation. This is managed

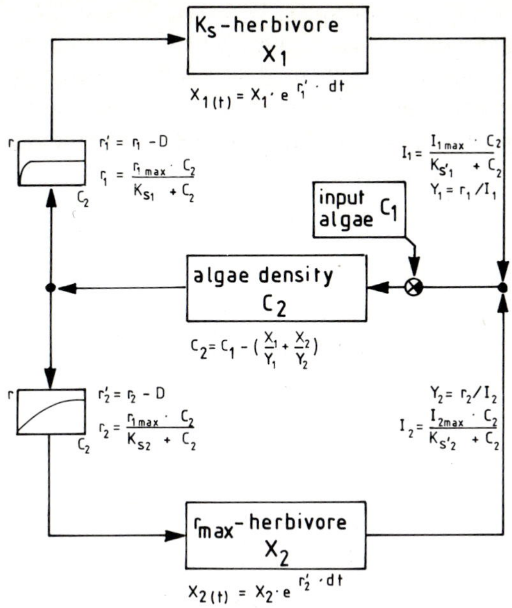

Fig. 5.1.6 Model for regulation loops with competing consumers

directly by an explicit declaration of the maximum number of individuals which will be achieved in the limited growth phase (= carrying capacity). This is no real feedback. In nature, however, this abundance is realized in a confrontation between the population and the feedback on the resources of the environment. The weakness of the logistic and derived models is, therefore, that the carrying capacity does not reflect the varying environment.

In contrast, the regulation model assumes a defined substrate limitation influenced by the growth activity of the population. It regards the substrate regulator as a connecting link for all other consumer populations which are dependent on this substrate. It assumes exploitative competition and no direct interference between the competitors. In competition of two populations (a, b) for a common substrate, the following equation is derived from Eq. (3.3.10) (Sect. 3.3.4, see also Eq. 7.2.1).

$$C_2 = C_1 - (X_a/Y_a - X_b/Y_b). \qquad (5.1.9)$$

Accordig to Fig. 3.3.5 the difference $C_1 - C_2$ is identified as analogous to the carrying capacity of the system (environment and populations). This is nothing else than the total ingestion of all populations sharing a common food, i.e., of all members of a specific trophic level. This difference is variable and is divided between the competitors acording to their grazing activities. As C_1 is generally much greater than C_2, the carrying capacity is determined by the algal input variable. The competitors take part by their different yield coefficients.

The control loops for competing species according to this model (Fig. 5.1.6) are an extension of the simple case with one herbivore (Fig. 5.1.1). This extended model additionally only accepts Eq. (5.1.9) as a formulation of the "regulated difference." This means that C_2 is determined by the population with the lowest K_s or the lowest Y, which could be shown by the outcome of the model simulations.

5.2 Comparison of Ecological and Microbial Growth Models

MARTIN E. BORAAS

5.2.1 The Problem

Models play a key role in defining the experimental design and theoretical interpretation for both field and laboratory work. The models themselves, however, reflect fundamental precepts of the importance of underlying ecological processes. The distinction between density-dependent and density-independent factors that affect population growth is emphasized in much of ecology. However, models of animal or microbial growth in continuous cultures require that explicit mechanisms of growth limitation be incorporated into the model structure. Such models tend to be mathematical analogues of the physical system they describe and explicitly use mass-balance equations for both the population and its resource base. Is there simply a procedural difference between these two approaches, or is the distinction more fundamental?

In ecology, for the past 70 years, the model of choice for describing animal growth has been the logistic model (Pearl 1925). This familiar model describes population growth as a smooth, S-shaped curve, which stabilizes at an upper limit, the carrying capacity (see any introductory biology or ecology text and further discussion in Chap. 5.1). Although the carrying capacity is often said to be determined in part by resource availability, the model has no explicit definition of how the population size and resources interact as populations grow. This has led to the uncomfortable situation in which a simple model is presented, particularly at the introductory level, that is said to represent population growth in general, but rarely can it be applied in any specific case.

In this chapter, I first give a verbal description of what I believe to be fundamental differences in these two classes of models. I then develop a microbial model that can be used to elucidate the growth regulation mechanisms of rotifers.

From this analysis, I conclude that density dependence is not a necessary or a useful concept in understanding population regulation in continuous cultures and, by extension, under many other circumstances.

5.2.2 Limits to Population Growth

The means by which the potential for an exponential increase of populations is curtailed is of fundamental interest in ecology. Two general approaches to this problem have been used, one in ecology and one in microbial ecology. In traditional ecology, factors influencing the growth dynamics have been classified as "density-dependent," reflecting a negative interaction between r (specific growth rate) and X (the population density), or "density-independent," indicating no relationship between the factor and X (e.g., Ehrlich and Roughgarden 1987). In microbial ecology, in contrast, population growth dynamics have typically been modeled from the perspective of an explicit limiting resource that originates from the environment.

Historically, the differences in the two approaches may have developed from the motivation of the investigator. The primary motivation for the "ecological" approach was a search for generality: a "law" of population growth (see Kingsland 1985). To characterize this generality, factors limiting population size were lumped together into a single parameter, the "carrying capacity." The developers of the microbial approach were more concerned with developing quantitative models of specific cases (e.g., Fredrickson and Tsuchiya 1977). The investigators wanted to be able to detect and model how changes in growth media, initial conditions, etc., would be expected to alter the sustainable yields of interesting microbes. Much of this work was pioneered by engineers, who assumed that population growth models should be designed to conform explicitly to physical and chemical laws of mass balance.

In the logistic, and related, models the depression in growth rate as a population approaches the carrying capacity is assumed to be a linear function of the population size. Hence, the population implicitly or explicitly is assumed to limit its own rate of growth by density dependence. To apply the model, it is not necessary to define the mechanism for the negative association or even to identify whether it affects fecundity or mortality, although a density-dependent mortality is frequently invoked (Stubbs 1977). Density-independent factors, in contrast, are usually considered to be sources of nonselective mortality (e.g., Ehrlich and Roughgarden 1987).

5.2.3 Experimental Analysis of Specific Mechanisms for Growth Regulation

For the analysis of specific cases, a clear distinction between mechanism and description is necessary. In its simplest form, the assumption of a density-dependent regulation of growth deliberately ignores the effects of specific mechanisms of growth regulation. It is a correlation that presumably reflects such mechanisms. Most authors distinguish between the statistical identification of density dependence and specific mechanisms of density dependence. However, this distinction is seldom extended to discriminate between density-dependent mechanisms that are properties of the population itself, e.g., cannibalism or hormonal suppression (Lomnicki 1988), and mechanisms that are properties of the environment, e.g., resource limitation (Tilman 1982). This distinction is critical and is, I believe, more basic than the classification of density-dependent vs. density-independent processes.

In a continuous culture, a population can show density dependence or not when precisely the same mechanisms are operating. For instance, after inoculation into a culture, the population biomass increases, initially exponentially, at the expense of food mass, while r declines until growth is balanced by washout, a nonlinear density dependence. However, if the input food concentration, C_1, is increased while the dilution rate, D, is held constant, then the population can increase independently of r. Finally, the steady-state concentration of rotifer biomass vs. specific growth rate (i.e., the dilution rate) is not monotonic (e.g., Fig. 3.1.4). In all three cases, the organisms are food limited, but density dependence is evident for only one, and in that case the relationship is not linear, i.e., not logistic.

Causation cannot be inferred from a correlation. A negative relationship between r and X is clearly a correlation. Without additional evidence, there can be no implication of causality. If this correlation is observed in the data, then further investigation is certainly warranted. However, once the causal relationships are known, then does the concept of density dependence continue to have utility? That is, is density dependence a useful basis for model development? Solomon (1958) pointed out that the term "density dependence" was misused and confusing at that time.

5.2.4 Points of Discrepancy Between the Logistic Model and Observed Population Growth

As a specific application of density dependence, the logistic equation assumes a linear decline in r with increasing X to the "carrying capacity of the environment" or K. The concept of an upper limit to population density is consistent with the First Law of Thermodynamics: Energy and mass cannot be created or destroyed. However, the value of the carrying capacity cannot be predicted without first observing the population growth. Nevertheless, an increase in population density requires that the population also increase in biomass. Since the logistic makes no explicit reference to a concomitant decrease in environmental mass, the logistic may be strictly interpreted to violate the First Law by allowing the de novo creation of biomass.

Since the logistic does not assume any specific mechanism of growth regulation, only that growth is ultimately limited, it cannot be tested, i.e., falsified (Williams 1972), with any specific data set. However, the assumption of a linear decline of r with X is experimentally testable. In any population growing exponentially, whether in a steady-state culture ($r = D$) or during exponential increase, r is clearly not related to X in a linear manner (e.g., Boraas et al. 1990; Chap. 5.1).

The simplest test is to examine r vs. X in a batch culture of a unicellular organism, such as bacteria or algae. In this case, as X increases during the exponential population growth of the organisms, r remains constant at r_{max} until just before the ultimate limit is reached. This gives a nonlinear, even rectangular, shape for r vs. X. Since these batch cultures of unicellular organisms mimic the major constraint of the logistic (the carrying capacity is determined in a batch culture), it would be expected that such experimental situations would fit the logistic. This is not the case. The batch culture system provides one of the best demonstrations of how the assumption of linearity of r with X can be falsified with data (Williams 1972).

A linear relationship of r with X is assumed in the logistic, but a convex relationship is usually observed in continuous and batch cultures. However, Smith (1963) observed a concave relationship between r and X in continuously expanding *Daphnia* cultures. Smith used a unique culture method, essentially a fed-batch culture with population harvesting. In a fed-batch culture, food is added and the population increases until there is sufficient biomass present that the metabolic losses balance the food input, so that the population enters a steady state (see Chap. 2.1 for additional details). When the population is also harvested, the steady-state density must decline since the harvesting represents an additional loss to the population. Therefore, the functional form observed for r and X by Smith (1963) is not a peculiar property of *Daphnia* but is a property of the culture method, i.e., the organism-environment interaction. In general, any relationship between r and X has little meaning unless the organism-environment interaction is well understood.

5.2.5 The Utility of Steady-State Continuous Cultures for Assessing Growth Models

A primary value of steady-state continuous cultures is the lack of ambiguity: All factors necessary for sustained growth and reproduction within specific environmental boundaries must be provided, or a steady-state population will not be obtained. Models of steady-state continuous cultures must also meet these rigorous constraints. This does not mean that either the data or the models of steady-state populations are peculiar or restricted in any sense. It does mean that a fuller understanding of environment-organism interactions is required than in most other laboratory or natural studies. Data are now available from steady-state populations of osmotrophic and phototrophic prokaryote and eukaryote unicells, as well as of holotrophic unicells and metazoans in both unispecific and multispecific culture. In general, the similarities between both experimental results and mathematical models of organisms of widely divergent phylogenies are more striking than the differences.

The assumption of density dependence as a causal factor for population growth can be falsified with data from continuous cultures. This is not a trivial circumstance. The logistic assumes that population growth is regulated as a function of itself, that is:

$$\frac{dX}{dt} = f(X). \tag{5.2.1}$$

In contrast, steady-state growth in continuous cultures requires that the growth be described as functions of the environment, that is:

$$\frac{dX}{dt} = f(X, Env) \tag{5.2.2a}$$

$$\frac{dEnv}{dt} = g(X, Env), \tag{5.2.2b}$$

where Env is environmental material that is potential biomass; f(X, Env) is the growth function of the organism; and g(X, Env) is the function describing the dynamics of Env. In continuous cultures, population growth cannot occur in the absence of environmental change, e.g., mass becomes biomass (or biomass becomes waste products). This is a condition that is necessary for life itself, and population models that do not reflect it are unlikely to be of much value for interpreting natural processes.

5.2.6 Models That Explicitly Express Specific Mechanisms for Population Regulation

An alternative to general density-dependent models are models with specific mechanisms of population regulation. Such models have the advantage of being falsifiable (Williams 1972). One such model was originally developed by Monod (1942). Another, related approach is the "cybernetic" model discussed by Walz in Chap. 5.1. Monod assumed that bacteria grew according to a hyperbola on a limiting substrate concentration. His model, subsequently modified by Herbert (1958) to account for metabolic losses, bears little resemblance to most models of animal population growth. This is an advantage, since all models are, at best, approximations, and different viewpoints can provide different insights into ecological processes.

The Monod-Herbert (M-H) model consists of 2 equations and 4 biological constants as follows:

$$\frac{dX}{Xdt} = \frac{r_{max} C_2}{K_s + C_2} - \mu_c - D, \quad \text{and} \tag{5.2.3a}$$

$$\frac{dC_2}{dt} = D(C_1 - C_2) - \frac{r_{max} C_2 X}{Y_g(K_s + C_2)}, \tag{5.2.3b}$$

where r_{max} is the maximum specific growth rate, K_s is the value of C_2 at $r_{max}/2$, μ_c is the maintenance coefficient (minimum specific metabolic loss rate), and Y_g is the maximum conversion efficiency of ingestion to X.

These "constants" are not, however, inherent properties of the organism but are a result of the organism-environment interaction. Figure 5.2.1 shows the specific growth rate of *Brachionus calyciflorus* on three food sources consisting of two algal species. Each food source has a growth threshold that differs by about a factor of 10, despite the fact that two of the foods are the same species at the same temperature and in the same inorganic medium. The foods are *Cryptomonas erosa* at 19 °C (triangles; Stemberger and Gilbert 1985) and *Chlorella pyrenoidosa* at 25 °C (closed and open circles). The sources of the data are explained in the figure legend. Clearly, rotifer growth is affected by food type and temperature (see also Rothhaupt, Sect. 7.1.3).

This model expresses both population and the limiting resource as mass (with Y_g as the conversion factor) and assumes that the population is carbon-energy limited and that the system is in a steady state. It is not, however, a complete bioenergetics model in that all metabolic loss processes are lumped into the terms μ_c and Y_g. The model assumes a hyperbolic growth function for

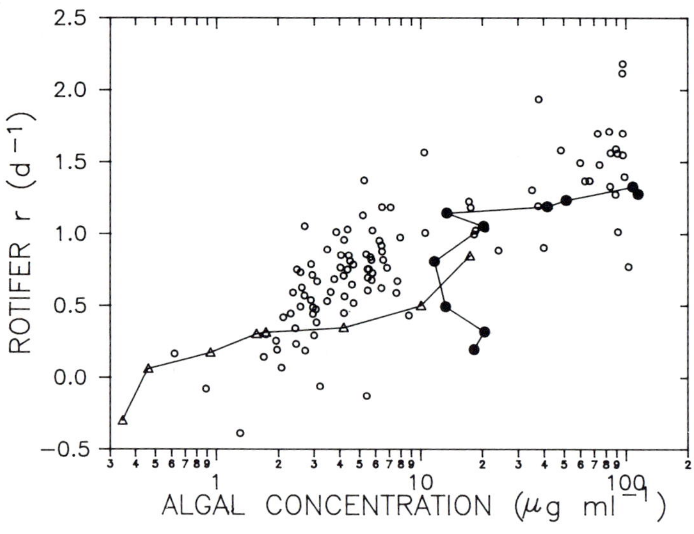

Fig. 5.2.1 Rotifer-specific growth rate as a function of algal dry mass. *Triangles* indiate the growth of *Brachionus calyciflorus* on *Cryptomonas erosa* at 19 °C. These data were digitized from Stemberger and Gilbert (1985, their Fig. 3). *Closed circles* represent residual algal biomass (*Chlorella pyrenoidosa*) in the darkened second stage of a two-stage chemostat vs. the dilution rate (= r in the steady state) at 25 °C (see Chap. 3.4). The algae were supplied by a steady-state alga chemostat under constant conditions. *Open circles* are calculated values of rotifer r vs. algal (also *C. pyrenoidosa* at 25 °C) biomass from a single-stage rotifer-algal culture (see Chap. 3.1, Fig. 3.1.1). The algae were constantly growing in this experiment. In this particular culture, the rotifer and algal populations fluctuated, allowing a wide range of r vs. C_2 values to be calculated

mathematical convenience (see Dabes et al. 1973). Given the carbon-energy limitation, the model predicts that the observed yield for a given D, Y, is:

$$Y = \frac{Y_g * D}{D + \mu_c}, \tag{5.2.4}$$

i.e., Y is a hyperbolic function of D (derivation in Boraas 1983). This appears to be a critical test for carbon-energy limitation for a given population on a given food and permits the values of Y_g and μ_c to be calculated from the steady-state concentrations of X and C_2. This relationship appears to be in good agreement with the data (e.g., Fig. 3.1.4; Boraas 1983).

Steady-state solutions for the differential equations 5.2.3a and b are, respectively:

$$C_2{}^* = \frac{K_s(\mu_c + D)}{r_{max} - \mu_c - D}, \quad \text{and} \tag{5.2.5a}$$

$$X^* = \frac{DY_g(C_1 - C_2{}^*)(K_s + C_2{}^*)}{r_{max} C_2{}^*}, \tag{5.2.5b}$$

where the asterisk signifies a steady-state concentration. A steady-state solution for X*, completely in terms of constants ad independent variables, is given in Boraas (1983) and Lassiter (1979).

The coefficient r_{max}, μ_c, and Y_g have reasonably clear physiological interpretations, but the value of K_s is more ambiguous. However, it should be noted that there is usually a lag before a population can achieve r_{max} when it is placed in an "optimal" environment after having been grown at a very slow r for several generations (e.g., Bennett and Boraas 1988). K_s, which is usually interpreted as the ability of the organism to acquire limiting resources at low resource concentrations, can be measured by fitting the model to observed data with a nonlinear regression algorithm. However, organisms growing at different rates may be in different physiological states, particularly when carbon-energy limited. One way to investigate this possibility is to calculate K_s from steady-state data assuming that the other coefficients of the model are, in fact, constants. That is, if the calculated K_s is constant for all D, then the physiological state of the organism does not vary under the experimental conditions. If, however, K_s varies with D, then the physiological state evidently does vary.

Both Eqs. (5.2.5a) and (5.2.5b) can be solved for K_s for a given value of D. The solution for Eq. (5.2.5b) expresses K_s in terms of X*, C_2*, D, Y_g, and r_{max}, whereas Eq. (5.2.5a) includes only C_2*, r_{max}, D, and μ_c. Since the derivation of K_s from Eq. (5.2.5b) uses more measurements than Eq. (5.2.5a), it is presented here as:

$$K_s = \frac{C_2{}^*(X^* r_{max} + DY_g)}{DY_g(C_1 - C_2{}^*)} - C_2{}^*. \tag{5.2.6}$$

Where this calculation has been carried out (Boraas 1983; Chap. 3.1, Fig. 3.1.4c), the K_s of rotifers has declined with D. This relationship leads to the hypothesis that the efficiency of resource acquisition increases with D, i.e., that the physiological state of rotifers "improves" with increasing r. The hypothesis seems reasonable and is testable from the measurement of swimming rates, feeding curves, etc. vs. D. That is, both swimming rates and feeding rates are predicted to decline for rotifers as their steady-state r declines.

Comparisons of the M-H model predictions to X* vs. D (Fig. 3.4.3) strongly indicate that rotifers grown in turbidostats and chemostats, are carbon-energy limited. In subsequent experiments, I have extended this hypothesis: Rotifers selected for extremely rapid growth in turbidostat culture (capable of about three population doublings per day) show clear cost-benefit trade-offs, which almost certainly are genetically determined (Bennett and Boraas 1989a, b). This hypothesis is consistent with a number of theoretical studies of the evolution of life history traits (e.g., Townsend and Calow 1981; Reznick 1985; Reznick et al. 1988).

5.2.7 Towards Testable Growth Models

Density dependence has often been elevated from a simple correlation that may indicate some type of environmental limitation to population growth to a causality. This is reflected in population growth models in which the only variable is the population itself. However, population growth always necessitates a transfer of mass from the environment to the population and from the population to the environment. Therefore, the environment is at least one other variable that accompanies any change in the population, as dictated by the law of the conservation of mass. Very general population models that ignore a variable environment, e.g., the logistic, are subject to severe misinterpretation. Furthermore, single-variable models are difficult to test since they rarely incorporate explicit growth mechanisms. Models that do incorporate environmental variability, e.g., the M-H model or the "cybernetic" model of Walz (Chap. 5.1), are less general but reflect biological processes much more accurately. These models incorporate explicit growth mechanisms almost by necessity and, therefore, can be tested by falsification.

There is a growing disparity between the sophistication of population models and the quality of data. The rigorous environment of continuous cultures provides an experimental platform on which the simplifying environmental assumptions of models can be met, together with high-quality data on population and environmental dynamics. The rapid give-and-take between experiment and theory in this framework may stimulate an equally rapid growth in our understanding of animal populations.

This is contribution no. 371 from the Center for Great Lakes Studies.

6 Competition and Life History Strategies of Rotifers

6.1 Use of Semicontinuous Culture Methods for Examining Competitive Outcome Between Two Freshwater Rotifers (Genus *Brachionus*) Growing on a Single Algal Resource

DIANNE B. SEALE, MARTIN E. BORAAS,
and J. BRIAN HORTON

6.1.1 Introduction

By definition, simple resource competition takes place when two or more populations grow at the expense of the same single limiting resource (Hsu et al. 1977). The limiting resource is extracted from the common environment, and the growth rates of the competitors decline as they deplete the resource (Fredrickson and Stephanopoulos 1981). In the simplest case, only one resource is assumed to determine the growth rates of both species. The competitive mechanism is differential consumption and growth on this limiting nutrient.

Several investigators have developed models which explicitly describe the mechanism for simple resource competition (e.g., Powell 1958; Megee et al. 1972; Fredrickson and Tsuchiya 1977; Hansen and Hubbell 1980; Fredrickson and Stephanopoulos 1981; Butler and Wolkowicz 1987; Chap. 5.1, this Vol.). Typically, mathematical expressions of these resource competition models (RCMs) have least three simultaneous equations: two for the growth rates of each potential competitor and one for the limiting resource. Hence, RCMs explicitly incorporate a reference to an environmental variable, in contrast to the classical Lotka-Volterra competition models (see Tilman 1982; Rothhaupt 1988; Boraas et al. 1990). The mechanistic approach of RCMs thus shifts the focus from descriptions of interspecies interactions, typical of the Lotka-Volterra approach, to analyses of mechanisms for differential growth by competing species as mediated through the environment. Unlike the Lotka-Volterra model, RCMs can be used a priori to predict the competitive dominant from preliminary examinations of the growth potentials of putative competitors under specific environmental conditions. Time-course simulations and steady-state solutions for RCMs in a chemostat predict that the competitive outcome will depend on the growth parameters (maximum growth rates and half-saturation constants) of the competing species and the turnover rate of the system, but not on the feed nutrient concentration (Megee et al. 1972; Chap. 5.1).

A simplified, graphical expression of resource competition models may be used to predict competitive outcomes. In this case, equations need not be defined nor mathematical parameters estimated. The RCMs are expressed simply by plotting the specific growth rate, r, of the organisms vs. the concentration or availability of the limiting resource (Jannasch 1967; Fredrickson and Stephanopoulos 1981; Tilman 1982; Rothhaupt 1988; Boraas et al. 1990; Chap. 6.2, this Vol.). In the plots, the empirical observations for growth vs. resource level for each competing species are depicted in the shapes of the corresponding curves. Loss rates may be overplotted. Predictions are made from a straightforward examination of the plots. The predicted competitive dominant is the species with the fastest net r (r = gross growth – loss rate) for a given set of conditions. The faster growing species will remove a disproportionate share of the limiting resource. As a result, the resource value will be reduced to a level below that required for sustained growth by its competitor, causing extinction of the slower-growing species. Plots may show one clear competitive dominant (net growth always higher) for all levels of a single, limiting resource. A more interesting case is where the plots of growth curves show a "crossover point" for net growth: One species has the higher r at low resource levels, but the other has a higher r with ample resources. The competitive outcome may be reversed simply by manipulating the resource availability.

For the special case of competition in a chemostat, both the resource supply rate and the loss rate are imposed by the system flow rate. However, the model may be applied to any experimental system in which resources, growth rates, and loss rates may be controlled, measured, or estimated (Tilman 1982). In addition to simple resource competition, the approach can be used to examine competition on multiple resources (Rothhaupt 1988) and a host of additional, more complex competitive interactions (Tilman 1982). This paper is concerned only with simple resource competition, where the single limiting resource is algal food.

In this paper, we expand our previous analyses (Boraas et al. 1990) of the usefulness of resource competition models and semicontinuous cultures (Chap. 2.1) for examining competition between two metazoans growing on a single limiting resource in an open system. First, we evaluate the maximum and resource-dependent growth rates of two species of freshwater rotifers within one genus, *Brachionus rubens* and *B. calyciflorus*, each growing in isolation on the alga *Chlorella pyrenoidosa* as the sole food supply. We then apply a graphical competition model (Jannasch 1967; Tilman 1982; Boraas et al. 1990; Chap. 6.2, this Vol.) for predicting the competitive outcome in mixed cultures of these species. Finally, the results from a competition experiment are compared with model predictions. The graphical RCMs predict two, mutually exclusive competitive outcomes: *B. rubens* should win at the slow D (10% replaced daily), but *B. calyciflorus* should win at the rapid D (50% replaced). This qualitative prediction of the model could be falsified with competition experiments showing either a sustained coexistence or displacement of the wrong species.

6.1.2 Special Methods

The study incorporated both semicontinuous cultures (SCs; Boraas et al. 1990; Chap. 2.1 and 3.1, this Vol.) and chemostat cultures (Boraas 1983; Chap. 2.2, this Vol.). Chemostat cultures were used for the rotifer *B. calyciflorus* and for all cultures of *Chlorella*, the algal food. Because of space and equipment limitations, SCs were used for unispecific cultures of *B. rubens* and for the *B. calyciflorus-B. rubens* competition trials. For all experiments, the measured dependent variables were the densities of the rotifers, X (X* for the steady state), and residual algae, C_2 (C_2* for the steady state). The independent variables, under experimental control, were the input (limiting) resource (here algal) concentration, C_1, and the system dilution rate, D. However, for purposes of constructing the competition models from unispecific rotifer cultures, the independent variable plotted was C_2, and the dependent variable was D (= r). Based on these results, independent competition trials were established for two critical D where (1) *B. rubens* would have the higher net specific growth rate (r), or (2) *B. calyciflorus* would have the higher r. Both steady-state and time-course data were evaluated for the competition trials.

The sole source of food for all rotifers was the alga *C. pyrenoidosa* Chick (UTEX No. 26), supplied by separate algal chemostats, maintained under constant light, temperature, and dilution rates with limiting nitrate at 0.5 mM (medium in Boraas 1983). The algal cultures (first stage) produced *C. pyrenoidosa* at a constant input concentration, C_1, of 114.0 ($\pm$ 8.8, SD) μg dry mass ml^{-1} to the rotifers (second stage). This C_1 was sufficiently high to allow nutrient-saturated growth of both rotifer species (see Boraas 1983, and Chaps. 2.1 and 3.1). The rotifer strains used in this experiment were provided by J. Gilbert (*B. calyciflorus* Pallas) and U. Halbach (*B. rubens* Ehrenberg). In these cultures, both rotifer species were parthenogenetic; resting eggs and males were not seen.

For the unispecific *B. calyciflorus* growth studies in two-stage continuous cultures, a peristaltic pump was used to replace a fraction of the medium at 30–60-min intervals (Boraas 1980, 1983). The *B. rubens* SCs were operated simply by collecting outflow from the algal chemostat and using this to replace a volume removed from the rotifer culture (see Chap. 2.1 and 3.1, this Vol.). The unispecific *B. rubens* cultures were held at 24.0 $\pm$ 1 °C, the *B. calyciflorus* cultures at 25.0 $\pm$ 0.1 °C, and the algal stage at 25 $\pm$ 0.2 °C. As required in a two-stage chemostat, algal growth was inhibited in all second-stage rotifer cultures by operating that culture in near darkness.

The *Chlorella* populations in the first stage, C_1, and their residual densities, C_2, in the second stages, were measured with a Coulter counter and a C-1000 Channelyzer, using a 50-μm aperture (Boraas 1980, 1983). These counts provided both a numerical count and a biovolume (see Seale and Beckvar 1980). The algal biovolume was converted to dry mass with the constant 0.57 g dry mass cm^{-3} live biovolume (Boraas 1983). Algal counts and biovolumes were taken from fresh samples for the algal chemostat and unispecific rotifer growth studies and from samples preserved in 2% glutaraldehyde for the competition study.

Preliminary studies showed a negligible change in cell size after preservation. The Coulter counter was also used to quantify rotifer populations in the unispecific two-stage cultures using a 560-μm aperture and an electrolyte consisting of 0.075% NaCl (mass/vol; see Chap. 3.1, this Vol.; Boraas 1983). The size distributions of the two rotifer species overlapped, particularly at the slower dilution rate, precluding use of the Coulter counter for the mixed-species competition cultures. Hence, the adults in a 10-ml preserved subsample were identified and counted under a stereoscopic microscope. Only adult counts are shown, because juveniles and eggs were not reliably identifiable to species.

The competition studies were carried out in duplicate, foil-covered, 250-ml polycarbonate Erlenmeyer flasks filled to 200 ml, at $23 \pm 2\,^{\circ}C$. All competition cultures were inoculated initially with 100 adults of both rotifer species. The cultures were mixed gently several times each day to resuspend the algae and rotifers. In addition, once daily cultures were mixed gently, then a predetermined fraction of the culture volume was poured off, preserved, and replaced immediately with an equal volume of fresh algal suspension. The cultures and replenished algae were poured immediately into a second, clean Erlenmeyer flask, to minimize biofilm accumulation.

For the chemostat cultures, D was measured as the medium pumping rate/culture volume. The D of an SC is computed from the fraction (F) of the culture replenished daily:

$$D = \ln[1/(1 - F)]. \tag{6.1.1}$$

For our SCs, the interval of dilution was 1 day, and the daily growth rates (dilution rates) were computed directly from Eq. 6.1.1 (see Chap. 2.1 for additional information about the operation of the SCs). The best method for measuring resource levels is somewhat ambiguous in SCs and may be a source of error (Chap. 2.1, this Vol.). We elected to monitor the value of C_2 just after mixing but before the daily dilution, for heuristic reasons (see Boraas et al. 1990). It should be noted that in the SCs, the algal levels actually experienced by the rotifers would have been at times higher (e.g., just after dilution) or lower (e.g., just before mixing).

6.1.3 Estimates of Maximum Growth Rates

Like Pirt (1975), we measured the maximum growth rate, r_{max}, from the sum of D and the slope of the exponential line:

$$r = [(\ln N_0 - \ln N_t)/t] + D \tag{6.1.2}$$

both during exponential increase (approach to a steady state) and during decrease (wash out) of the population (Boraas et al. 1990). Exponential increase of the *B. calyciflorus* population was used to compute r_{max} after inoculation into a chemostat at $D < r_{max}$. After a steady state had been achieved, D was increased to a value above r_{max}, causing wash out, and a second estimate of

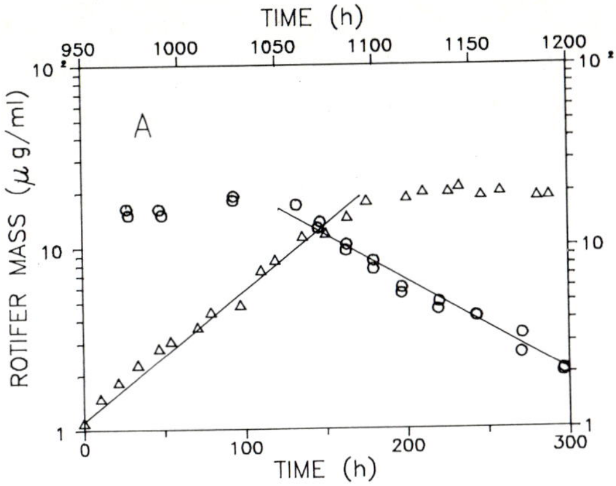

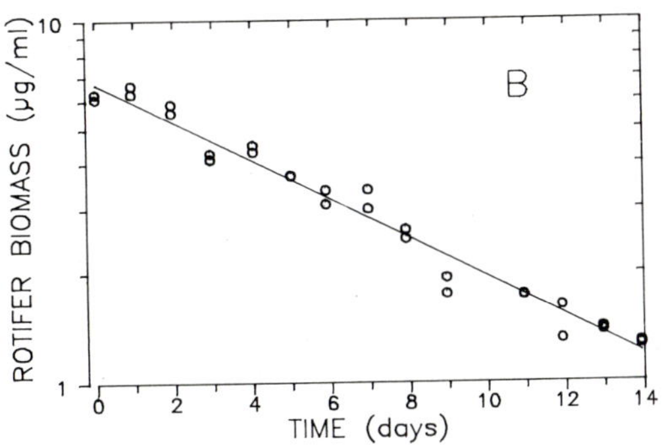

Fig. 6.1.1 A, B Exponential growth (*triangles*) and decay (*circles*) of population densities of rotifers in continuous culture. **A** Data for *Brachionus calyciflorus* growing at $D < r_{max}$ (*triangles*) and at $D > r_{max}$ (*circles*). **B** *B. rubens* in semicontinuous culture when $D > r_{max}$ (*circles*). Both populations were measured using a Coulter counter. See text for numerical values

r_{max} was computed from the subsequent exponential decay of the population. The r_{max} of *B. rubens* was computed from the exponential decay of a population when $D > r_{max}$ (see Boraas et al. 1990).

The nutrient-saturated maximum growth rate, r_{max}, of *B. calyciflorus* (Fig. 6.1.1A) exceeded that of *B. rubens* (Fig. 6.1.1B) by about 60%. For *B. calyciflorus*, with D at 1.62 day^{-1}, the exponential decay rate was -0.31 day^{-1} for a r_{max} of 1.31 day^{-1}; and at $D = 0.92$ day^{-1}, the exponential growth rate was 0.36 day^{-1} for a r_{max} of 1.28 day^{-1}. For *B. rubens* at $D = 0.92$ day^{-1}, the slope for exponential decay was 0.12 day^{-1} (Fig. 6.1.1A); hence, r_{max} was 0.80 day^{-1}.

6.1.4 Resource-Dependent Growth

For the preliminary analyses of the growth potential of each species in isolation, steady-state rotifer densities (X^*) and residual algal densities (C_2^*) were determined over a range of D for both *B. rubens* (14 values for D between 0.048

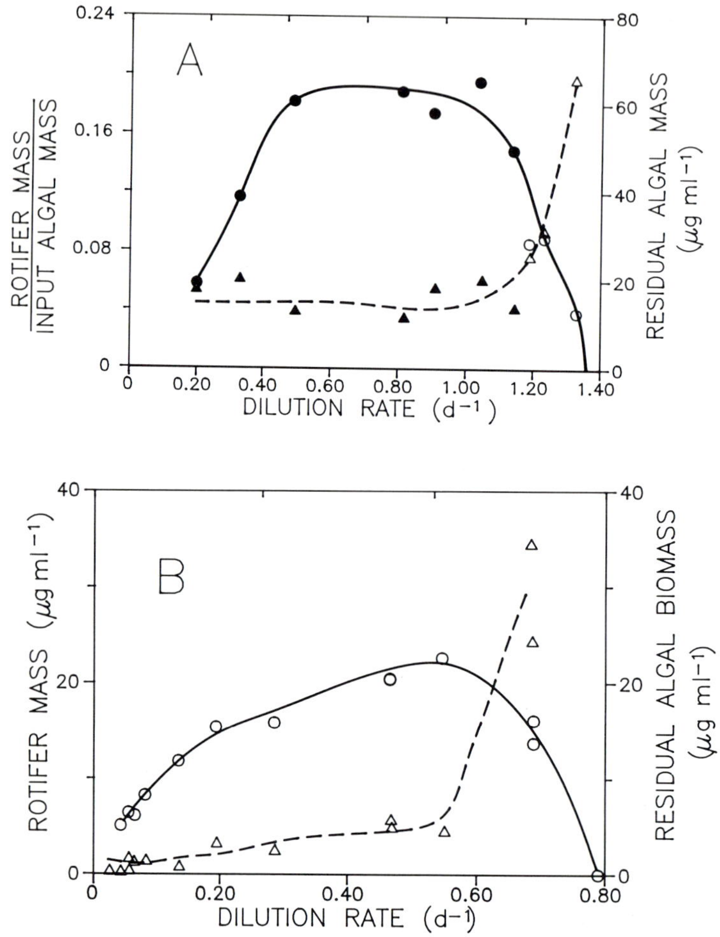

Fig. 6.1.2 A, B Steady-state data for rotifer (*circles*) and residual algal biomass (*triangles*) observed over the full range of dilution rates (D = r, specific growth rate). Lines drawn by eye. **A** Data from continuous *B. calyciflorus* cultures. *Closed circles* redrawn from Boraas (1983). *Open symbols* show independent estimates for r and corresponding residual algal concentrations (C_2), collected with a modified turbidostat culture (see text). Because of variations in input algal concentrations with the different studies, rotifer biomass was normalized by dividing by C_1 before plotting. **B**. Data for semicontinuous *B. rubens* cultures. Rotifer data not normalized, because of constant C_1 in the study

and 0.69 day^{-1}) and *B. calyciflorus* (7 values for D between 0.17 and 1.06 day^{-1}). These experiments provided the data for plotting r vs. C_2* (resource levels) for graphical RCMs (Jannasch 1967; Tilman 1982; Boraas et al. 1990).

We measured X* and C_2* as functions of D for both rotifer species in independent studies (Fig. 6.1.2). Previously published data from a chemostat (Boraas 1983) and turbidostat (Bennett and Boraas 1988; Boraas and Bennett

1988; Chap. 2.3, this Vol.) were used for *B. calyciflorus*, and SC data were used for *B. rubens* (Boraas et al. 1990; Chap. 3.1, this Vol.). In both *B. calyciflorus* and *B. rubens*, X* was an unimodal ("inverted U") function of D, and C_2* was a positive nonlinear function of D (Fig. 6.1.2). In plots for graphical RCMs (Tilman 1982), the dependent variable is r and the independent variable is resource availability. Therefore, we replotted our data to show rotifer growth rates vs. the standing crop concentrations of the algae, C_2* (Fig. 6.1.3). Although the C_2* varied little over much of the range of D for both rotifer species, the general shapes of r vs. C_2* were similar to the rectangular hyperbolae used for previous constructions of mathematical RCMs (e.g., Powell 1958; Tilman 1982). As the data were obtained for graphical RCMs, we did not estimate parameters for the model but rather drew the "best-fit" line "by eye" for our plots (Fig. 6.1.2).

The graphical model (Fig. 6.1.3) reveals a clear "crossover point" (at approximately $r = 0.62\ day^{-1}$, $C_2 = 16\ \mu g\ ml^{-1}$). *B. rubens* has a higher r at low C_2*, and *B. calyciflorus* grows faster at higher C_2*. Hence, the model predicts that *B. calyciflorus* will win at the rapid D and *B. rubens*, at the slow D (Fig. 6.1.3). The broken horizontal lines show the density-independent mortality rates imposed on each species by the dilution rates in SCs. These horizontal lines intersect the growth function at the point where the species' r balances these imposed mortality rates. The dotted vertical lines show the critical C_2* for each

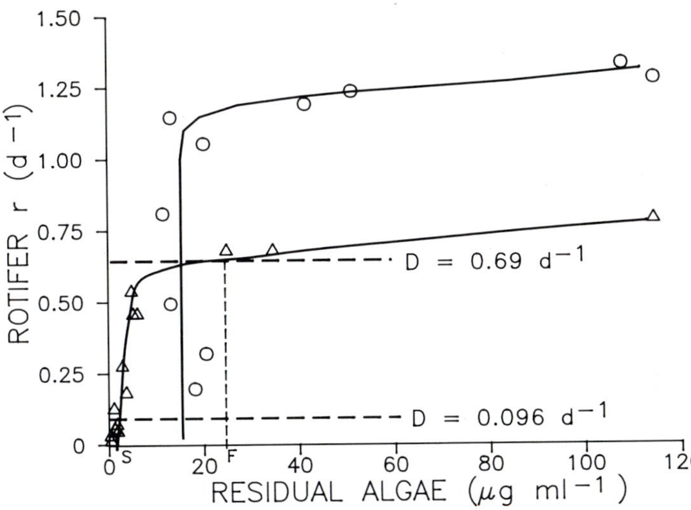

Fig. 6.1.3 Resource-dependent rotifer-specific growth rates (r), using data for rotifers and C_2*; *circles* are *B. calyciflorus* r, and *triangles* are *B. rubens* r. *Horizontal dashed lines* show the two dilution rates (mortality rates) used: rapid ($D = 0.69\ day^{-1}$) and slow ($D = 0.096\ day^{-1}$). *Vertical lines* indicate the critical residual food concentrations for the graphical competition model. Points of intersection are projected to the x-axis: S and F are the critical concentrations for *B. rubens* at the slow and rapid D, respectively. The point of intersection of *B. calyciflorus* is apparent from the solid line. According to the graphical model, these results indicate that *B. calyciflorus* should displace *B. rubens* at rapid D, and the reverse should occur at slow D

species at this point of intersection. Based on this plot, we made two predictions concerning the predicted C_2* at rapid and slow D: (1) At rapid D, *B. calyciflorus* can grow faster and can reduce C_2 to about half that needed by *B. rubens* to maintain balanced growth. (2) At the slow D, *B. rubens* grows faster and can reduce C_2* to a level $< 20\%$ of that needed by *B. calyciflorus* to sustain growth.

6.1.5 Competition Experiments

Competition studies were conducted at two flow rates, slow and rapid. For the slow D cultures, 10% of the volume was removed and replaced with fresh algae each day (from Eq. 6.1.1, $D = 0.096\ \text{day}^{-1}$); in rapid D cultures, 50% of the volume was similarly replenished ($D = 0.69\ \text{day}^{-1}$). Two replicates were maintained for each D, and all four cultures were maintained simultaneously. After 30 days, the D was reversed for each culture vessel, and then time-course data were obtained for X and C_2; due to technical problems, data for residual algae during the second 30-day period were incomplete.

For the time-course plots (Figs. 6.1.4, 6.1.5), the population densities are shown as numbers ml^{-1} for *B. calyciflorus*. The corresponding *B. rubens* numbers were multiplied by a factor of 4 before plotting, to compensate for the difference between the relative body sizes of the two rotifer species (see Boraas et al. 1990). Thus, the rotifer densities shown (Figs. 6.1.4, 6.1.5) roughly approximate the relative biomass estimates. The algal populations are shown as dry mass.

In slow D cultures, *B. calyciflorus* first increased and then steadily declined with increasing *B. rubens* populations; after the initial surge, *B. calyciflorus* seemed to be displaced to exclusion (Fig. 6.1.4). In rapid D cultures, only *B. calyciflorus* was counted, even though *B. rubens* grew in isolation at the rapid D (Fig. 6.1.2). While we saw qualitative agreement between prediction and observation in this study, the actual C_2* in competition studies were higher than predicted from unispecific studies. In both slow D cultures, C_2* averaged about $12\ \mu g\ ml^{-1}$ (Fig. 6.1.3), rather than the predicted $2\ \mu g\ ml^{-1}$. At rapid D, with *B. calyciflorus* dominant, C_2* averaged about $35\ \mu g\ ml^{-1}$, rather than the predicted $17\ \mu g\ ml^{-1}$ (Fig. 6.1.5). This discrepancy is further discussed in Sect. 6.1.9.

Thirty days after inoculation, the rapid or slow D was reversed for all cultures (vertical arrows in Figs. 6.1.4, 6.1.5). When D was shifted up (Fig. 6.1.4), *B. rubens* initially decreased, with a concurrent increase in C_2, and then persisted with large fluctuations. *B. calyciflorus* never reappeared in these cultures, indicating its competitive displacement during the previous slow D period. In the cases in which D was shifted down (Fig. 6.1.5), two different results were seen in the two replicates: (1) *B. calyciflorus* initially increased and then declined; (2) a small *B. rubens* population increased rapidly, concurrently with a decline (apparently to extinction) in the *B. calyciflorus* population. The appearance of *B. rubens* in one replicate indicates that a few individuals (uncounted) had maintained themselves in the second culture throughout the rapid D time course.

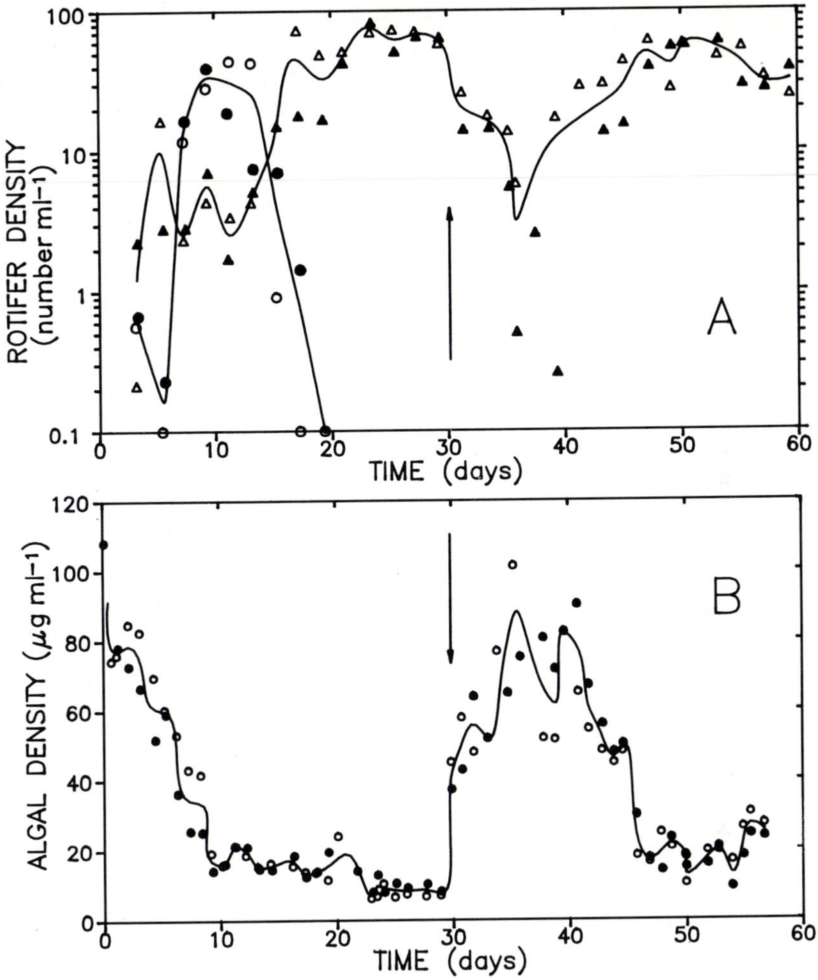

Fig. 6.1.4 A, B Time course for densities of rotifers (**A**) and residual algal biomass (**B**) in competition experiments. *Closed* (replicate 1) vs. *open* (replicate 2) *symbols* indicate means for rotifers and algae in each replicate of the competition study for each flow rate. *Circles* indicate *B. calyciflorus* and *triangles*, *B. rubens*. *Solid lines* are drawn by eye through the medians for the two replicates. The *vertical line* in each panel indicates the time at which the dilution rate was reversed for the cultures. In this experiment, the dilution rate was shifted up. *Left blocks* show data at the slow flow rate ($D = 0.096\ day^{-1}$). *Right blocks* show data at the rapid flow rate ($D = 0.69\ day^{-1}$)

6.1.6 Application of Competition Studies and Models to Analyses of Aquatic Systems

An ability to predict zooplankton population growth rates and densities with experimentally tested models is critical to understanding freshwater ecosystem dynamics. Zooplankton populations require a source of particulate food, mostly

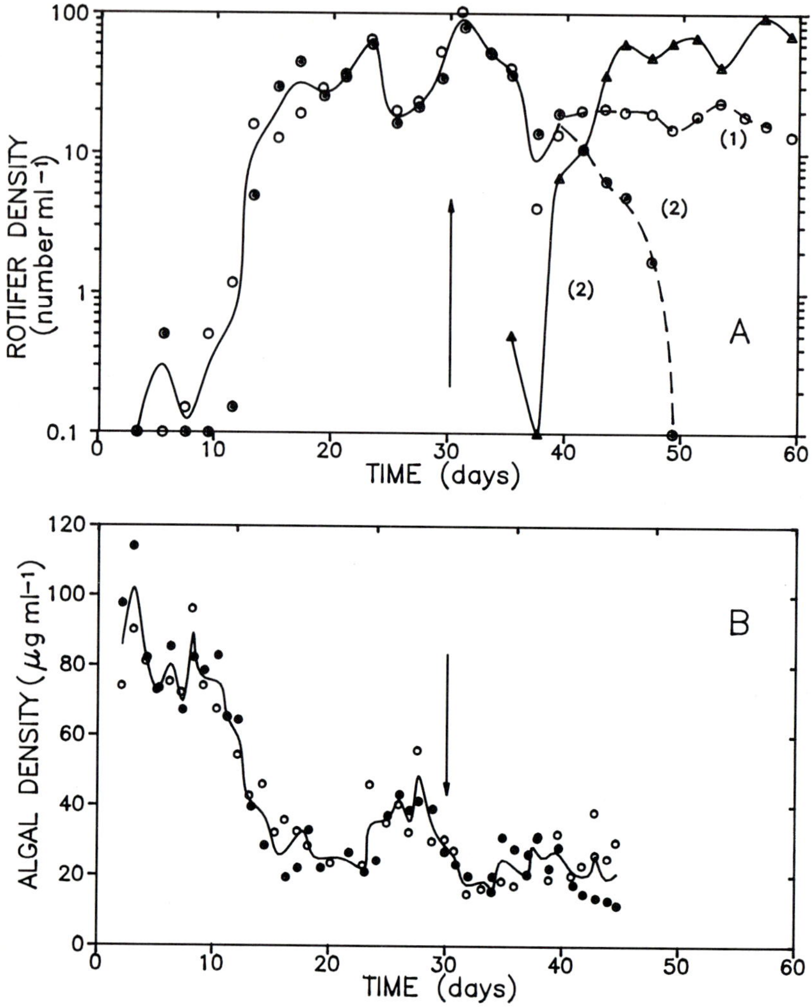

Fig. 6.1.5 A, B symbols as in Fig. 6.1.4. In this experiment, the dilution rate was shifted down. *Left blocks* are for D = 0.69 day^{-1}; *right blocks* are for D = 0.096 day^{-1}. Different results for the two replicates [(1), (2)] were seen after the shift down (see text for further explanation)

microscopic algae and bacteria, to provide essential resources for growth and proliferation. Resource limitation can be as important as predation in determining zooplankton species abundance and population dynamics (Neill 1975; Lampert 1978; Smith and Cooper 1982; Hoenicke and Goldman 1987). Consumption rates, which place an upper limit on the growth rates, can depend on the food concentration in filter-feeding cladocerans and rotifers (Mullin et al. 1975; Starkweather and Gilbert 1977b; Porter et al. 1982; Boraas 1983; Stemberger and Gilbert 1985; Yúfera and Pascual 1985). Under controlled conditions, the rotifer densities are determined in large part by food quantity (Halbach et al. 1983). Filter-feeding zooplankton have similar food preferences; hence, their resource requirements often overlap extensively (Lane 1975). However, they have some ability to select food particles, for example, on the basis of particle size (Gliwicz 1969) or algal species (DeMott and Kerfoot 1982; Leibold 1989), perhaps allowing species to avoid direct competition in some field situations. For a more complete review of this question, see Chap. 6.3, this Vol., and Sommer (1989a), particularly the chapters by DeMott (1989) and Gliwicz and Pijanowska (1989).

Because of the difficulties in demonstrating competitive displacement in the field, some ecologists have expressed skepticism concerning the role of competition in regulating population densities in nature (den Boer 1986). It can be extremely difficult to demonstrate an exact correspondence between resource levels and differential growth rates of competing populations in nature. Resource competition among zooplankton species is moderated by many factors, including variation in the ability to maintain energy reserves (Goulden et al. 1982), different survivorships of juveniles and adults under food limitation (Romanovsky and Feniova 1985), and temporal and spatial differences in habitat use or nutritional needs (Hoenicke and Goldman 1987). The existence of interference competition, e.g., allelopathy in zooplankton (Folt and Goldman 1981), can also obscure the resource-competition relationship. With interference competition, a population's competitors can be denied equal access to limiting resources through some physical, behavioral, or chemical means.

The most fundamental function of a model in science is to describe formally the relationships that best reflect the processes under investigation. In the case of competition, the process involved is the differential use of at least one limiting resource by at least two populations, causing depression of the growth rates in both. When bared to their fundamental structure (Fig. 6.1.5), the resource-consumer and the Lotka-Volterra (Lotka 1925; Volterra 1931) competition models differ in one key feature: The resource-based models explicitly include a measurement of the environment, and the Lotka-Volterra models do not. In traditional applications of the Lotka-Volterra competition model, all variables affecting the "competitive abilities" of the putative competitors are incorporated into a "competition coefficient" for each species. The traditional Lotka-Volterra model includes neither a direct reference to the ecological mechanism for competition or coexistence nor any explicit term for an environmental variable. Implicitly, the Lotka-Volterra models assume a constant or random environment, with the competing species interacting independently of the environment (Williams 1972). In contrast, the resource competition models assume that (1) the environment, as expressed by a mutually limiting resource, is variable and (2) the "competitive interaction" is mediated only by the consumption of this resource and the subsequent growth of the competing populations (Fig. 6.1.5).

In the natural sciences, models are used to present a formal hypothesis for experimental evaluation. A test of a model is nothing more than a test of its assumptions (Williams 1980). The predictions of the model can be compared to real data to show discrepancies. Such discrepancies are to be expected, since all models are abstractions, and the real world always has some noise. Some discrepancies indicate errors in the assumptions forming the framework of the model. If these errors are large, the model must be adjusted or discarded. Rather than being a difficulty with model development, this stepwise approach reveals the power of models to clarify hypotheses. Descriptive models, which do not include a representation of the mechanisms under investigation, can mimic natural processes. However, a useful model must be vulnerable to rejection on the basis of comparison with data. Our experimental system was designed to hold environmental conditions, including the rate of supply of the limiting resource, as near to constant as possible. Hence, the experimental design, as much as possible, satisfied the major assumptions of both the Lotka-Volterra models and the resource-consumer models. Hence, we were in a position to judge the relative value of the two classes of models for understanding competitive interactions between these two rotifers.

6.1.7 Limitations of the Lotka-Volterra Model for Competition

The Lotka-Volterra and resource approaches to competition can be used interchangeably to describe competitive interactions at or near a steady state (May 1975). However, the Lotka-Volterra competition models are descriptive models which cannot be used to make predictions or to elucidate mechanisms for competition. For example, they do not include, and hence cannot predict, the relationships among competing species and their resource levels. It is impossible to use the models to predict the potential impact of changing resource levels on competing species, without first observing the competitive interactions. All interactions between competing populations are expressed as a simple competition coefficient. These competition coefficients cannot be predicted a priori from previous data, such as relative growth rates or food levels; they can only be calculated after the presumptive competitors have been grown together. The Lotka-Volterra "carrying capacity" cannot be related a priori to the resource needs of the organisms. The Lotka-Volterra model fits all possible contingencies and, therefore, cannot be falsified or rejected on the basis of output alone. As long as descriptive models of competition are accepted as sufficient, Lotka-Volterra models cannot be discarded. The intractability of Lotka-Volterra models has become a liability for the development of ecology as a predictive science, regardless of their historical appeal and usefulness as descriptive tools.

The inadequacy of the Lotka-Volterra competition models has long been apparent. For example, they assume that r declines linearly with population density. As previously discussed (Williams 1972; Smith-Gill and Gill 1978; and others), typical data do not support this assumption. In our study, both rotifer populations entered periods of exponential increase in which r was constant (r_{max}), while the population density varied. Clearly, r was not a linearly decreasing function of X (also see Chap. 5.1).

6.1.8 Mechanisms for Competition

One key role that could be played by models in competition theory is to evaluate the relative importance of various mechanisms for competition. For example, interference competition, sometimes mediated through aggressive interactions, could be modeled and tested. MacArthur (1972) clearly explained why the existence of interference competition necessarily implies the existence of resource competition. "First we note that there is a natural priority: competition for resources can exist by itself, with no aggressive overtones, while aggressive competition would be pointless unless, in its absence, there would be resource competition; there would be no reward for the fighting if some new resource were not gained" (MacArthur 1972). The simplest representation of competition mediated through a limiting resource is illustrated in Fig. 6.1.5. The resource competition models can be modified easily to include explicit terms for interference, if desired (e.g., Boraas 1974). However, Lotka-Volterra models subsume all negative species interactions in a single "competition coefficient" and cannot be used to distinguish mechanisms for competition.

6.1.9 Predicting the Competitive Outcome from A Priori Growth Data

The growth parameters of potential competitors under specific environmental conditions, determined from measurements made on pure cultures, can be used to predict the competitive dominant of two metazoans, *B. rubens* and *B. calyciflorus*, growing on a single limiting resource, *C. pyrenoidosa*. In our studies with rotifers, explicit resource competition models provided an excellent qualitative and semiquantitative prediction of competition outcome for a single limiting resource (Boraas et al. 1990). Essentially the same RCMs and the experimental systems (continuous and semicontinuous cultures) have been used to examine competition among bacteria (Jannasch 1967; Fredrickson and Stephanopoulos 1981), as well as algae and plants (Tilman 1982).

As predicted by the resource competition models, the competitive outcome can depend on system turnover rate. In previous studies of single-resource competition between these two rotifer species using a fed-batch culture, where food is added but organisms are not removed, *B. rubens* has always replaced *B. calyciflorus* (Halbach 1984). (See Chap. 3.1 for further discussion of the fed-batch culture system.) This experimental design and the result are both consistent with our SC competition trial at low D (Fig. 6.1.3): *B. rubens* can grow more rapidly than *B. calyciflorus* at low resource levels. This may be interpreted to mean the *B. rubens* has a lower half-saturation value (see Powell 1958). However, *B. calyciflorus* has a higher r_{max} and therefore is the projected and observed competitive dominant at the higher resource levels seen at the rapid D (Fig. 6.1.3).

We favor a stepwise examination of the different assumptions and predictions of resource competition models. Here, we minimized complicating factors by placing the organisms into open laboratory culture with a controlled environment, restricted to a single food source. In this manner, we were able to isolate the biological interaction of interest, simple resource competition, from environmental noise. If a more complex model is desired, for example seasonal variations in food levels or temperature, the models (and the experimental systems to evaluate them) can be modified accordingly. For example, the availability of alternative foods can modify the competitive outcome. Rothhaupt (1988) used the RCM-SC approach to examine competition between *B. rubens* and *B. calyciflorus* growing on varying proportions of two alternative food sources. Both rotifers used both algae as prey, but *B. rubens* preferred and grew better on *Monoraphidium*, and *B. calyciflorus* preferred *Chlamydomonas* (Rothhaupt 1988). Tilman's (1982) graphical model for a two-consumer, two-resource system predicted the outcome, either coexistence of the two species or competition to exclusion by one rotifer species, in 11 of 12 cases (Rothhaupt 1988; Sect. 6.2.5, this Vol.).

With models and experimental protocol emphasizing mechanisms for interaction, this approach could test more complex predictions of microbial RCMs: the outcome for consumers competing for a growing prey population (Baltzis and Fredrickson 1984), the effects of parasites or predators (Fredrickson and Stephanopoulos 1981), the coexistence of competitors growing on periodically varying inputs (Stephanopoulos et al. 1979), complex interactions among species which can display commensalism and mutualism as well as competition (Megee et al. 1972), and multiple-species food webs (Jost et al. 1973b).

The prediction and observation of the competitive displacement of *B. rubens* at high D may appear to be counter-intuitive. How can nutrient limitation cause competitive displacement at such high residual algal levels? According to the RCMs and our experimental results for rapid D conditions, competitive displacement can take place even though the residual algal concentrations remain relatively high. The initial conditions in all competition trials were the same: Both populations were inoculated at very low concentrations. The prediction's simplicity can be illustrated by computing the predicted net growth rates in mixed cultures. The maximum net rate of population increase achievable for each species under nutrient-saturated conditions is $r_{max} - D$. Thus, at the rapid D (0.69 day^{-1}), *B. calyciflorus* can increase at a rate of 0.60 day^{-1}, while at best *B. rubens* could only grow at 0.12 day^{-1}. Clearly, the initial conditions gave the edge to *B. calyciflorus*, which was therefore able to consume algae and decrease C_2 at a high rate, so that *B. rubens* was never able to become established. These results are consistent with the theoretical assumptions of the RCM: if C_2 is depressed below a level necessary to maintain growth at a given turnover rate, the population will decline. While maintaining growth in isolation at rapid D, the C_2* remaining in the presence of the steady-state *B. rubens* populations was about 33 $\mu g\,ml^{-1}$, which was close to the C_2* remaining in the competition chambers. This may have been sufficient to maintain *B. rubens* in isolation (but see discussion concerning possible measurement discrepancies below) but not to allow population increase in the face of a faster-growing competitor.

The quantitative model predictions for residual algal concentrations were not achieved accurately in the competition trials (see also Boraas et al. 1990), for three possible reasons. (1) The model may be in error, particularly when one of the species is grown near its r_{max}. (2) One rotifer species may have adapted, either physiologically or genetically; that is, either *B. rubens* became more efficient or *B. calyciflorus* became less efficient, as previously observed for marine rotifers (Droop and Scott 1978). (3) The predilution C_2*, which represents a value somewhere between the highest (postdilution) and lowest (premixing) values for residual algae, does not accurately assess the actual C_2*. We have no supporting evidence for hypothesis (2) in our system. There are several potential sources of experimental error to support hypothesis (3). Probably most important, the competition cultures were agitated intermittently, which could have allowed *Chlorella* to settle out. By agitating samples just prior to sampling, settled algae were resuspended, perhaps resulting in overestimates for C_2* in the competition studies. Hence, the algal concentrations actually experienced by the rotifers were, very likely, lower than the C_2* we measured (also see comments in methods above). Rothhaupt (Sect. 7.1.3) assumes that the accumulation of feces in continuous cultures has the consequence of a lower nutritional value of the measured "food" concentration. Therefore, the residual concentration is higher.

From our data (Fig. 6.1.3), we believe it should be more difficult to assess growth as a function of food concentration rather than as food supply rates. In both species, a large range of variation in r was associated with a very small range of C_2* (Fig. 6.1.2; also see Chap. 7.1, this Vol.). Increased food supply rates resulted in higher rotifer densities, but C_2* remained nearly constant except when D began to approach r_{max}. We contend that the small variation in C_2* virtually precludes the major alternative method for measuring r as a function of food: setting C_2* constant, measuring the rotifer abundance in time, and calculating r from the exponential growth equation (e.g., Pilarska 1977b). In this measurement, the residual food concentration would be assumed to be an independent variable. Slight errors in regulating this C_2* would result in large errors in r. By setting D (and hence r in the steady state) as an independent variable in a continuous culture, allowing C_2 to be a dependent variable, these experimental errors are greatly reduced. In fact, slight variations in C_2 probably have little effect on r. The rotifer growth rate is limited by the rate that the food is supplied, rather than its instantaneous concentration. This dependence of growth rate on food supply rate is implicitly assumed in mathematical RCMs (e.g., Powell 1958).

6.1.10 Application of the Resource Competition Model in Aquatic Ecology

Although relatively new to theoretical ecology, resource-consumer approaches to evaluating comptition models have been commonplace in microbial ecology for some time. Their usefulness in microbial ecology has been tied to analyzing practical problems. The key question is: Which microbial species dominate under well-defined culture conditions? For example, Powell (1958) developed competition models to explain why single-species chemostats rarely became contaminated with other bacterial species. Once the mathematics of growth in chemostats had been developed, numerical solutions for independent growth parameters could be used to predict and to manipulate the dominant species in a continuously renewed culture. Jannasch (1967) used this method to separate two bacterial species, obtaining cultures of *Aerobacter* at $D = 0.6\ h^{-1}$ and *Achromobacter* at $D = 0.4\,h^{-1}$. Our results are a simple extension of this approach to separating rotifer species. The Lotka-Volterra models cannot be adapted to these pragmatic uses. Hence, Lotka-Volterra models have not become prevalent in applied microbial ecology.

Our primary purpose was to reveal the mechanisms determining the relative growth rates of two rotifer species and ultimately to understand the population dynamics of zooplankton in aquatic ecosystems. Simple descriptive models, with no predictive utility, are not adequate for this purpose. Although they require additional work, RCMs resolve some of the problems associated with Lotka-Volterra models (Rothhaupt 1988; Tilman 1982). They are predictive and testable; the assumptions are realistic; the mechanisms are stated explicitly and can be distinguished. Our approach is to evaluate the usefulness of RCMs for describing the simplest case: a model for single-resource competition. We have shown that the fundamental qualitative prediction of the model was satisfied: The dominant species were predicted from data on resource requirements and mortality rates. As predicted by simple graphical analyses (see Jannasch 1967), the competitive outcome was reversed by changing the dilution rate of the system. Although we did not obtain accurate quantitative predictions for $C_2{}^*$ from the competition studies, the shifts in $C_2{}^*$ in competition vs. unispecific studies were in the predicted direction. In our view, simple resource competition is the most likely mechanism for explaining our results. However, other factors, such as stochastic processes and behavioral or chemical interference, cannot be excluded without further study.

There are some heuristic problems associated with the models that may have limited their popularity with ecologists. A known resource must be quantifiable and limiting to the organisms of interest. The resource dynamics, including supply and loss rates, must be known and measurable. These restrictions would seem to constrain a quantitatively rigorous study of competition to the laboratory. Our goal is understanding the process of competition by a step-by-step comparison of data to models, to improve our understanding of the process of competition. By comparing model predictions to observations, we can modify or discard incorrect assumptions. Additional data will be needed to account for competitive interactions in the field (see Seale 1980), but initial efforts to apply mechanistic approaches to understanding "real-world" competition are encouraging (Tilman 1982). We expect a general theory of competition to emerge from these efforts. The prospect is sufficient to continue the tedious, but rewarding process of measurement, prediction, and comparison.

Acknowledgements. Support for the experimental phase of study was provided by National Science Foundation grant BMS 75–18749 to D. B. Seale and F. M. Williams at the Pennsylvania State University. Funds for data analysis and manuscript preparation were provided by Environmental Protection Agency grant R-810871-01-0 to M. E. Boraas and D. B. Seale at the University of Wisconsin Milwaukee. We gratefully acknowledge the stimulating seminars on ecological competition at these two institutions, particularly the contributions and comments of F. M. Williams, Brian Dennis, James Barrett, and William Bennett.

J. Brian Horton, the third autor of this manuscript, died in 1986, after a lingering illness and paralysis, before completing his doctoral thesis. The clarity of his insight into theoretical ecology and his love of life and its beauty will always be an inspiration to those who knew him and had the honor of working with him.

This is contribution no. 372 from the Center for Great Lakes Studies.

6.2 Rotifers and Continuous Culture Techniques: Model Systems for Testing Mechanistic Concepts of Consumer-Resource Interactions

KARL O. ROTHHAUPT

6.2.1 Introduction

The causal understanding of the mechanisms governing plankton communities has increased during the last decades, probably more than in any other field of ecology (e.g., Kerfoot 1980; Lampert 1985; Sommer et al. 1986; Sommer 1989a). It is the joint effort of descriptive field studies, field and laboratory experiments on the community and population level, and ecophysiological research on the organism level that has advanced this field of science. This progress would not have been possible, however, without the development of integrating concepts. A very successful concept has been predation. It provided an explanation for puzzles like cyclomorphosis (Havel 1987) and the vertical migration of zooplankton (Lampert 1989), as well as for whole community effects between trophic levels (McQueen et al. 1986).

The understanding of resource-related processes, in contrast, and of how they interact with predation, has lagged behind. An example is the size efficiency hypothesis (Brooks and Dodson 1965) that predicts zooplankton community structure from body size. This hypothesis assumes that large-bodied species are superior resource competitors but more vulnerable to vertebrate predation. While the predation assumptions have been generally confirmed (e.g., Zaret 1980; Kerfoot 1987), there has been much debate about the competition side (e.g., Dodson 1974; Romanovsky 1984; Romanovsky and Feniova 1985). Understanding consumer-resource interactions requires insight into the mechanisms involved (Tilman 1987; Chap. 6.1). Under this premise, considerable advances have been achieved concerning direct and in direct nutrient-mediated interactions of the phytoplankton (e.g., Tilman et al. 1982; Sommer 1989b). In doing so, phytoplankton ecologists first tested theoretical concepts in relatively simple laboratory systems, before they increased the complexity and successively approached the field situation. It is exactly this procedure that Kareiva (1989) has in mind when he writes "we need more bottle experiments," in an attempt to renew the dialogue between theory and experiment in ecology. Bottle experiments, in a broad sense, have never been completely out of vogue in plankton ecology, probably because the organisms we are working with are particularly amenable to this kind of experimentation.

Among the herbivorous zooplankton, rotifers are especially appropriate for testing theoretical concepts that predict the outcome of direct or indirect interactions from ecophysiological traits. Many rotifers have a cosmopolitan distribution and are easy to culture. Most importantly, due to their small size and short generation times, they are suited for continuous culture techniques. In the following, I will try to specify some elementary and more complex current concepts of resource-related processes of the herbivorous zooplankton and discuss how rotifers and continuous culture techniques can be applied for approaching them experimentally.

6.2.2 Numerical and Functional Responses

A resource can be defined as any factor or substance that can lead to enhanced population growth as its availability increases and that is consumed by the organisms (Tilman 1982). When speaking about herbivorous zooplankton, usually algae of an appropriate size are regarded as resources in the above sense. However, protozoa, bacteria, or detritus may also contribute at times. The resource dependence of population processes can be divided into two aspects: How food intake depends on food supply, and how this food supply is converted into the production of offspring and biomass ("functional response" and "numerical response" sensu Holling 1959a). Both partial aspects are important, and in their combination they describe how an organism interacts with its material environment.

Several methods exist to estimate the functional and numerical responses of planktonic rotifers. Theoretically, it should be possible to estimate both from steady states in the second stage of rotifer chemostats. In practice, however, this approach is only of limited applicability, due to the ill-defined food quality of these systems (see Chap. 7.1 for a detailed discussion). Functional responses are most easily determined using short-term radiotracer methods or closed bottle incubations with cell counts before and after incubation (see Peters 1984 for methods). Both methods yield consistent and comparable results (Rothhaupt 1990a). The use of artificial particles (e.g., plastic beads) as "tracers" may give erroneous results, since rotifers may ingest these particles at different rates than the food algae (DeMott 1986; Rothhaupt 1990a).

When numerical responses are determined, it is important to supply a qualitatively and quantitatively constant food environment. A predetermined food concentration may be realized by applying "rotating cultures," in which the animals are regularly transferred into fresh food suspensions (see Chaps 4.1, 4.2, 7.1) and densities are kept low so as to prevent food depletion due to grazing losses (Stemberger and Gilbert 1985; Rothhaupt 1990b; Walz et al. 1989). Another possibility are turbidostat-controlled systems that keep food densities constant by adjusting the inflow rate of the food suspension to balance grazing (Bennett and Boraas 1988; Chap. 3.4). In both cases, the food concentration is the independent variable, and the resulting growth rate is the dependent variable. This is in contrast to chemostat systems, in which the steady-state growth rate is determined beforehand by the dilution rate, and food quantity (or quality) is the resulting variable. Both methods, especially intermittently fed cultures, probably give poor results in the range of the Monod curve above the threshold concentration (Eq. 6.2.1; Fig. 6.2.2) if the slope is very steep, where minor changes in the food concentration result in major changes of the growth rate.

Rotifer ingestion rates usually increase with food density (Starkweather 1980) and saturate, reaching some maximal ingestion rate, at high food concentrations. For *Brachionus* species, the feeding behavior below a certain critical food particle size appears to be best described by a rectilinear model (Rothhaupt 1990c). This model assumes constant maximal clearance rates up to a certain food concentration (incipient limiting level, ILL) and maximal ingestion rates above this concentration. Consequently, below the ILL, ingestion rates increase linearly with food concentration, the proportionality factor being the filtering rate. Bigger particles mechanically interfere with the feeding process, and this interference increases with the particle concentration. The resulting functional response curves are therefore better described by a curvilinear type 2 model with continuously decreasing filtering rates (Rothhaupt 1990c). In some cases, ingestion rates may even decrease at high food concentrations of large particles.

Numerical responses are usually characterized by a minimum food concentration for zero population growth and a curvilinear increase with food concentration, asymptotically approaching a maximal growth rate. This can be well described by a modified Monod equation, including a minimum food concentration for zero population growth.

$$r = r_{max}(C - C_0)/(C - C_0 + K_s), \tag{6.2.1}$$

where r = population growth rate (day^{-1}), r_{max} = maximum population growth rate, C = food concentration, C_0 = minimum food concentration for zero population growth, K_s = Monod constant. Alternatively, the Monod model as modified by Herbert (1958) for maintenance metabolism has many of the same properties (Boraas 1983).

Among the rotifers, the characteristics of the Monod equation appear to be body size related. Larger species have higher maximal growth rates but require more food for zero population growth; i.e., r_{max} and C_0 both increase with body size (Stemberger and Gilbert 1985; Chap. 6.3).

For nutritionally similar foods, the numerical responses strongly depend on the ingestability of the food items. Two examples are the numerical responses of a strain of *B. rubens* to the coccal green algae *Monoraphidium minutum* and *Chlorella minutissima*. *B. rubens* feeds less efficiently on the very small *Chlorella* (2 μm in diameter) than on *Monoraphidium* (5 μm in longest linear dimension). Maximal filtering rates, measured with radiotracers, are about 11 $\mu l\,h^{-1}$ when *Monoraphidium* is fed, and 4 $\mu l\,h^{-1}$ for *Chlorella* (Rothhaupt 1990c). Both numerical response curves are well described by the same Monod-type equation,

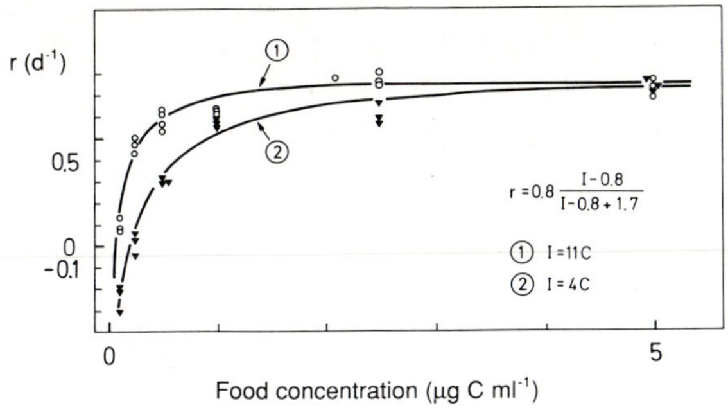

Fig. 6.2.1 Population growth rates of the rotifer *Brachionus rubens* when fed the green algae *Monoraphidium minutum* (1) or *Chlorella minutissima* (2). Both numerical response curves are described by the same equation, when a term representing the clearance efficiencies for the respective algae is introduced

modified to include a term that reflects the different ingestabilities of both algae (Fig. 6.2.1). Similarly, the decreasing ingestion rates with high concentrations of large algae are reflected by decreasing population growth rates, resulting in inversely U-shaped numerical response curves (see Fig. 6.2.9b).

As stated above, the chemostat is not the ideal experimental system to determine rotifer functional and numerical responses. Chemostats are very instructive, however, to illustrate how an imposed mortality (a "top-down" influence) and resource supply (a "bottom-up" influence) interact. In the field, mortality may be due to predation or displacement by currents. In the chemostat, the imposed mortality is given by the dilution rate (D). This model does not include the different nature of either mortalities, which is density dependent and size selective for predation and density independent and unselective for wash out (Chap. 7.2). The resource level required to balance the losses due to dilution are given by the intersection point of the numerical response curve and the dilution line (Fig. 6.2.2). If the numerical response curve is an increasing function of food concentration around the intersection point, and if the food supply concentration is sufficient, the system will stabilize at this point. As long as D does not surpass r_{max} and the food supply concentration is high enough, the resulting steady-state food concentration will increase as the externally imposed mortality rate increases. The food concentration will always be reduced to the level that permits the mortality to be balanced by reproduction.

In the case of decreasing growth rates at high food concentrations, there can be two intersection points of the numerical response curve with the dilution

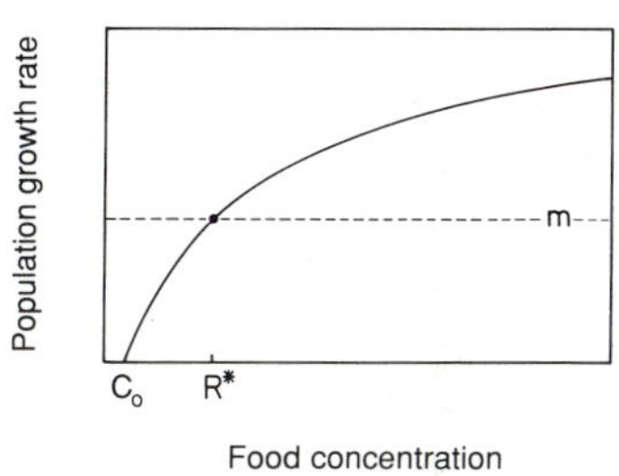

Fig. 6.2.2 Graphic illustration of the threshold resource level required to persist (R*). R* is the resource concentration where the growth rate equals the mortality rate (m). In a chemostat, m is equivalent to D

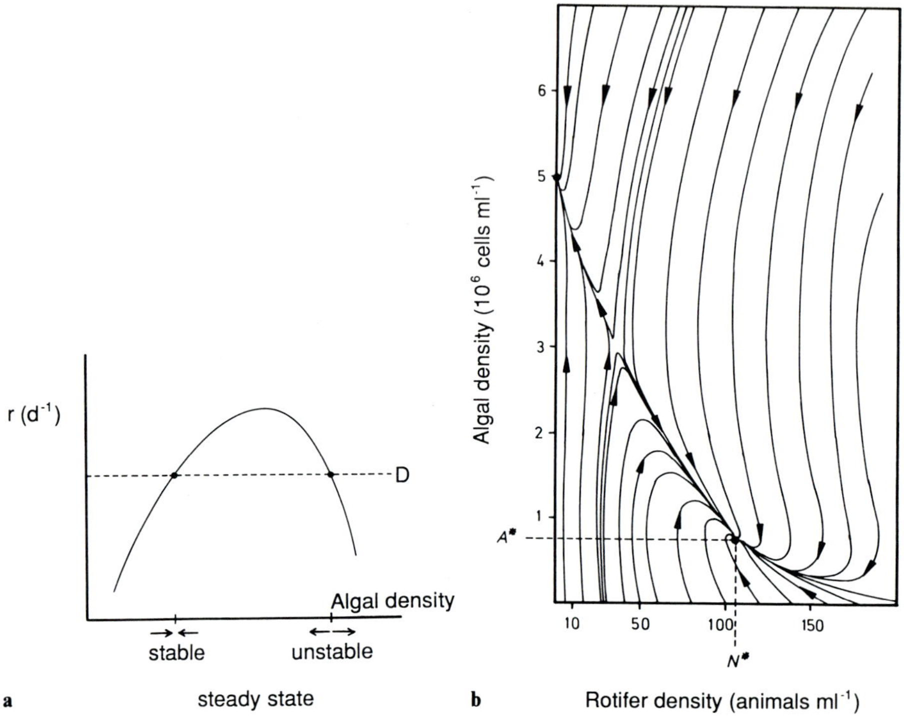

Fig. 6.2.3 a, b Schematic representation of the effect of substrate inhibition (decreasing growth rates at high food concentrations) in a chemostat. **a** There are two intersection points of the numerical response curve with the dilution line. Only the left point represents a stable equilibrium. **b** Phase plane representation of the simulated system behavior. Depending on the starting conditions, trajectories lead either to the wash out of the animals or to the stable equilibrium (After Rothhaupt 1985)

line (Fig. 6.2.3a). Only the left point, however, fulfills the condition that the growth rate increases with food concentration and hence is a stable equilibrium point. The other equilibrium point is an unstable one. Even small perturbations from it lead either to a transition to the stable equilibrium or to the wash out of the animals. The simulated behavior of such a system is shown in Fig. 6.2.3b.

6.2.3 Food Quality Effects

It is well-known that algal species differ in their biochemical composition and hence in their nutritional quality for the herbivorous zooplankton (Ahlgren et al. 1990). In particular, blue-green algae have often been found to be relatively

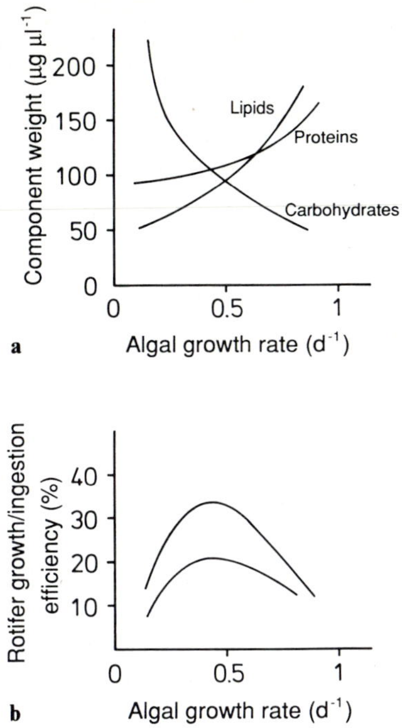

Fig. 6.2.4 a Biochemical composition of the food alga *Brachiomonas submarina*, and **b** calorific growth/ingestion efficiencies of two strains of the marine rotifer *Brachionus plicatilis* as functions of the algal growth rate (After Scott 1980)

low quality foods (e.g., Arnold 1971; Lampert 1981; Rothhaupt 1991). According to this view, the nutritional quality of the resource base is expected to depend on the algal species composition. There is additional evidence, however, that biochemical changes caused by different algal growth conditions can have a profound effect on the consumers, even if the food species composition does not change (Smith 1991). An early example of this kind was given by Scott (1980). Using a two-stage chemostat system, he studied the influence of the growth rate of the food alga, *Brachiomonas submarina*, on the growth/ingestion efficiency of the marine rotifer *Brachionus plicatilis* (Fig. 6.2.4). Although the author did not state it explicitly, the food algae in this system were most probably light limited. The proportion of carbohydrates in the food algae decreased with algal growth rate, whereas lipids and proteins increased (Fig. 6.2.4a). The calorific growth/ingestion efficiencies of two strains of *B. plicatilis* were highest when algae grown at medium rates were fed (Fig. 6.2.4b). Scott (1980) attributed this to the balanced proportions of the major biochemicals in these algae.

Recent studies with crustacean zooplankton have shown that mineral nutrient limitation of the food algae can have pronounced effects, too. Phosphorus deficiency (Sterner in press) as well as nitrogen deficiency (Giani 1991) have been found to decrease the nutritional value of algal foods for

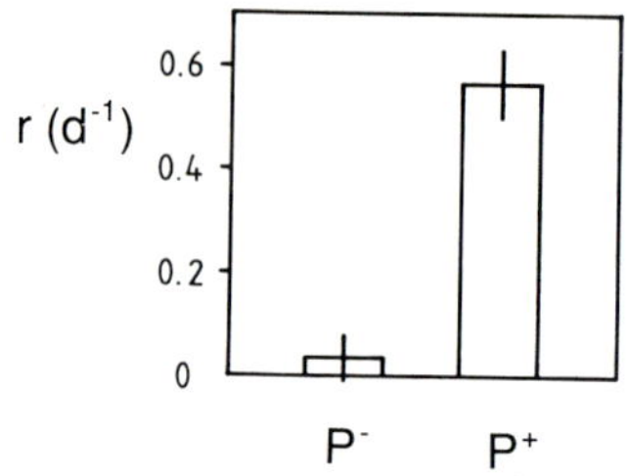

Fig. 6.2.5 Population growth rates of the rotifer *Brachionus rubens* when fed the food alga *Monoraphidium minutum* from a severely phosphorus-limited (P^-) or from a not nutrientlimited (P^+) culture. The food concentration was $0.5\,\mu g\,C\,ml^{-1}$ in both cases

Daphnia. Severely phosphorus-limited algae have been found to be a poor food for *B. rubens*, too (Fig. 6.2.5). It is not yet clear whether these effects are due to varying algal stoichiometry, i.e., N or P contents per se, or to shifts in the proportions of biochemical constituents. Two-stage rotifer chemostats appear to be especially suited to studying the effects of algal nutrient limitation on algal stoichiometry, biochemical composition, and nutritional quality for herbivorous zooplankton. In the first stage of such a system, algae are easily grown under defined conditions with regard to the nature and the severity of the limitation.

Animals in the field will always have a variety of food particels available. Even if the dominant food type is biochemically insufficient, they may obtain the essential component from a few additional bits. The value of low quality foods, like some blue-green algae, can be improved when they are combined with other foods (Lampert 1981). A mixture of a low and a high quality alga may even be superior to the higher quality alga alone (Fig. 6.2.6). *B. rubens* cannot exist when it is fed only the filamentous blue-green alga *Anabaena flos-aquae* ($1\,\mu g\,C\,ml^{-1}$). It grows well with the green alga *Monoraphidium minutum* as food. Growth rates are highest, however, when equal proportions of both algae are offered (Fig. 6.2.6).

Similar complementary effects are conceivable, even expected, concerning the nutritional quality of nutrient-limited algae. Sommer (1988b, 1989c) found phytoplankton at times to be nutrient limited in situ, but never were all species strongly limited by the same nutrient at the same time. However, simultaneous

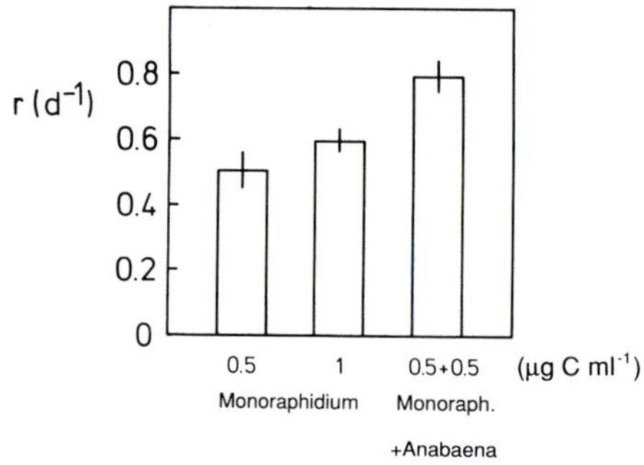

Fig. 6.2.6 Population growth rates of the rotifer *Brachionus rubens* when fed different concentrations of *Monoraphidium minutum* or a mixture of *Monoraphidium* and *Anabaena*. *Anabaena* alone ($1\,\mu g\,C\,ml^{-1}$) did not permit population growth (After Rothhaupt 1991)

N and P limitation did occur, and this is consistent with the theory of resource limitation (Tilman 1982). If nutrient-limited and nonlimited algae or simultaneous limitation by different nutrients co-occur in nature, the question arises as to whether these potential zooplankton foods can balance each other's nutritional qualities and therefore represent complementary resources. It is possible to run continuous algal cultures in which species pairs coexist and are limited by different mineral nutrients (e.g., Holm and Armstrong 1981; Tilman 1981). This could easily be used to investigate the occurrence of such complementary effects in two-stage rotifer chemostats. The first algal stage would have to consist of such an algal species pair, either grown singly (then two separate algal stage are required) or in combination, with each alga limited by a different mineral nutrient.

Another possibility of complementary resources involves the role of the "microbial loop." When algae are severely nutrient-limited and hence (usually?) of low nutritional quality, they may excrete excess photosynthetic products in the form of dissolved organic compounds thus fuelling bacteria and their consumers (Bratbak and Thingstad 1985). Bacterial growth rates are thought to be mainly limited by the availability of organic carbon sources, and the stoichometry of the heterotrophic complex would thus tend towards higher nutrient contents. This is corroborated by findings that bacteria generally contain more nutrients and are less variable in their elemental composition than algae (Jürgens and Güde 1990). The "microbial loop" could thus act as a complementary food source for some zooplankton. The question would not be how much carbon or energy is transferred to the traditional food chain, but whether the microheterotrophs improve the food quality for the zooplankton, especially when the phytoplankton are nutrient limited. Zooplankton have

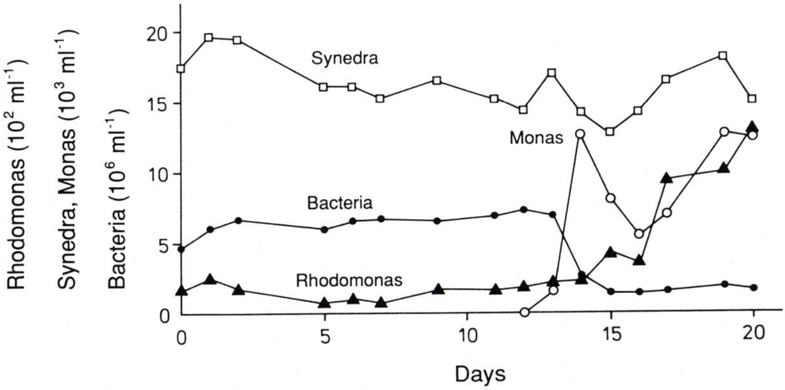

Fig. 6.2.7 A laboratory experiment that demonstrates some possibilities of constructing complex "resource communities" in the laboratory. During the first part, the diatom *Synedra*, the cryptophyte *Rhodomonas*, and bacteria approach a steady state. *Synedra* is Si-limited, *Rhodmonas* is P-limited, and the bacteria are probably limited by algal exudates. On day 12, the bacterivorous flagellate *Monas* is successfully introduced into the system

been found to feed and reproduce on heterotrophic flagellates and ciliates (DeBiase et al. 1990; Sanders and Porter 1990), and qualitatively additive effects of microheterotrophs in zooplankton diets are reported (Stoecker and Capuzzo 1990).

Again, these ideas are amenable to laboratory testing. Nutrient-limited algae, bacteria, and bacterivorous flagellates can coexist in chemostat culture (Güde 1985), thus offering the opportunity to use such a system as the food base for rotifers. It would be possible to test for complementary nutritive effects by comparing algae, subject to different degrees of nutrient limitation, with and without the heterotrophic organisms. Figure 6.2.7 illustrates how both the coexistence of algae limited by different nutrients and the inclusion of heterotrophs might be used to construct complex "resource communities" (see also Arndt et al. 1992).

6.2.4 The Threshold Concept

Thus far, we have been considering the direct interactions of one consumer with its resource base. If we now consider competition for shared resources, we have to keep in mind that this is an indirect interaction and that direct consumer-resource interactions are basic to it. In a previous section, we have seen and exemplified for a chemostat situation (Fig. 6.2.2) that there is a certain food requirement that is needed to allow population losses to be compensated by the production of offspring under the given circumstances. This has been termed "resource requirement at equilibrium" (R^*; Tilman 1982) and is the point at steady state where the apparent growth rate is zero ($r' = 0$; see Sect. 4.2.5). This threshold food concentration must not be confused with the minimum food requirement (C_0) for zero population growth ($r = 0$) from the modified Monod equation (e.g., 6.2.1). C_0 is an estimate of the threshold food concentration when population losses are only given by natural mortality. Threshold food requirements are higher if a mortality rate exerted, e.g., by predation in the field or by dilution in the chemostat, has to be compensated (cf. Fig. 6.2.2).

The threshold concept states that if species compete for a common resource base, the species with the lowest threshold food requirements under the given circumstances will win and exclude its competitors. This species is able to lower the food concentration to a level at which its competitors are no longer able to achieve positive growth or to withstand mortality. Figure 6.2.8 examines two hypothetical cases. The two species involved are assumed to suffer identical mortality rates, corresponding to the conditions of continuous or semicontinuous culture. In the first case (Fig. 6.2.8a), one species is superior over the whole range of resource concentrations and is thus expected to be generally the superior competitor. In the second, more interesting case (Fig. 6.2.8b), the numerical response curves intersect. In this system, the outcome of competition depends on the mortality imposed (= dilution rate in a chemostat). If a growth rate

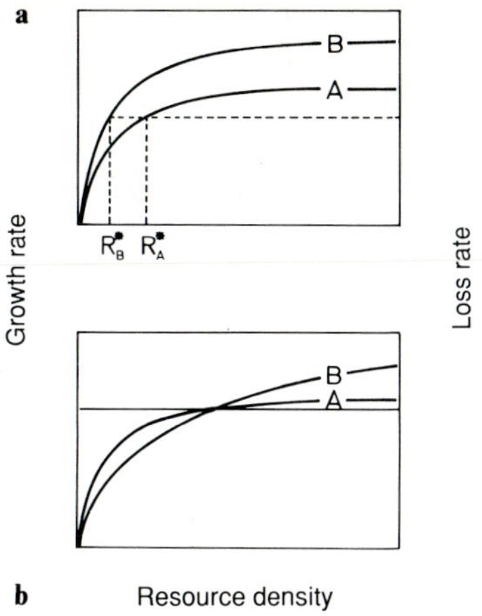

Fig. 6.2.8 a, b Schematic representation of the competition for a limiting resource. The species with the lower threshold resource requirement to compensate the losses (R*) is superior. **a** Species B is always superior. **b** Below a critical mortality rate species A, but above that rate species B, is superior

below the intersection point is required, species A is the superior competitor, having the lower food requirements. For growth rates above the intersection point, the opposite is true, and the outcome of competition is reversed. The applicability of this approach was first shown in microbial studies (Jannasch 1967) and has recently been confirmed for competing rotifers (Rothhaupt 1988; Boraas et al. 1990; Chap. 6.1).

6.2.5 Resource Partitioning

Herbivorous zooplankton differ in their feeding strategies and modes of food selection. This results in the differential use of components of the food spectrum and hence implies the possibility of food partitioning. Rotifers of the genus *Brachionus* are a good model system for studying these effects, as they have been shown to differ in their food size preferences (Rothhaupt 1990a). When the green algae *Monoraphidium minutum* (5 μm in longest linear dimension) or *Chlamydomonas sphaeroides* (12 μm in diameter) are offered, *B. rubens* more efficiently ingests the smaller alga, while *B. calyciflorus* is better able to feed on the bigger one. This is reflected by the growth rates attained in unmixed food (Fig. 6.2.9). According to the threshold concept, *B. rubens* is expected to be the superior competitor when *Monoraphidium* is fed, and *B. calyciflorus* when *Chlamydomonas* is fed. Coexistence seems to be intuitively possible when both foods are mixed.

The outcome of competition for pairs of resources can be predicted by a graphical method (Tilman 1982) that involves the division of a resource plane – a two-dimensional space bounded by the densities of the resource species – into a zone in which coexistence, and complementary zones in which superiority of one of the competitors, is predicted. The relevant statistics are both competitors'

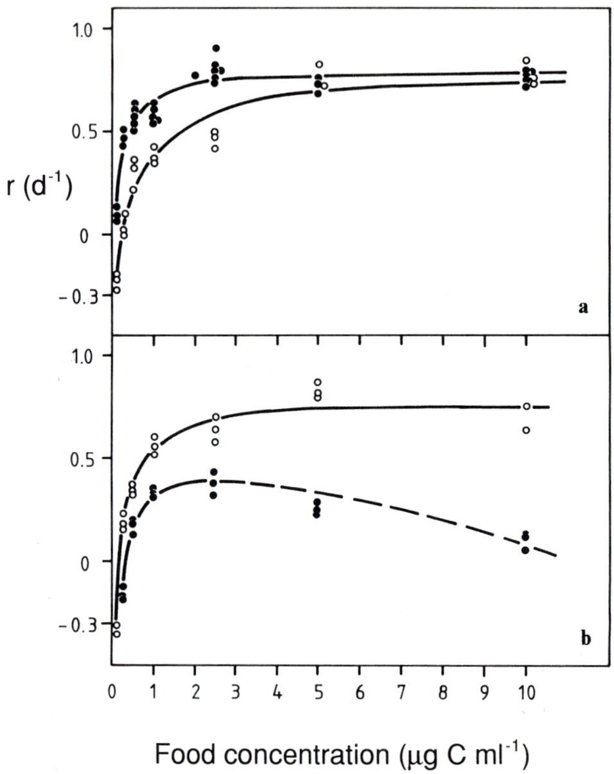

Fig. 6.2.9 a, b Relationship between environmental food concentration and population growth rate for *B. rubens* (*closed circles*) and *B. calyciflorus* (*open circles*). *B. rubens* is the superior competitor for *Monoraphidium* (**a**), whereas *B. calyciflorus* is always superior for *Chlamydomonas* (**b**). The decline in growth rate for *B. rubens* at high densities of *Chlamydomonas* is caused by mechanical disturbance of the feeding process (After Rothhaupt 1988b)

"zero net growth isoclines" and "consumption vectors". Zero net growth isoclines join up combinations of resource densities that permit zero population growth. They are thus equivalent to the threshold food concentrations for single resources. Consumption vectors indicate the effects that the feeding activities have on the resource densities, i.e., they describe the proportions in which the resources are removed from the system (see Tilman 1982 for a detailed description; Rothhaupt 1990c for a brief outline).

Applied to the example of both *Brachionus* species, prediction and experimental outcome can be compared for two series of competition experiments that were done in semicontinuous cultures (Chap. 2.1) at two dilution rates with different mixtures of both foods (Fig. 6.2.10). There is a high degree of correspondence between prediction and observation. In 11 of 12 cases, the theory accurately predicts the experimental outcome. The stable coexistence of both

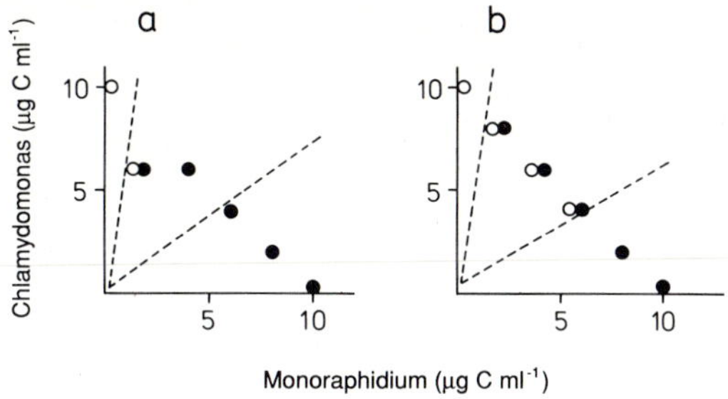

Fig. 6.2.10 a, b Comparison of prediction and experimental outcome in the competition experiments with the rotifers *B. rubens* and *B. calyciflorus*. Food mixtures in the *middle sector between the broken lines* are expected to lead to coexistence (Br + Bc). Mixtures in the *outer sectors* should lead to the exclusion of one species. The species predicted to persist is indicated (Br or Bc). *Full circles* mark experiments in which only *B. rubens* persisted, *open circles*, only *B. calyciflorus*. Both symbols side by side indicate coexistence (After Rothhaupt 1988b)

rotifer species is unequivocal under certain conditions with mixed foods (Rothhaupt 1988).

6.2.6 Resource Variance

Variability of resource availabilities appears to be an important factor influencing herbivorous zooplankton communities (DeMott 1989; Rothhaupt 1990d). There is theoretical evidence (see review by Chesson 1986) as well as empirical evidence from laboratory studies of phytoplankton competition (Sommer 1985; Gaedeke and Sommer 1986) that resource variations can promote the coexistence of competitors and thus enhance the diversity of coexisting species.

In the field, variations of the zooplankton food resources may be caused by external factors as well as by internally driven consumer-resource interactions. Externally triggered variations range from dial patterns to seasonal events. On the intermediate time scale, the more erratic pattern of meteorological events dominates. Internally driven resource fluctuations are usually considered to be analogous to predator-prey cycles, and in the field they are often caused by herbivorous cladocerans (McCauley and Murdoch 1987).

Culture techniques involving rotifers offer interesting possibilities of simulating resource variance effects in the laboratory. Internally driven consumer-resource fluctuations are often observed in one-stage chemostats (Fig. 6.2.11). These cycles have either been observed to dampen and stabilize (Boraas 1980) or to continue for longer periods in an apparently chaotic fashion (Schlüter 1984). In any case, resource availabilities change considerably in response to the population dynamics, leading to alternating phases of over- and

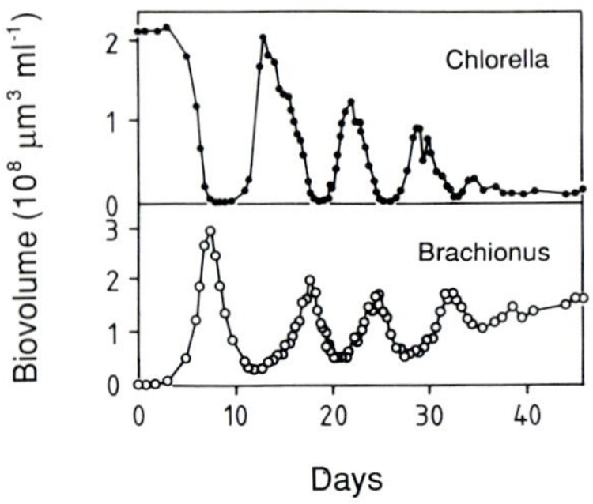

Fig. 6.2.11 Resource variations caused by rotifer population oscillations in a one-stage chemostat (After Boraas 1980)

underexploitation of the food resources. Competitive interactions are thought to be influenced by this kind of food resource variation. The idea, originally developed for crustacean zooplankton, is that some species are better able to store energy and to survive phases of low food density (Romanovsky and Feniova 1985; Matveev 1985, 1987).

The effects of externally forced, short-term variation can be studied using semicontinuous cultures (Chaps. 2.1, 6.1). Similar to a chemostat, the populations are allowed to increase and to exploit resources. At certain time intervals, a part of the medium, including animals, is removed and replaced with fresh food suspension. Each dilution event thus is a twofold disturbance: reduction of the population size and addition of new resources. The long-term average of the dilution rate is given by (Gaedeke and Sommer 1986):

$$D = -(\ln V_t - \ln V_0)/t \tag{6.2.2}$$

where t is the time interval between dilution events, V_0 the culture volume, and V_t the volume of the culture remaining after dilution.

Figure 6.2.12 schematically shows the changes of the relative food concentrations resulting from different experimental protocols. The average dilution rate is identical in all cases shown ($D = 0.5\,\text{day}^{-1}$). It is assumed that the cultures have reached a steady state, i.e., that the average rotifer densities are at their carrying capacity and that the carrying capacity is proportional to the difference $C_1 - C_2$ (see Sect. 3.3.4). The actual food concentrations were calculated according to the following equation, taking into account that rotifer populations increase during a dilution interval:

$$C = C_0 * \exp - [FR * t * No * \exp(D * t)], \tag{6.2.3}$$

where C_0 and N_0 are the resource concentration and the animal density after a dilution event, FR is the clearance rate per animal, and t is the time.

By varying the dilution pattern or the input resource concentration (C_1), it is possible to run cultures that are identical in their long-term average dilution rate, and thus in their average degree of resource limitation, but that differ in their degree of resource variance. If the input algal concentration is held constant, the temporal heterogeneity increases with the length of the time interval between dilution events (Fig. 6.2.12a): Food pulses are higher with increasing time

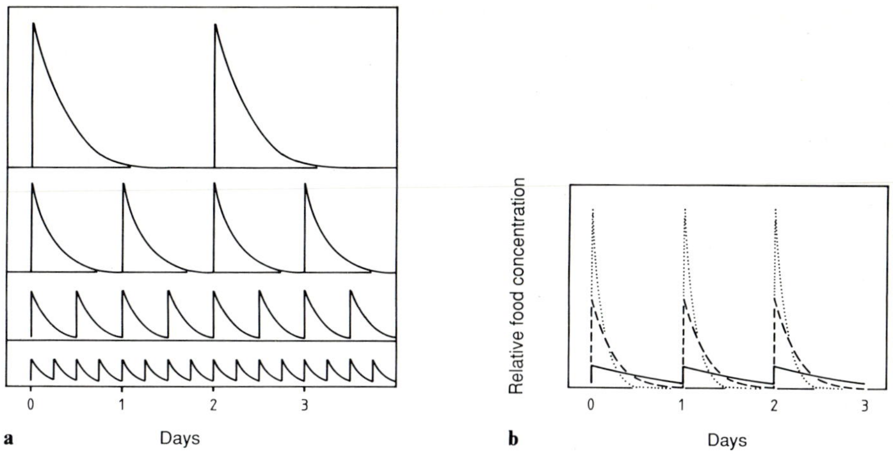

Fig. 6.2.12 a, b Simulated fluctuations of relative resource densities in semicontinuous culture when (**a**) the input food concentration is held constant and the time interval between dilutions is varied, and when (**b**) the time pattern of dilution is held constant and the input food concentration is varied. The long-term average of the dilution rate is identical in all cases ($D = 0.5\,day^{-1}$)

interval, but there are also increasing periods of little or no food. Consequently, both the frequency and the amplitude of resource variation change with the time pattern of dilution. If the time pattern of dilution is held constant, only the amplitude of temporal heterogeneity increases with the input resource concentration (Fig. 6.2.12b). Again, high food pulses are bound up with subsequent phases of food depletion.

The short-term resource variations caused by semicontinuous techniques can have a pronounced effect on the consumer species. Depending on their frequency and amplitude, the fluctuations around the long-term mean may be perceived or not. High food peaks may favor species that efficiently collect and store resources under rich conditions, while pronounced periods of food depletion will require some degree of starvation resistance. It was shown by a graphical model that short-term resource variance without serious starvation can promote the coexistence of zooplankton species that differ in the characteristics of their functional response curves (Rothhaupt 1990d).

6.2.7 Suitability of Rotifer Continuous Cultures for Testing Mechanistic Models

Rotifers are ideal model organisms for studying pelagic consumer-resource interactions. There are methods to determine numerical and functional responses to resource availabilities. The knowledge of these responses is a basic requirement for constructing mechanistic consumer-resource models. Of special interest are mechanistic competition models that predict the outcome of indirect interactions from previous ecophysiological knowledge. Such models are virtually unique in ecological studies.

Continuous and semicontinuous techniques offer a wide spectrum of possibilities for investigating various aspects and modifications of such models. Influences like food quality effects, possibility of complementary foods, resource partitioning due to different feeding preferences, or differential response to variations in the resource availability can be approached experimentally. If a certain effect or mechanism is to be studied, however, the experimental methods must be carefully chosen to match the underlying assumptions. When semicontinuous methods are applied, for example, it has to be kept in mind that this necessarily leads to fluctuations of the food availability which may influence the experimental outcome, even if this is not the primary aim of the study.

Carefully applied, the methods described thus hold the potential to match various theoretical approaches with the corresponding experiments. In this manner, models can be tested, and the gap between theory and observation can be narrowed by identifying mechanisms and causalities.

6.3 Life History Strategies of Rotifers

Norbert Walz

6.3.1 Introduction

Life history strategy depends on the conflict between different metabolic demands of the organism and the biotic and abiotic conditions of the environment. Abiotic situations often change very quickly and hinder the stable adjustment of steady-state phases (Chap. 7.2). The influence of only three, but among the most important abiotic parameters is studied here. The relationship of organisms to biotic parameters can be better predicted. Organisms with low energetic demands have advantages in environments with low food concentrations. On the other hand, they easily come under the control of a predator ("top-down" control in the well-known energy pyramid). Other organisms with a higher energetic metabolism require a larger food supply and are outcompeted at lower concentrations. They are easily controlled by "bottom-up" factors. On the other hand, they are able to avoid "top-down" controls by predators by their high kinetic features of their metabolism. Both types represent the extremes of different strategies selected to high r_{max} or low K_s metabolic characteristics (Sect. 6.3.7). An excellent introduction to basic rotifer ecology was given by Wallace and Snell (1991).

6.3.2 Response of Rotifer Populations to Abiotic Conditions

Abiotic conditions do not have regulatory effects on populations, but they influence many metabolic functions. Although this book emphasizes the role of biotic interactions in defining the life history strategies of rotifers, abiotic situations supply the boundary conditions for individuals and population development. Temperature, for example, has an accelerating influence on energy metabolism and, therefore, on the level of food necessary to fulfill energy demands (Walz et al. 1989). Embryonic development depends nearly exclusively on temperature as an abiotic condition (Duncan 1983; Herzig 1983). Rotifers, as all other organisms, are adapted to different temperature levels and to a larger or narrower temperature range. Different species present different temperature optima for their population growth rate. In laboratory cultures, for example, *Brachionus angularis* grows at a maximum rate at 20 °C and *Keratella cochlearis* at 15 °C (Walz 1987a). Much information comes from observations of natural communities with an ecological classification into eurythermal and stenothermal forms. According to such studies, *K. cochlearis* is eurythermal (Koste 1978a; May 1983) with on maximum abundance found at about 14 °C (Bērziņš and Pejler 1989a), and *B. angularis* is warm-stenothermal and most abundant at about 16 ° (Bērziņš and Pejler 1989a). *K. quadrata* (May 1983), *Asplanchna priodonta* (Bērziņš and Pejler 1989a), *Polyarthra vulgaris* (Stenson 1983), *Ascomorpha ecaudis* (Koste 1978a), and *Ascomorpha ovalis* are prominent examples for other eurythermal planktonic species. *Kellicottia longispina* is eurythermal also (Nauwerck 1963; Pourriot 1965; Zimmermann 1974; Herzig 1987). According to Koste (1978a), this species is "cold-eurythermal," i.e., with a distribution only below 15 °C. Threlkeld (1976) considers it to be a winter form.

The temperature range of stenothermal species is restricted to either cold or warm temperatures. Warm-stenothermal planktonic rotifer species include, for example, *Trichocerca similis* (Pourriot 1965, May 1983) and other *Trichocerca* species in Lake Constance (Walz 1987b) and *Pomphoyx sulcata* (May 1983). *Polyarthra dolichoptera* (May 1983, Stenson 1983), *Filinia terminalis* (Bērziņš and Pejler 1989a), *Notholca squamula* (Koste 1978a; May 1983), *Notholca caudata* (Laxhuber and Hartmann 1988), and *Keratella hiemalis* (Bērziņš and Pejler 1989a), on the other hand, have been classified as cold-stenothermal, although *P. dolichoptera* was also found by Bērziņš and Pejler (1989a) in smaller lakes and ponds at higher temperatures.

Oxygen tolerance is a most important factor limiting the distribution of populations. Stenson (1983) designated *Polyarthra dolichoptera* as tolerant to low oxygen concentrations, whereas *P. vulgaris* does not tolerate such conditions. *Filinia terminalis* (Guerrero et al. 1978), *F. hofmanni* (Mikschi 1989), and *Keratella hiemalis* (Bérzinš and Pejler 1989b; Mikschi 1989) are other prominent examples of planktonic rotifers which are able to tolerate an oxygen deficit. In a comprehensive list, Bērziņš and Pejler (1989b) could not find any true anoxybiont rotifer species.

Another parameter influencing rotifer distribution is pH. As a consequence of the problems of "acid rain," this question has gained much interest in lakes situated in unbuffered grounds. Most of the 227 rotifer species in a list of Bērziņš and Pejler (1987) tolerate a broad range of pH values. Several species were found only in acidic environments but none with their peak abundance < pH 5.5. The populations of most common species were reduced in acidic lakes: *Kellicottia longispina*, *Filinia terminalis*, *Polyarthra euryptera*, *P. vulgaris*, *Asplanchna priodonta*, *Conochilus unicornis*, and *Trichocerca cylindrica* (Havens and DeCosta 1988). Rotifer communities in many acidic lakes of northeastern USA are dominated by *Keratella taurocephala*, which replaced *K. cochlearis* (Siegfried et al. 1984; Siegfried 1991; Brett 1989). As the whole living community was changed in acid-stressed lakes, other biotic interactions and trophic states developed (Siegfried 1991). A nearer inspection showed that *K. cochlearis* per se is not acid sensitive (Havens and DeCosta 1988). This reveals that rotifer abundance is strongly related to the trophic state and availability of algae, evidently by depression of their crustacean competitors.

6.3.3 Niche Separation According to Food Particle Size

For a long time, the partitioning of resources according to particle size has been in the foreground of interest for scientists working on zooplankton. Since Hutchinson's (1961) formulation of the "paradox of the plankton," many efforts were undertaken to measure size-selective feeding by zooplankton. The size efficiency hypothesis looked for a relationship between the body size of the consumer and the size of its food particles (Brooks and Dodson 1965; Hall et al. 1976). Smaller species, especially rotifers, should be able to ingest smaller particles and vice versa (Gliwicz 1969). A linear relationship was demonstrated by Burns (1968) between the length of cladocerans and food size. Such early experiments were made with plastic beads which do not necessarily represent the characteristics of living cells. Tracer experiments with natural algae often come to other results, especially in the smaller size range (e.g., Gilbert and Bogdan 1984).

According to analyses of the space between their filter setulae and of their ingestion and growth rates, many cladocerans filter unselectively over a wide size range from bacteria, picoplankton, to nanoplankton of about 30 μm. To this group belong *Diaphanosoma*, *Chydorus*, *Ceriodaphnia*, and some species of the genus *Daphnia*. Another group filters bacteria and picoplankton ($< 3\ \mu$m) very inefficiently, if at all: *Bosmina*, *Simocephalus*, *Sida*, and other *Daphnia* species (Geller and Müller 1981; Pace et al. 1983; DeMott 1982, 1985; Brendelberger and Geller 1985; Bogdan and Gilbert 1987). In those species capable of ingesting bacteria at all, the capacity drops with increasing setulae space of the feeding appendages, which is correlated with body size (Porter et al. 1983a; Brendelberger 1985). Cladocerans, especially daphnids, are held to feed unselectively in a broad range of nanoplankton from 3 to 20 μm (Sterner 1989). Beyond the ingestion of larger particles with larger body size, the upper size of ingestible particles seems to be variable for different morphological types (Gliwicz 1980; Gliwicz and Siedler 1980; Fulton and Pearl 1987).

Herbivorous copepod adults and copepodite stages are also able to ingest nanoplankton algae, but never bacteria (Bogdan and Gilbert 1987). In general, they feed on larger algal cells (Frost 1972).

By no means are rotifers the dominant bacteria and picoplankton algae feeders in plankton (e.g., Pace et al. 1990) as previously supposed (Nauwerck 1963). In many cases, rotifers ingest bacteria very inefficiently, and they require a very high particle density (Starkweather et al. 1979). Only for the species group with *Keratella cochlearis*, *Kellicottia longispina*, and *Conochilus unicornis* there is an efficient consumption of bacteria and small aflagellate algae (Gilbert and Bogdan 1981; Ross and Munawar 1981; Bogdan and Gilbert 1984, 1987). These species also feed unselectively on larger algae. They may be characterized as "generalists" (Gilbert and Bogdan 1984). The food preference of *Keratella quadrata* is not much different from *K. cochlearis. K. quadrata* is a polyphagous species feeding on particles smaller than 10–12 μm, especially Chlorococcales, Volvocales, Chrysomonadales, *Euglena, Rhodomonas*, and central diatoms, but also on detritus, with a higher portion of nanoplankton (Nauwerck 1963; Pourriot 1977). Similar feeding conditions are typical for *K. hiemalis* (Zimmermann 1974).

Rothhaupt (1988) studied the particle size spectrum for ingestion of three *Brachionus* species. The filtration rate and incipient limiting level depended on the optimum particle size specifically, which was correlated to the body size of the rotifers. The optimum for the studied *Brachionus* species was 12 μm and smaller. The ingestion rate of smaller particles at higher concentrations was as high as for optimally sized particles. Larger particles were ingested at a somewhat lower rate, but the rate remained high. In demonstrating the general strategy of rotifers in comparison with other zooplankton groups, these differences seem to be too low for a selective advantage, but perhaps for the succession of rotifers some species may be favored by these mechanisms. For that reason, *Brachionus* species may belong to these generalists, as well. This corresponds to the general feeding morphology in the family Brachionidae (Edmondson 1965; Gilbert and Starkweather 1977; Starkweather 1980).

Filinia species are perhaps exceptional since they are often found in bacterial horizons of eutrophic or merolimnic lakes (Guerrero et al. 1978). They feed on bacteria and minute detritus particles (Nauwerck 1963). In larger lakes, these species are very rare.

The above-mentioned cladocera feeding on bacteria do not have an important grazing effect on the bacteria (Platzek 1986; Børsheim and Andersen 1987), and their nutritional value for the cladocera is very low (Platzek 1986). In other cases, cladocera are significantly involved in bacterial grazing (see review of Güde 1989), but today it is highly evident that the heterotrophic flagellates are the most important bacteria grazers in limnic biotopes (Bern 1987; Arndt 1992).

Other rotifer species belong to the "specialists" (Gilbert and Bogdan 1984). *Polyarthra* species are unable to feed on bacteria, and their ingestion efficiency decreases for small aflagellate algae. They prefer larger flagellate *Cryptomonas* species (Gilbert and Bogdan 1981; Bogdan and Gilbert 1984, 1987) and ingest *Euglena* up to 45 μm (Bogdan and Gilbert 1982). Some *Synchaeta* species are pronounced specialists for larger particles, especially *S. pectinata*, sucking out large *Cryptomonas*, and *Trichocerca* species, feeding on larger filamentous algae (Pourriot 1970). *Ascomorpha ovalis* and *A. saltans* live exclusively by sucking out dinoflagellates (Pourriot 1977; Koste 1978a), and *A. ecaudis* catch green algae and Chrysomonades (Pourriot 1965; Ruttner-Kolisko 1972a). *A. ecaudis* captured *Cryptomonas erosa* in experiments by means of its gelatinous envelope (Stemberger 1987). Other *Synchaeta* and *Trichocerca* species are carnivorous.

To this group belongs the omnivorous *Asplanchna*, feeding on large algae and rotifers (Ejsmont-Karabin 1974; Guiset 1977).

As cladocera hardly graze on particles larger than 20 μm (Gliwicz and Siedlar 1980; Sterner 1989), some of these specialist rotifers find their niche in feeding on bulky algal cells ("inedible algae"). Partly, this group may be in competition with *Eudiaptomus gracilis* and *Cyclops vicinus* (Horn 1985a, b) and generally with other copepods, preferring larger algal cells. However, copepods show a highly selective feeding behavior, grasping individual cells separately. Physical and chemical characteristics ("taste," DeMott 1986) of the particles play a dominant role. In this range of particles ("netplankton"), different specialist consumers may find specific niches.

In the presence of cyanobacteria, especially filamentous forms, different reactions are shown. Larger cladocerans seem to be more inhibited than smaller cladocerans, copepods, and rotifers (Dumont 1977; Richman and Dodson 1983; Orcutt and Pace 1984). The inhibition may consist of mechanical interference, poor food quality, or toxicity of the cyanobacteria. Rotifers, including generalists like *Brachionus* species, are remarkably uninhibited and may profit in such situations. *B. calyciflorus* feeds on filamentous *Aphanizomenon* spp. by nibbling the ends of the filaments (Dumont 1977). *B. plicatilis* grows with a mixed diet of *Chlorella* and the blue-green *Schizothrix calcicola* better than on either unialgal diet alone (Snell et al. 1983). In Lake Nakuru, an East African soda lake, *B. plicatilis* and *B. dimidiatus* nourish exclusively on the blue-green *Spirulina platensis* (Varesci and Jacobs 1984). Whereas some strains of blue-green algae are also toxic for rotifers (e.g., *Anabaena flos-aquae* for *Asplanchna girodi*; Snell 1980), most "toxic" strains do not affect rotifers. *B. calyciflorus* grows with toxic strains of *Anabaena flos-aquae* (Starkweather 1981; Starkweather and Kellar 1983) and with *Microcystis aeruginosa* (Fulton and Pearl 1987) in contrast to cladocerans which are not able to do this. In competition experiments done by Fulton and Pearl (1988), however, blooms of nontoxic *M. aeruginosa* did not improve the competitive ability of *B. calyciflorus* as the rotifers were outcompeted by *Daphnia ambigua* at low food concentrations of the accompanying algae (see Sect. 6.3.4). The advantage of rotifers seems to lie solely in their resistance to toxicity.

Niche separation on the basis of particle size and morphological characteristics of algae cells is functional only for the rotifer specialists. In contrast, rotifer generalists are in sharp competition with cladocera. This is the reason why in the "energetic mainstream" of the pelagial, other principles of niche separation must be active.

6.3.4 Niche Separation by Food Quantity

A further diversification of possible niches is achieved by different pretensions of animals on the quantity of food. In aquatic herbivorous animals, this is expressed by different K_s values in the Monod equation (Eq. 3.3.7). Species with low K_s values and high k values ($k = r_{max}/K_s$ = slope of the first order reaction; see Sect. 4.1.2) demonstrate a higher affinity to food. At low food concentrations, their rates of ingestion and growth are higher. According to Table 4.1.3, for example, *Keratella cochlearis* is adapted to low food concentrations. On the other hand, *Brachionus angularis* is superior at higher concentrations.

Rothhaupt (1988) found additional K_s values for *B. rubens* and *B. calyciflorus*. The food affinity of *B. rubens* is higher (higher k values, lower K_s values) than for *B. calyciflorus*. Inside the range of particle size between 2 and 12 μm, the food affinity was further dependent on the food particle size; the k values were higher at the optimal food size.

These findings correspond well with the above-mentioned K_s relations of *B. angularis* and *K. cochlearis*. The smaller species attains the higher food affinity. This relationship is supplemented by Stemberger and Gilbert (1985) by an analysis of 8 rotifer species (Fig. 6.3.1). This generalization also holds within the genus *Asplanchna* (Stemberger and Gilbert 1984).

Parallel to K_s values, yield coefficients ($Y = K_1$, gross growth efficiency; appendix) also indicate different energetic preferences (see Sects 3.3.6, 4.1.7, and 4.2.6).

Equation (3.3.6) in the following form:

$$C_2 = C_1 - X/Y \qquad (6.3.1)$$

shows the relationship between Y and the food concentration required for growth. In steady states of chemostats, this is the self-adjusted "residual" con-

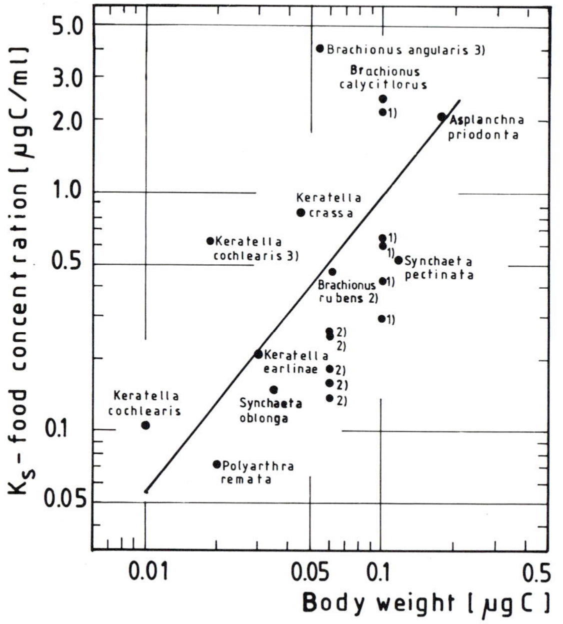

Fig. 6.3.1 Food concentration for half-maximum growth (K_s) in relation to rotifer body wight: *1*, *Brachionus calyciflorus* according to Rothaupt (1988), different algae; *2*, *B. rubens* according to Rothhaupt (1988), different algae; *3*, *B. angularis* and *Keratella cochlearis* according to Walz (Chap. 4.1) (After Stemberger and Gilbert 1985)

centration of algae. From this it follows that as the efficiency for growth is higher, the required food concentration must increase. On the other hand, organisms with low K_s values adapted to low food concentrations favor their maintenance functions and survival by a lower Y. Yield for growth, therefore, is lower. Table 6.3.1 demonstrates the pattern: protozoa – rotifers – cladocera have decreasing food concentrations necessary for growth.

A comparison of Y or K_1 values in the literature is only possible with caution, as in many cases it is not known which portion of growth was measured: total, somatic, or reproduction. Especially for higher organisms, the energy metabolism of individuals is usually considered without concerning all life aspects of a population (e.g., reproduction, mortality). Additionally, Y is not constant but dependent on food concentration (Chap. 4.1) and on food ration and temperature (Chap. 4.2). The efficiency of the energy metabolism increases for nearly all ectothermic organism (especially fishes, mollusks, and crustacea; see literature survey, Walz 1978a) with the food ration and temperaure up to a maximum value. Above the temperature optimum, or at higher food rations, Y drops considerably. On other cases, this relationship is not clear. For example, Kersting (1978) did not find a relationship between the K_1 and the food concentration in *Daphnia magna*. The efficiency was constant at 0.19.

Table 6.3.1 shows a comparison of Y and K_1 values for protozoans and a selection of ectothermic metazoans. Bacteria reach efficiencies of 0.41–0.53 (Table 2 of Bergter 1983) or 0.5–0.64 (Table 2 of Calow 1977a).

Other Y and K_1 values show that the efficiencies for poikilothermic metazoans can increase over 0.5, but in most cases they remain below 0.35 (Calow 1977a; Brett and Groves 1979; Schroeder 1981). Poikilothermic animals are able to convert more food to biomass than homothermic ones (Calow 1977b).

Table 6.3.1 demonstrates, in general, that protozoans have higher efficiencies than metazoans, they do not have to spend energy for mortality. Cladocerans seem to have lower efficiencies than rotifers.

Calow (1977a) emphasized that the more efficiently an organism transfers energy under normal conditions, the lower is its adaptive value under poor food conditions. Indeed, the high yields of rotifers are coupled with high K_s values. This underscores that rotifers, and even more so the protozoans, are users of high food concentrations.

The rotifers reach their high efficiencies at high concentrations only. The food ration of *B. angularis* required to reach Y_{max} is 70% C day^{-1}, i.e., they have to ingest 70% of their body weight daily (Chap. 4.2). The maximum food ration in chemostats was 300% day^{-1}. *B. rubens* ingests a daily maximum ration of 360% day^{-1} (Pilarska 1977c). Protozoans take in even more than ten-fold their body weight per day (Fenchel 1982). Higher aquatic metazoa need much lower food rations. *Daphnia* spp. have maximum specific ingestion rates of about 80% day^{-1} (Lampert and Muck 1985). Fish consume a daily maximum ration of about 20% (Brett 1979).

Cladocerans are not so much dependent on high food resources as are rotifers. From experimental studies, "incipient limiting levels" (ILL) of the

Table 6.3.1 Yield coefficients (Y) = gross growth efficiency (K_1) for ectothermic organisms; if possible Y_{max} values are given

Species	Substrate	Temperature (°C)	Y	Reference
Protozoa, Ciliata:				
Colpidium campylum	*Alcaligenes faecalis*	20	0.11	Sudo et al. (1975)
Tetrahymena pyriformis	*Aerobacter aerogenes*	25	0.73	Canale et al. (1973)
T. pyriformis	*Klebsiella aerogenes*	25	0.41	Curds and Cockburn (1968) (batch culture)
T. pyriformis	*K. aerogenes*	25	0.54	Curds and Cockburn (1971) (continuous culture)
Colpoda steinii	*Escherichia coli*	25	0.45	Drake and Tsuchiya (1977)
C. steinii	*E. coli*		0.78	Proper and Garver (1966)
Stentor coeruleus	*T. pyriformis*	20	0.63–0.82	Laybourn (1976)
Protozoa, Flagellata:				
Ochromonas spp.	*Pseudomonas* spp.	20	0.34	Fenchel (1982)
Pleuromonas jacularis	*Pseudomonas* spp.	20	0.43	Fenchel (1982)
Rotatoria:				
Brachionus plicatilis	*Dunaliella salina*	20	0.11	Doohan (1973)
B. plicatilis (slow strain)	*Brachiomonas submarina*	21	0.23	Droop and Scott (1978)
B. plicatilis (fast strain)	*B. submarina*	21	0.38	Droop and Scott (1978)
B. calyciflorus	*Monoraphidium minutum*	20	0.29	Leimeroth (1980)
B. calyciflorus	*Chlorella pyrenoidosa*	20	0.26	Boraas (1983)
B. calyciflorus	*Scenedesmus obliguus*	20	0.36	Galkovskaya (1963)
B. calyciflorus	*Monoraphidium minutum*	20	0.35	Rothhaupt (Chap. 4.3)
B. rubens	*M. minutum*	20	0.45	Rothhaupt (Chap. 4.3)
B. rubens	*Scenedesmus costato-granulatus*	20	0.43	Schlüter et al. (1987)
B. rubens	*Chlorella vulgaris*	20	0.24	Pilarska (1977c)
B. angularis	*Stichoccoccus bacillaris*	20	0.34	Walz (Chap. 4.1)
Encentrum linnhei	*Brachiomonas submarina*	21	0.30	Scott (1988)
Crustacea, Cladocera:				
Daphnia pulex	*Chlamydomonas reinhardii*	20	0.17	Richman (1958)
D. magna	*Chlorella vulgaris*	18	0.19	Kersting (1978)

ingestion rate can be compared. All authors who studied this and gave food concentrations in comparable units are in correspondence: Daphnids reach maximum ingestion rates at about 0.3 μg C ml^{-1} (e.g., Geller 1975; Porter et al. 1982; Lampert and Muck 1985). Among the cladocerans, the ILL and the ingestion rates seem to be higher for the larger species as the comparison between the smaller *Bosmina longirostris* and the larger *Daphnia rosea* (DeMott 1982) shows.

In contrast, the ILL for rotifers are generally higher. The ILL for *Keratella cochlearis* is $>$ 1.5 μg C ml^{-1} (Walz, Chap. 4.1) and 2 μg C ml^{-1} for *B. angularis* for both *Stichococcus bacillaris* and *Coccomyxa* spp. as food algae. A smaller form of *B. angularis* has an ILL of 1.3 μg C ml^{-1} with food particle sizes comparable with these algae (Duncan and Gulati 1983). The ILL of rotifers seems to be dependent on particle size. It is lowest for optimum sizes and increases especially for smaller particles in an unproportional measure (Rothhaupt 1988). Yúfera and Pascual (1985) report the ILL for *B. plicatilis* for different algae, which are above 5 μg C ml^{-1} (if C = 50% dry weight). However, Korstad et al. (1989) found ILL of 1.46 and 1.90 μg C ml^{-1} for this species. The ILL for *B. calyciflorus* is 2.5 μg C ml^{-1} with the green alga *Euglena gracilis* as food (Starkweather and Gilbert 1977c) and 1.5 μg C ml^{-1} with the cyanobacterium, *Anabaena flos-aquae* (Starkweather 1981).

Another parameter that demonstrates that rotifers are dependent on a high food supply is the threshold food concentration. There are different thresholds for the metabolism of individuals and populations (Sect. 4.1.4). The population-based thresholds include the effort to overcome losses due to mortality (r = 0; see Sect. 4.1.5, 4.2.8; Fig. 4.2.3). From experiments of Stemberger and Gilbert (1985), combined with the results of Rothhaupt (1988) and Hartmann (1987), it is evident that threshold concentrations increase with the body size of rotifers (Fig. 6.3.2) and are higher than related values of cladocerans (Lampert and Schober 1980; Gliwicz 1990), which again tend to surpass those of calanoid copepods (Lehman 1988). A similar relationship to body weight is obvious inside the genus *Asplanchna* (Stemberger and Gilbert 1984). Additionally, Snell and Boyer (1988) found higher threshold food concentrations for the mictic than for the amictic reproduction of *B. plicatilis*.

Other zooplankton groups do not follow this positive correlation. Thresholds in cladocerans decrease with increasing body size (Gliwicz 1990; Gliwicz and Lampert 1990). Thresholds for the large *Daphnia magna* are about 0.015 μg C ml^{-1}, and for the smallest tested cladoceran, *Ceriodaphnia reticulata*, about 0.04 μg C ml^{-1}. Schiemer (1985) arranged aquatic pelagic animals according to increasing threshold concentrations, with cladocerans having 0.04–0.12 μg C ml^{-1}, calanoids 0.07–0.11 μg C ml^{-1}, and ciliates 0.6–4.0 μg C ml^{-1}. With converting factors from Sect. 2.2.13 (this Vol.) threshold bacterial densities for ciliates (Taylor 1978) give similar POC concentrations. The thresholds of copepods are even lower, about 0.05 μg C ml^{-1}, for the larger *Calanus finnmarchicus* and 0.01 μg C ml^{-1} for the smaller *Pseudocalanus* spp. (Frost 1985).

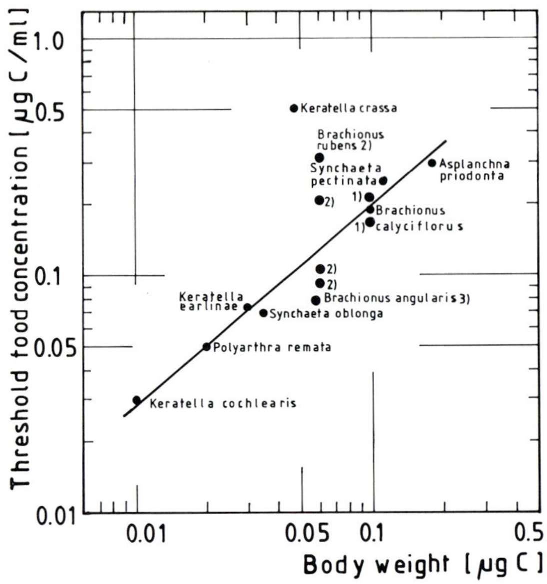

Fig. 6.3.2 Threshold food concentrations as related to rotifer body mass: *1*, *B. calyciflorus* according to Rothhaupt (1988), different algae; *2*, *B. rubens* according to Rothhaupt (1988), different algae; *3*, *B. angularis* according to Hartmann (1987) (After Stemberger and Gilbert 1985)

As an overall trend, as body size increases between the taxonomical groups, the thresholds decrease.

Below the metabolic-based threshold concentration, which is slightly lower than the threshold that includes the reproduction of populations (Fig. 4.2.3), individuals starve and die when the reserve material has been consumed. At low food concentrations, cladocerans have twofold superior abilities, first by having lower thresholds, second by possessing more energy reserves. The starvation resistance of cladocerans depends on age and lipid reserves (Tesssier et al. 1983; Goulden et al. 1987). Threlkeld (1976) found a positive relationship between the times until death of the different cladoceran species and body size, but in any case these times are considerably longer in comparison with rotifers. These have little ability to survive scarce food conditions. Gilbert (1985) reports on "LT_{50}" of about 2 days for *Keratella cochlearis*.

6.3.5 Impact of Predators on Rotifer Communities

The interference between invertebrate predators and rotifers is relatively well-known. Many species, especially of the crustacean plankton, are evidently predators. The copepod *Cyclops vicinus* (older copepodite stages > stage 4 and adults) prefers, according to Brandl and Fernando (1978), *Polyarthra*, spp., *Keratella quadrata*, *K. cochlearis*, *Synchaeta* spp. and *Asplanchna* spp. *Polyarthra* has the highest coefficient of selection for *C. vicinus*. On the other hand, Zankai

(1984) showed that *C. vicinus* did not select rotifers positively but consumed them in great quantities. Recently, it became clear that adults of *C. vicinus* are omnivorous and feed at least 50% on algae (Toth and Zankai 1985; Toth et al. 1987).

In Lake Constance, *C. vicinus* is the key factor for rotifer development at the end of the spring phytoplankton development (Walz et al. 1987). The abundance of rotifers begins to increase only after *C. vicinus* goes into dormancy and disappears. Then populations of *Polyarthra vulgaris/dolichoptera*, *Asplanchna priodonta*, *Keratella cochlearis* and *K. quadrata* start their development, but even when *C. vicinus* was present, they have high birth rates. Additionally, at that time the *Synchaeta* populations are totally depressed.

Similar conditions are found between *Cyclops bicuspidatus* and *Synchaeta* spp. in Lake Michigan (Stemberger and Evans 1984). This copepod prefers soft-bodied rotifers like *Synchaeta* spp. and *Polyarthra major*. *P. vulgaris* (perhaps because of its fugitive movements) and loricate rotifers like *Keratella cochlearis* and *Kellicottia longispina* are avoided.

Mesocycops leuckarti feeds on *Asplanchna* spp. and *Synchaeta pectinata* (Gophen 1977) but prefers *Polyarthra dolichoptera* and *Synchaeta oblonga* and avoids *Keratella cochlearis* (Karabin 1978), whereas rotifers are its main food. In Lake Constance, a comparison of rotifer numbers by Walz et al. (1987) and of *Mesocyclops leuckarti* (Stich 1985) in the same year leads to the conclusion that this copepod controls the population of *Polyarthra vulgaris/dolichoptera* in autumn. *Mesocyclops edax* preys on *Asplanchna girodi* and *Polyarthra vulgaris* but not on *Keratella cochlearis* (Gilbert and Williamson 1978).

Recently, cyclopoid copepods considered to be obligate predators were found to be omnivorous (see above), and calanoid copepods believed to be filter feeders were observed to be raptors. *Diaptomus pallidus* feeds on rotifers, improving its chances of survival and reproduction. It feeds on *Synchaeta oblonga*, which is defenseless, less on *Polyarthra ramata*, which escapes by its jumping response, and not at all on the hard loricate *Keratella cochlearis* (Williamson and Butler 1986).

Asplanchna priodonta feeds on *Polyathra* spp. *Keratella* spp., *Synchaeta* spp., and large algae (Salt et al. 1978). It prefers *Keratella cochlearis* and *K. quadrata* (Ejsmont-Karabin 1974; Guiset 1977). According to Gilbert and Williamson (1978) it selects *Keratella cochlearis* but avoids *Polyarthra vulgaris*. Further experiments (Gilbert 1980b) with *Asplanchna girodi* demonstrate the easy capture of *Synchaeta pectinata*, *K. cochlearis*, and *Conochilus* spp. individuals. It rarely captured *Kellicottia bostonensis* and intact *Conochilus* colonies and rejected the peritriche ciliate *Rhabdostylis* spp. and the dinoflagellate *Peridinium*.

In Lake Constance, after the reduction of *Cyclops vicinus* in late spring, a second predator on rotifers appears: *Asplanchna priodonta* (Walz et al. 1987). It executes a strong pressure on *K. cochlearis*, which develops only after the total reduction of *Asplanchna*. *K. quadrata* is not affected to such an extent and appears in the presence of *Asplanchna*. The same situation, but on a spatial scale, was observed in a small oligomesotrophic pre-Alpine lake, but *Asplanchna* and *K. quadrata/hiemalis* were at different depths (Höffgen 1988). In the Lake Fasaneriesee, a eutrophic gravel pit lake near Munich, FRG, *K. cochlearis* evades the predators *Asplanchna girodi* and *A. priodonta* by entering deeper depths (Huber 1982) where *Asplanchna* spp. could not follow, demanding higher oxygen contents (Mikschi 1989). Bosch and Ringelberg (1985) report high death rates of *K. cochlearis* in the presence of high *Asplanchna* densities.

Synchaeta species have a "virgate" mastax. Several species are carnivorous (Naumann 1923, according to Nauwerck 1963; Koste 1978b). Similar conditions are found for the *Trichocerca* spp. They suck on large algae, but the same species are carnivorous on other rotifers (Naumann 1923; Williamson 1983, in Nauwerck 1963). *T. capucina* sucks out the eggs of other rotifers (Pourriot 1970, 1977), especially *Keratella* spp. (Naumann 1923, in Nauwerck 1963).

Instars of *Chaoborus* larvae live as predators on zooplankton species. They prefer smaller prey sizes (Swift and Fedorenko 1975). For this reason, rotifers are the main prey of early instars (Moore and Gilbert 1987), but *Chaoborus* does not prefer any particular species (Dumont 1977). According to Williamson (1983), *Chaoborus* feeds on *Asplanchna*. The first two instars of *C. trivittatus* feed to a high degree on *K. cochlearis* and *Kellicottia longispina* when they are present in high numbers after the elimination of their main competitor, *Daphnia rosea* (Neill 1985). As *Chaoborus* is mostly active in the pelagial at night, being near the sediment at daytime, it may be that the influence of this species was often neglected.

Of particular interest is the interference of rotifers with cladocerans. The competitive superiority of *Daphnia* to *Keratella cochlearis* has been known for a long time (Dieffenbach and Sachse 1911). Suppression is managed by the up to 1000-fold greater grazing rate of *Daphnia* (Sect. 7.2.5), lowering phytoplankton densities under the threshold of rotifers (Sect. 6.3.4). Beneath this exploitative competition (Vanni 1986a), in the case of *Daphnia* interference competition is also active (Gilbert 1985). Rotifers, especially *Keratella cochlearis* and *Brachionus calyciflorus*, are damaged when swept into the filtering apparatus of *Daphnia* (Burns and Gilbert 1986a, b). Rotifers have even been found in the intestine of *Daphnia*. In seminatural enclosures, *K. cochlearis* was only suppressed by the resource exploitation of *Daphnia* (Threlkeld and Choinski 1987). On the other hand, even at low (1 – 5/l) *Daphnia* densities, this mechanical interference should also play a role in natural communities (Gilbert 1988a), even in Loch Leven, Scotland (MacIsaac and Gilbert 1990), where May and Jones (1989) first held that interference competition was unlikely.

Because they are of smaller size (*Synchaeta oblonga*) or delicate (*Ascomorpha ecaudis*), other rotifer species are vulnerable to *Daphnia* damage (Gilbert 1988b). In contrast to large *Daphnia*, smaller cladocerans (< 1.2 mm) do not seem to inhibit rotifers. They often co-occur at high densities and do not mechanically interfere with them (Gilbert 1988a). Exploitative competition is also much weaker as the food depletion by smaller cladocerans is much lower (MacIsaac and Gilbert 1989).

Rotifers have evolved many defense mechanisms against invertebrate predators, such as quick escape responses (e.g., *Polyarthra*, *Filinia*), hard lorica (e.g., *Keratella*), spines (e.g., *Filinia, Kellicottia, Brachionus*), and others listed by Stemberger and Gilbert (1987b). Another strategy is to evolve a high maximum growth rate, r_{max}. By that means, populations are able to withstand predation pressure (Sect. 6.3.6).

Because of their optical orientation, fish are not direct predators of rotifers, whose body size is too small. Large *Asplanchna* and colonies of *Conochilus unicornis* have been found in the stomachs of pelagial fishes (Stenson 1982). The influence of fish is indirect as they feed on the competitors of rotifers (Neill 1984; Gliwicz and Pijanowska 1989). On the other hand, fish larvae, after having consumed their yolk sac, like to feed on rotifers as their first food. This was shown for perch, carp, and roach (Filatov 1972; Guma'a 1978; Hammer 1985). However, this predatory activity is of no ecological importance in the pelagial as the young fishes live in the littoral or near the benthos of the lakes. In freshwater aquaculture, *Brachionus calyciflorus* is used as the first food of carp larvae (Widigdo 1988). Marine aquaculture relies on *B. plicatilis* as food for marine fish, shrimp, and crab larvae (Sulkin 1975; Emmerson 1984; Lubzens et al. 1989).

6.3.6 Kinetic Implications of Development

For a given zooplankton species, the predation pressure is withstood by avoidance mechanisms (spines, hard lorica, etc.) or by high growth rates. The latter mechanism allows the organisms to withstand flushing rates by short residence times of the water within the lake. Population growth rates can be described by the Lotka-Euler formula (Birch 1948):

$$\sum_{x=J+D}^{x_{max}} e^{-r \cdot x} * l_x * m_x = 1. \qquad (6.3.2)$$

The coefficients are the age (x)-dependent survivorship (survivors/cohort, l_x), the age-dependent birth rate (birth$*$female$^{-1}*$day^{-1}, m_x), and the boundary condition J + D (juvenile period + embryonic period or age to the first reproduction = "alpha" according to Cole 1954).

This equation looks only at the kinetic aspects of population development. Reproduction and survival cannot be maximized simultaneously (Snell and King 1977; Calow 1984). Likewise, energetic aspects must be considered (Sect. 6.3.4).

To maximize growth rates, the birth rate must be high, and the development time must be short. Both are variables and dependent on food conditions and temperature as well as body size and egg size.

6.3.6.1 Temperature

A temperature increase accelerates the bioenergetic rates (Fig. 4.2.10) and reduces the duration of development. In a pairwise comparison, *Brachionus angularis* behaves like a r_{max} strategist (Sect. 6.3.7). Durations of all life history

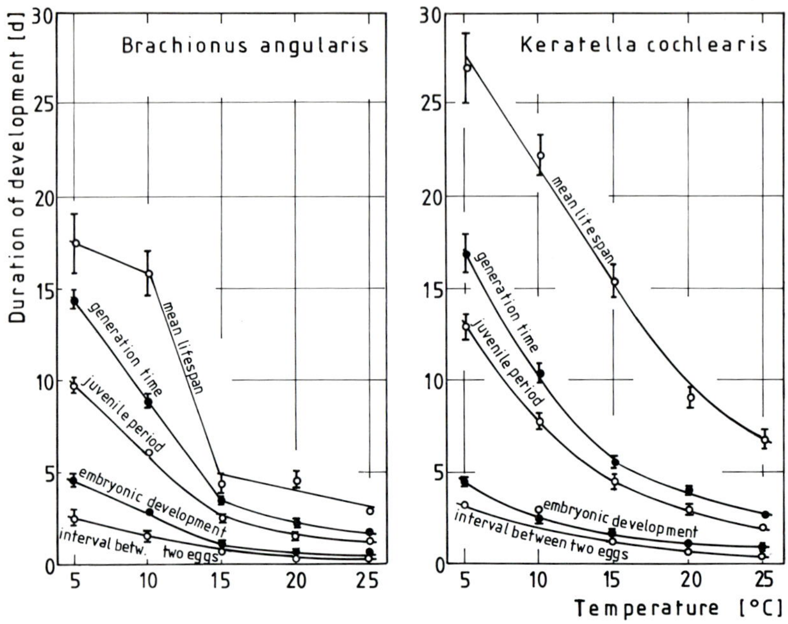

Fig. 6.3.3 Duration of different life periods and time of development at different temperatures (±95% confidence intervals of the means; *K. cochlearis* data after Walz 1983a)

periods are shorter than for the K_s strategist *Keratella cochlearis* (Fig. 6.3.3), but as in all other rotifers (Herzig 1983), durations of those periods decrease in both species with increasing temperature (Walz 1983a, 1987a).

The lifespan is especially important for the rates of development. Both species do not differ significantly between the number of eggs/female produced in a lifetime (Table 6.3.2.). The same is true for the number of young hatched (Mann-Whitney U-test, $P > 0.1$). But, as a result, egg-laying and hatching rates per female and per day is higher for *B. angularis* (Walz 1987a). This is also shown by the age-specific birth rate (m_x) and fecundity rate (offspring per surviving female per day, $l_x m_x$; Fig. 6.3.4). Throughout, but to a lesser extent at lower temperatures, m_x is higher for *B. angularis.* The lifespan is longer and survival better in *K. cochlearis* (Walz 1987a); therefore, l_x is higher in *K. cochlearis.* In this species, 50% of a cohort is dead after 15 days, whereas with *B. angularis* this percentage died after only 4 days. The product $l_x m_x$ (R_0) is, however, the most important factor determining the growth rate. This product is higher for *B. angularis* (Table 6.3.2; Fig. 6.3.4). For both species, with higher temperatures this fecundity rate is higher and skewed to younger age classes, which by itself accelerates the growth rate.

This is even more clearly seen in Fig. 6.3.5, where the growth rate is plotted against the age of the cohort. The growth rate after Eq. (6.3.2) was not calculated

Table 6.3.2 Number of eggs and surviving offspring (= R_0) per female and lifetime, egg-laying rate and hatching rate per day and relationship to temperature

Parameter	Temperature (°C)				
	5	10	15	20	25
Eggs/female					
B. angularis	1.8	3.0	1.9	4.3	2.3
K. cochlearis	1.8	2.4	3.1	2.1	2.3
Offspring/female					
B. angularis	0.9	2.5	1.6	3.6	1.5
K. cochlearis	1.1	1.6	2.2	1.5	1.3
Egg-laying rate					
B. angularis	0.10	0.19	0.42	0.92	0.77
K. cochlearis	0.07	0.11	0.21	0.23	0.37
Hatching rate					
B. angularis	0.05	0.16	0.36	0.78	0.51
K. cochlearis	0.04	0.08	0.16	0.18	0.25

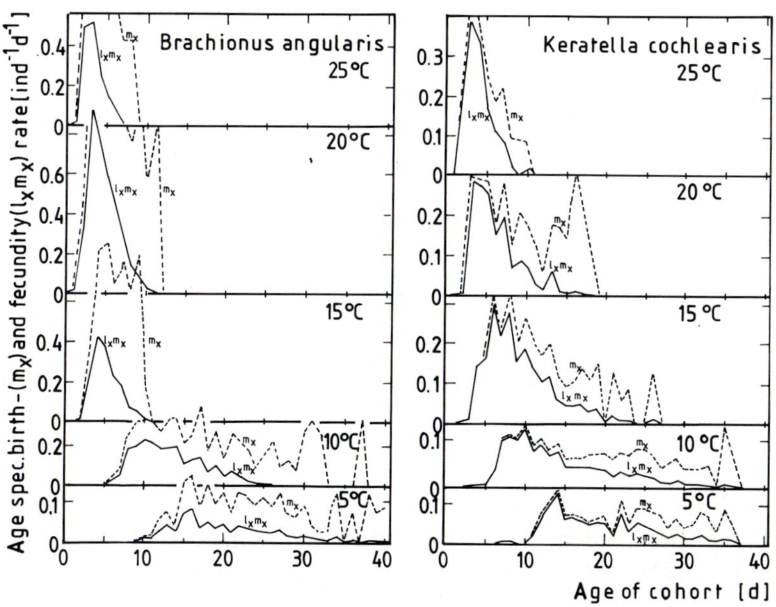

Fig. 6.3.4 Age-specific birth (m_x) and fertility rate (l_xm_x) at different temperatures (*K. cochlearis* data after Walz 1986a)

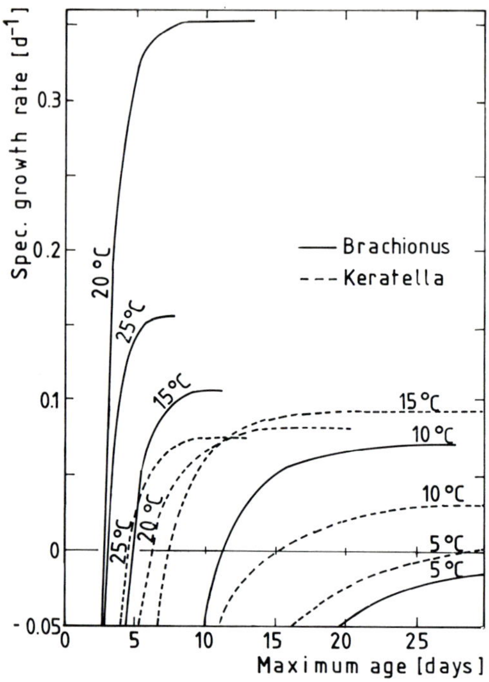

Fig. 6.3.5 Specific growth rate of *B. angularis* and *K. cochlearis* populations in dependence on the maximum age of the cohort (after Walz 1987a)

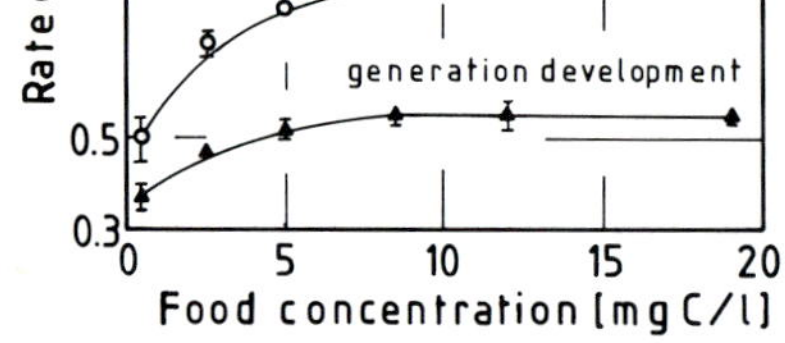

Fig. 6.3.6 Relationship between the food concentration and the rate of embryonic development ($1/D_e$), rate of juvenile development ($1/D_p$), and rate of generation development ($1/D_g$) of *B. angularis* ($\pm 95\%$ confidence interval of the means; after Walz and Rothbucher 1991)

for the whole cohort, but only for members up to a distinct age. This age is also equivalent to the residence time in continuous cultures. At $5\,\mu g\,C\,ml^{-1}$, this figure holds for populations of *B. angularis* which reach growth rates $> 0.35\,(day^{-1})$. If the residence time is lowered, the growth rate diminishes abruptly, and a point is reached at which the growth rate exactly balances losses by mortality ($r' = 0$). This is the "age of zero growth." In metazoans, this time cannot be less than the period of embryonic and juvenile development (the boundary condition in Eq. 6.3.2). In *B. angularis*, this period is 2.8 days (according to Fig. 6.3.5), whereas the sum of periods for embryonic and juvenile development is 2.3 days at 20 °C.

In a comparison of growth rates in Lake Constance (unpubl. data) compared with crustaceans of the same years (Schober 1980) and in the laboratory, it was established that maximum rotifer growth rates are higher for specialists such as *Asplanchna*, *Synchaeta*, and *Trichocerca* (Fig. 6.3.7) than for cladocerans, especially *Daphnia*. But it is exactly this "age of zero growth" which gives the rotifers the decisive time advantage, when the food conditions for development are favorable. Cladocerans do not reach the zone of positive growth before 8–14 days of population development (Jacobs 1978; Porter et al. 1983b).

6.3.6.2 Food Concentration

With increasing food concentration, the duration of juvenile phase (D_p, prereproductive period between hatching and laying of the first egg) shortens. In Fig. 6.3.6, the rate of juvenile development ($1/D_p$) is shown to increase with food concentration (Walz and Rothbucher 1991). This shortening of the prereproductive phase by higher food concentrations is established for many rotifers (King 1967; Pilarska 1977b) and planktonic crustaceans (Weglenska 1971). For *B. angularis*, the rate of embryonic development ($1/D_e$) is highest in a medium range of food concentrations, where the relative egg size is lowest. Increasing the juvenile development compensates for the prolongation of D_e. The rate of generation development ($1/D_g = 1/D_e + 1/D_p$), as an analogue to the population growth rate, therefore, reaches a pleateau acccording to a saturation function of the Monod model (Fig. 6.3.6).

6.3.6.3 Body Size

In Sect. 4.2.7, a positive relationship was shown between body size and maximum specific growth rate of rotifer populations. For the rotifers and planktonic crustaceans of Lake Constance in Fig. 6.3.7, a further relationship is demonstrated between r_{max} and body weight. For every population, the mean r_{max} values of the 5 maximum growth rates of the growing season were calculated. Both rotifers and crustaceans had higher mean growth rates with increasing body size. Rotifers at the lower end of this size spectrum (*Polyarthra dolichoptera/vulgaris*, *Kellicottia longispina*, *Keratella cochlearis*, and *K. quadrata*) are more limited by their kinetic constraints than *Synchaeta* spp. and *Asplanchna priodonta*, which are limited by energetic requirements. Also, *Trichocerca* spp. belong to the r_{max} group with high growth rates and a large body size, but an exact value for this was not available. It is interesting to see that the most pronounced r_{max} types are viviparous (as *Asplanchna*) or do not carry their eggs (*Synchaeta*), whereas all K_s strategists (see Sect. 6.3.7) bear their eggs until hatching.

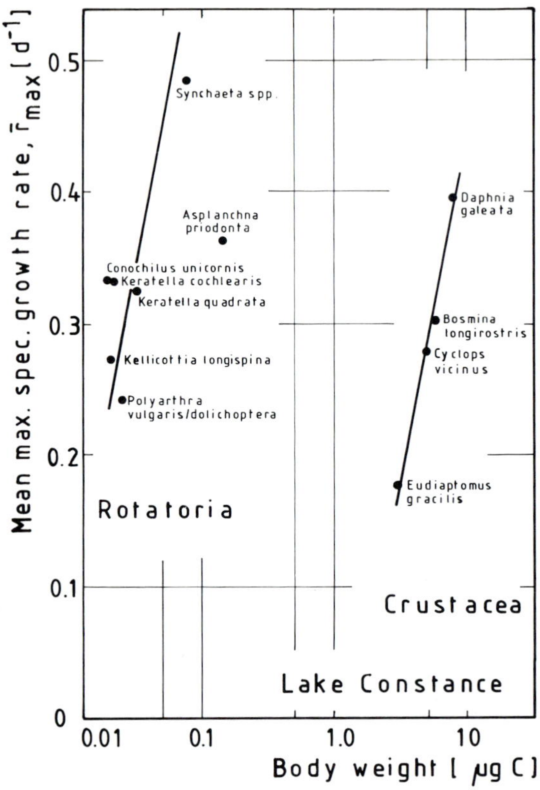

Fig. 6.3.7 Maximum specific growth rates of rotifers and crustaceans in Lake Constance in 1977/1978 as related to body size. Body sizes of most rotifers were recorded according to Walz (1987), for *Asplanchna priodonta* according to Dumont et al. (1975), for *Daphnia* spp., *Eudiaptomus gracilis*, and *Cyclops vicinus* according to Schober (1980), for *Bosmina longirostris* with length measurement of Schober (1980) according to regressions of Bottrell et al. (1976) (After Walz 1986a).

Similar positive relations between maximum growth rate and body size were found in laboratory rotifer cultures by Stemberger and Gilbert (1985; Fig. 4.2.6). Larger strains of *Brachionus plicatilis* are able to grow faster than smaller strains of the same species (Endo and Mochizuki 1979). In Lake Lanao (Philippines), a positive trade-off between birth rate and body size of rotifers and planktonic crustaceans was demonstrated by Lewis 1979). Stemberger and Gilbert (1985) could not decide between a double log plot of log r_{max} vs. log body mass and a semilogarithmic presentation with linear r_{max} as in Fig. 6.3.6, since both regressions gave the same level of significance and determination.

6.3.6.4 Relative Egg Size

Species with low relative egg sizes (egg volume/body volume) in Lake Constance and in chemostat cultures have higher maximum birth rates than species with larger relative egg sizes (Fig. 6.3.8). Further kinetic features of these species are that the rates of embryonic development are higher, and the periods of generation times are lower. Of course, only egg-bearing species can be compared, so *Asplanchna* and *Synchaeta*, placed at the upper end of the r_{max} group

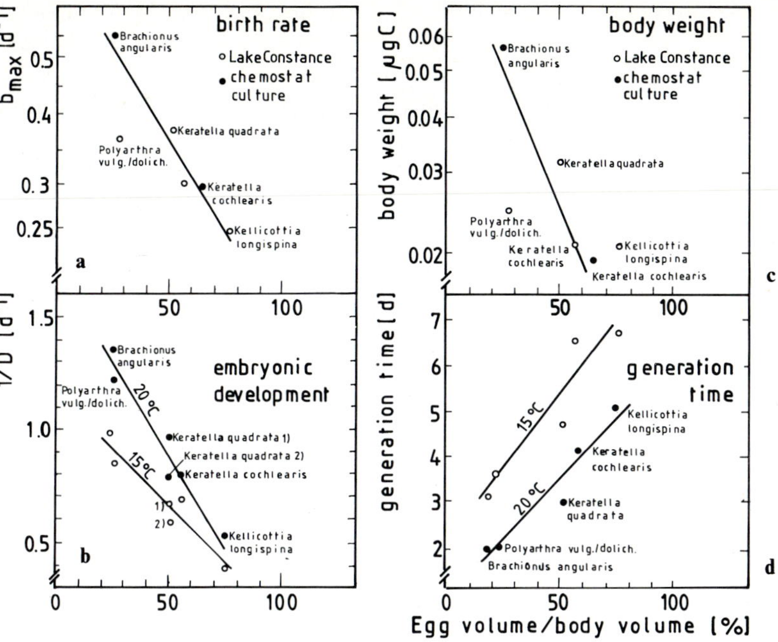

Fig. 6.3.8 a-d Relationship of different rotifer population dynamic parameters to the relative egg size: **a** maximum birth rate, **b** rate of embryonic development, **c** body weight, **d** generation time **a**, **b** after Walz 1986a

(Fig. 6.3.7) could not be considered. On the lower end, *Kellicottia longispina* has an egg that is 75% of the adult body volume.

Embryonic development which largely determined the birth rate was gathered from the following literature: *K. cochlearis* Walz (1983a); *B. angularis*, Walz (1987a); *K. quadrata*, Pourriot and Deluzarches (1971), Amren (1964); *P. dolichoptera*, Pourriot and Deluzarches (1971); *Kellicottia longispina*, Edmondson (1060). Generation times were taken from: *K. cochlearis*, Walz (1983a); *B. angularis*, Walz (1987a); *Kellicottia longispina*, (Winberg 1971); other species, Pourriot and Deluzarches (1971).

Figure 6.3.8 shows an inverse relationship between body weight and relative egg size. Larger rotifers have relatively smaller eggs with higher rates of embryonic development and maximum birth rates (r_{max} strategy). The same trade-offs between relative egg size and body weight, maximum birth rate, and rate of juvenile growth as in rotifers hold for cladocerans (Romanovsky 1985). Large cladocerans have, like the rotifers, relatively small eggs (Lynch 1980).

For many rotifer species there is no intraspecific variation of the relative egg size. In laboratory culture, *Keratella cochlearis* had always one egg with 65.4% of the body volume. Only occasionally under very good conditions were two eggs observed. This species had exactly the same relative egg size (66%) in

Lake Lanao (Philippines; Lewis 1979). In Lake Constance, *K. cochlearis* had a constant egg size of 55.7%. This constancy is reported also for other species, e.g., for *Asplanchna girodi* (Robertson and Salt 1981).

In contrast, *Brachionus* species seem to have variable egg sizes. Variation of the relative egg size in *B. angularis* is between 0.24 and 0.33. Absolute egg sizes increase with the body size of the mother and with the food concentration. Maximum relative egg sizes are found at intermediate absolute sizes (Walz and Rothbucher 1991). Duncan (1983), Sarma and Rao (1987), and Yúfera (1987) report positive correlations between the body size of the mother and the egg volume. Guisande and Mazuelos (1991) communicate for *B. calyciflorus* maximum absolute egg volumes at intermediate food concentrations.

The egg volume seems to have an influence on the duration of embryonic development. In *B. calyciflorus* and *B. caudatus*, the embryonic development is prolonged with egg size (Pourriot 1973a; Duncan 1983). In *B. angularis*, no continuous relationship between absolute egg volume and D_e was found (see above), but D_e shortens with greater relative egg size (Walz and Rothbucher 1991).

6.3.7 The r_{max} and K_s Strategies of Rotifers

Energetic and kinetic features of rotifers allow an ecological diversification within this taxonomic group. The hypothesis is as follows: Within any zooplankton taxonomic unity, small species are K_s and larger ones r_{max} selected. In other words there is a r_{max}/K_s continuum within any group. Species with pronounced differentiations of their metabolism maximizing r_{max} or minimizing K_s of the Monod function (Eq. 3.3.7) are not separated by a dichotomy (Christiansen and Fenchel 1977) but are connected in an continuum (Pianka 1970, 1972). Dugdale (1967) in studying the nutrient-limited growth of marine phytoplankton emphasized an evolutionary trade-off between energetic and kinetic parameters. Crowley (1975) found a relationship of the size of K_s to the variability of the substrate concentration. Species with particularly high r_{max} are species with special feeding modes and food types (Sect. 6.3.3) e.g., *Asplanchna*, *Synchaeta*, and *Trichocerca*. They require high food concentrations. On the other hand, the suspension-feeding generalists feed relative unselectively on all phytoplankton particles up to a size of about 30 μm. They do not reach such high r_{max} values and have a lower food demand. The transitions between the two types are conntinuous. Therefore, in a pairwise comparison *B. angularis* may be characterized as a r_{max}, and *K. cochlearis* as a K_s strategist (Walz 1983b, 1987a).

This r_{max}/K_s concept is parallel to the well-known r/K selection theory (MacArthur and Wilson 1967; Pianka 1978), which was founded on the terms of the logistic equation. K strategists are selected "in equilibrium" by density-dependent factors, and r strategists as opportunistic species by density-independent factors. In contrast, this r_{max}/K_s model is derived from the Monod model with kinetic (r_{max}) and energetic (K_s) selection factors (Chap. 5.1). In many cases, both models describe similar phenomena annd relationships, particularly those that do not require the strict derivation of a specific model (Pianka 1972).

In the Lotka-Euler equation (Eq. 6.3.2), only the kinetic aspects of the population growth are represented. Together with energetic arguments, they determine the position of a population in the r_{max}/K_s continuum. The r_{max} strategists are distinguished by the kinetic aspects of their growth and birth rates. Deciding factors are the age-dependent growth rate (m_x) and age at the first reproduction. The energetic aspect is not manifested in this equation, but there must be a high energy supply available. Since this is described by a high K_s value, r_{max} strategists are energy limited and likely controlled from "bottom up" in the food chain. With enough food, their growth rate is sufficient to withstand the predation pressure from a predator.

The kinetic aspect of the K_s strategy depends on the age-dependent survival rate, l_x. As a lower effort of energy is needed for reproduction, reproductive activity begins late in the lifetime because more energy is used for maintenance of the structure. Since survival is more secure and mortality is lower, reproduction may take place late in the lifetime, be distributed over a longer period, and, therefore, has a lower rate. Such a strategy does not require a high food concentration. In the lower range of food concentration, the actual value of the growth rate is higher for a K_s strategist than for a r_{max} strategist. Both strategies belong to different energetic domains. As populations of K_s strategists never attain high r_{max} values, they are likely to succumb to predator pressure and become controlled from the "top down".

Species with larger eggs, which should guarantee a higher individual fitness of the offspring, should be successful in times of shortage. This higher fitness of the offspring should be apparent as a lower mortality. In an interspecific comparison between *B. angularis* and *K. cochlearis*, this was shown for a saturation food concentration of $5\,\mu g\,C\,ml^{-1}$ (Walz 1987a). Therefore, a relationship between relatively large eggs and low K_s values is probable. The survival of copepod nauplii increased when the egg size was larger (Burns 1988). This pattern is also found in birds. Clutch size and survival of the young, which may be better fed individually at lower clutch sizes, are negatively correlated (Lack 1954; Cody 1966).

Only in the genus *Brachionus* is a temporally dynamic strategy found (Nichols et al. 1976). Such a variable strategy allows an optimum adaptation to the presently available resources, an important factor since the pelagial has a fluctuating food supply. Such temporal differences are more or less pronounced in fish (Ware 1975) and copepods, especially in *Eudiaptomus gracilis* with many small eggs in the spring and a few large ones in the summer (Czeczuga 1959), as the food concentrations are highest in spring. Hutchinson (1951) cited Wesenberg-Lund (1904) who made similar observations with *E. gracilis*.

6.3.8 Niche Differences Between Rotifers and Cladocerans

The r_{max}/K_s hypothesis also explains the niche separation between rotifers and their main competitors, the cladocerans. With food size as the only criterion, this was not successful (Sect. 6.3.3) as far as the generalists are concerned. Rotifer specialists avoid competition by feeding on larger food particles. In general, rotifers are r_{max} and cladocerans, K_s strategists. Cladocerans are likely to be under pressure from fish predators and are often controlled "top down" (Hrbaček et al. 1961; and others).

Problems occur for the small, slower growing rotifers (see above), which do not achieve higher growth rates than the cladocerans in many lakes. This is the case for *Kellicottia*, *Keratella*, and *Polyarthra* spp. in Lake Constance (Walz 1987b) and in Lake Fasaneriesee (Gilch 1981; Steiff 1981; Huber 1982, Precht 1983). Also, Banse and Mosher (1980) found, contrary to Allan (1976), that rotifers do not have the high growth rates expected from an interspecific allometric regression. They assumed that rotifers have a lower mortality. This is the case for the young of rotifers with relatively large egg sizes. But the mortality of cladocerans is even lower, since they are able to starve for longer times (Threlkeld 1976). The low mortality in the young stages does not give the rotifers a competitive advantage.

Additionally, the rotifers as r_{max} strategists need higher food concentrations (Sect. 6.3.4). This shows that they depend on "pioneer" conditions when the cladocera with their greater filtering efficiency are not present. In such situations they use their advantage: their short juvenile and embryonic period (Eq. 6.3.2). More exactly, this is the "age of zero growth" (Sect. 6.3.6). The distinction from the "generation time," for which more data are available, is low. Compared with the longer time delays of cladocerans, rotifer populations are present earlier and reach high numbers very quickly.

There is no parallel to the rotifer pattern in the cladocerans. According to Goulden et al. (1978) and Lynch (1980), the times to reproduction are longer for smaller species, in contrast to fecundity and b_{max}, which increase with body size (Goulden et al. 1978; Romanovsky 1985). In contrast to rotifers which stop or reduce their somatic growth at an advanced age (Walz and Rothbucher 1991), cladocerans allocate considerable amounts of energy for body growth throughout their life. The larger they grow, the larger is the brood chamber for more eggs (Taylor 1985). This pays off at a later time by a higher specific population growth rate.

Cladocerans, therefore, are able to react to higher food concentrations with a greater time delay, even though they finally reach similar or higher actual growth rates. This explains an often observed phenomenon (Herzig 1987) that rotifers increase at the begining of a phytoplankton bloom but later are followed by the cladocerans which then graze down the food concentration below the thresholds for rotifers. For the rotifers, a second criterion applies to r_{max} strategy: "being first is often a better strategy than trying often" and "precocious maturation may play an equally important role as fecundity" (Gould 1977).

7 Ecological, Ecotoxicological, and Aquacultural Significance of Continuous Culture Experiments

7.1 Critical Consideration of Chemostat Experiments

Karl O. Rothhaupt

7.1.1 Introduction

The use of chemostats in rotifer culture was first proposed by Droop (in Conover 1970) who regarded this technique as a promising approach for eco-physiological studies. He referred to, and later described in more detail (Droop 1976), a two-stage system in which the second-stage rotifer chemostat is supplied by a steady-state algal culture. This has been the most widely used system since then, and the following considerations essentially apply to it. Other continuous culture techniques, such as one-stage chemostats, turbidostats, or semi-continuous cultures (see other chapters in this volume), differ in important aspects.

In theory, the chemostat is thought to hold important advantages: It receives a known input of food algae in a defined physiological state, and as a self-regulating system it sustains time-invariant populations with a constant age structure, that essentially have lost their previous history. Most importantly, due to these characteristics, metabolic rates and efficiencies can be deduced from state variables that are easily measured (see Part 4).

The aim of this chapter is to check critically the above ideas. The assumption of time-invariant populations will be examind theoretically and compared with empirical data. Furthermore, I will give evidence that the assumption of a defined nutritional base may be erroneous and that this may lead to certain misinterpretations when determining functional and numerical responses from chemostat steady states.

7.1.2 The Occurrence of Oscillations

Periodic oscillations are known for laboratory populations of *Daphnia* (Pratt 1943) and of rotifers (Halbach 1970). Population cycles have also been reported for zooplankton in the field (McCauley and Murdoch 1987). They may thus be a widespread phenomenon rather than a mere laboratory artifact. This kind of population dynamics can be modeled by modifying the logistic equation to incorporate a time lag in the density-dependent term (Hutchinson 1948):

$$\frac{dN_{(t)}}{dt} = r_{max} N_{(t)}[1 - N_{(t-T)}/K], \tag{7.1.1}$$

where

N = population density
K = equilibrium population density
r_{max} = maximal growth rate
T = time lag.

The dynamic behavior of the model depends on the parameters r_{max} and T (May 1981):

$r_{max} * T < e^{-1}$	exponential approach to K
$e^{-1} < r_{max} * T < \pi/2$	damped oscillations
$r_{max} * T > \pi/2$	stable limit cycles

Generally speaking, if the expression $r_{max} * T$ surpasses a certain value, oscillations can be expected, the magnitude of which increases with $r_{max} * T$.

The applicability of the time-lagged logistic has especially been demonstrated for rotifer populations (Halbach and Burkhardt 1972). There was a reasonable agreement between the dynamic behavior of simulations using empirically determined parameters and laboratory population dynamics.

Time lags are due to the fact that it necessarily takes a certain time to convert food taken up into offspring. The length of this time span is largely temperature dependent (Halbach and Burkhardt 1972) and can therefore be regarded as invariable for a given clone of rotifers growing under stable laboratory conditions. The probability and magnitude of oscillations under these circumstances depends on r_{max}. When considering the stability of steady states in a rotifer chemostat, however, r_{max} has to be substituted by the "highest apparent growth rate" r'_{max}.

$$r'_{max} = r_{max} - D. \tag{7.1.2}$$

From this, it follows that oscillations can particularly be expected in rotifer chemostats running at low dilution rates and that the potential for fluctuations decreases as the dilution rate, D, approaches r_{max}. The occurrence of oscillations, however, means that the assumption of time-invariant populations with a constant age structure may not be given.

Since most work with rotifer chemostats aims at information deducible from steady states, the population dynamics are seldom described in the literature. However, some examples for the occurrence of oscillations in continuous

Fig. 7.1.1 a–d Examples of population dynamics that show oscillations in continuous culture: **a** *Brachionus calyciflorus* (after Boraas 1980); **b** *B. angularis* (after Walz 1983b); **c** *B. rubens* (Rothhaupt, unpublished in data); **d** *B. rubens* (*dots*) and *B. calyciflorus* (*circles*) in "rotating culture" competition experiments (After Rothhaupt 1988)

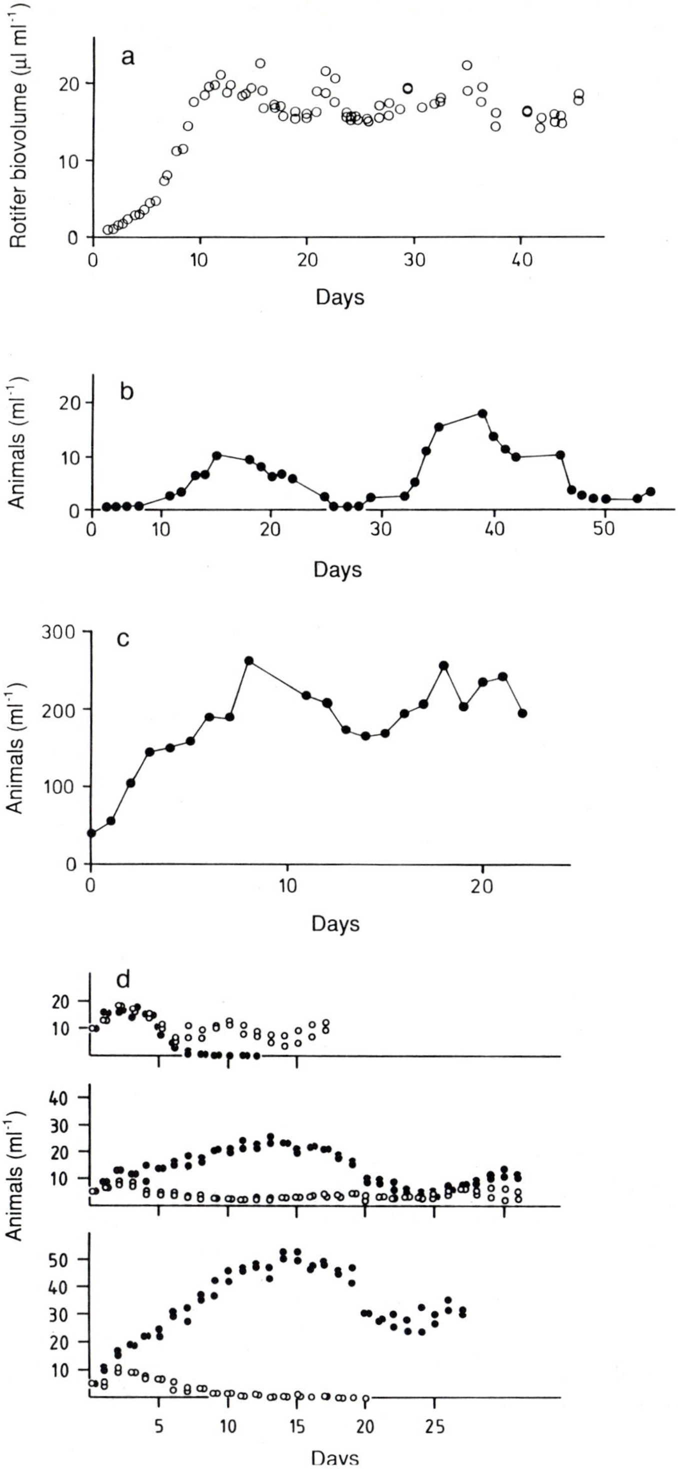
a
Rotifer biovolume (µl ml⁻¹)
Days
b
Animals (ml⁻¹)
Days
c
Animals (ml⁻¹)
Days
d
Animals (ml⁻¹)
Days

Table 7.1.1 Growth rate characteristics and estimated time lags from oscillating rotifer populations in continuous culture

Species	Temperature	r_{max}	D	r'_{max}	T	$r'_{max}*T$	Source (see Fig. 7.1.1)
B. calyciflorus	25	1.332	0.924	0.408	1.50	0.612*	a
B. angularis	20	0.579	0.250	0.329	5.75	1.893**	b
B. rubens	20	0.794	0.114	0.680	2.50	1.699**	c
B. calyciflorus	20	0.772	0.200	0.572	1.75	1.004*	d

T, time lag; D, dilution rate; temperature in °C; r_{max}, D, r'_{max} in day^{-1}; T in day
$r_{max}*T$ is a predictor of the dynamic behavior: * damped oscillations, ** undamped oscillations

rotifer cultures can be found (Fig. 7.1.1). Maximal growth rates and dilution rates are usually given. The time lag can be estimated as approximately 1/4 of the oscillation period (May 1981). An examination of the growth parameters shows that in all these cases oscillations could be predicted (Table 7.1.1), mostly damped oscillations. However, in real systems, whether in the laboratory or in the field, density-independent disturbances caused by stochastic fluctuations may sufficiently perturb the damped oscillations to yield persistent "quasi cycles" (Nisbet and Gurney 1976). When population fluctuations are not shown in the literature, steady-state rotifer densities are usually reported with a coefficient of variation of up to 15% (e.g., Boraas 1983; Schlüter et al. 1987; Rothhaupt, Chap. 4.3). This is further evidence that often there appear to be substantial fluctuations around the mean.

7.1.3 The Nutritional Quality of the Reactor Algae concentration (C_2)

If residual algal concentrations in the second stage of rotifer chemostats (Chap. 4.3) are compared with food requirements of the same rotifer clones for the same food alga determined using a "rotating culture" technique with daily transfer of the rotifers into fresh suspensions of algae of defined concentrations (Walz et al. 1989; Rothhaupt 1990a), the chemostat food concentrations appear to be very high (Fig. 7.1.2). The "rotating culture" method was designed to keep the food quantity constant and food quality as close as possible to the original algal culture. Defined food suspensions were prepared and changed daily, and at the same time rotifer densities were reduced to ca. 1 ml^{-1}. Food concentration being the independent, and rotifer growth rate the dependent variable, this method is similar to the turbidostat approach (Chap. 3.4; Bennett and Boraas 1988; Boraas and Bennett 1988). From Fig. 7.1.2 it may be concluded that the reactor algal concentration (C_2), measured in the second stage of rotifer chemostats as, e.g., organic carbon, does not consist of pure algae. It probably represents a mixture of algae, feces, detritus, and bacteria. The crux of a

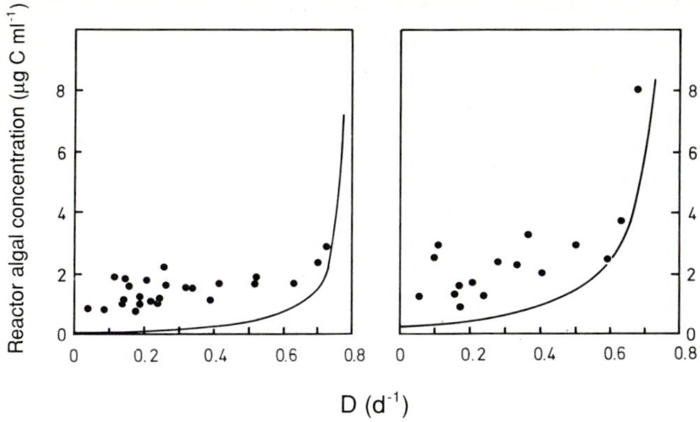

Fig. 7.1.2 Reactor algae concentration (C_2) at various dilution rates (*D*): *left panel, B. rubens; right panel, B. calyciflorus. Line* shows the expected residual algal concentration obtained in culture experiments with pure algal food

qualitatively ill-defined reactor algal concentration probably does not depend on whether it is determined by filtration and determination of a biomass component (e.g., Walz 1983b; Rothhaupt, Chap. 4.3) or by electronic particle counts (e.g., Boraas 1980, 1983). This problem has been recognized before (Droop and Scott 1978) but, I believe, has not yet been sufficiently considered when chemostat results have been interpreted. It means that the assertion of a well-defined food base is not valid for rotifer chemostats, and this has implications if certain aspects of functional and numerical responses are determined from steady states.

7.1.3.1 Numerical Responses

As a consequence of the lower average food quality, maximal growth rates attainable in chemostat steady states are usually lower than those in pure algal food. In the chemostat (Chap. 4.3) *Brachionus rubens* established a steady state at a highest growth rate of 0.725 day^{-1} and was washed out at $D = 0.77$ (day^{-1}); *B. calyciflorus* established a steady state at $D = 0.68$ day^{-1} and was washed out at $D = 0.72$ (day^{-1}). With the same food alga (*Monoraphidium minutum*) in "rotating" cultures with fixed food concentrations, both rotifers reached and maintained maximal growth rates of about 0.8 day^{-1} (Rothhaupt 1990). Boraas (1983) reported chemostat steady states with *B. calyciflorus* that reached up to ca. 90% of the maximal growth rates estimated from transient phases, but this species reached $r_{max} = 1.27$ day^{-1} when grown in chemostats. Walz (1983b), working with *B. angularis*, reached steady states up to dilution rates of 0.32 day^{-1}, whereas in transient states the highest growth rate was 0.58 day^{-1} (Table 4.1.3, see also Sect. 3.3.8). Working close to or at r_{max} obviously requires other techniques than two-stage chemostats. Alternatives for this purpose that keep the algal food quality more constant are turbidostat-controlled systems (Chap. 3.4) or the fed-batch culture technique mentioned (Rothhaupt 1990a). Bennet and Boraas (1988) reported the highest growth rate ever found for metazoans for *B. calyciflorus* in the turbidostat ($r_{max} = 1.92\,day^{-1}$).

But there are also problems in the range below the maximal growth rates. The kinetics of the growth rate with the food concentration ("numerical response") is in a way the inversion of the plot of the reactor algal concentration shown in Fig. 7.1.2. Numerical responses are usually described as Monod curves. Consequently, due to the high C_2 concentrations, estimates of the saturation constant (K_s) from chemostat steady states tend to be very high. Walz (1983b) determined a K_s of 4.05 $\mu g\,C\,ml^{-1}$ for *B. angularis*, Schlüter et al. (1987) report a K_s of 45 $\mu g\,dw\,ml^{-1}$ for *B. rubens*, and Boraas (1983) found K_s values between ca. 4 and 60 $\mu g\,dw\,ml^{-1}$ in chemostat experiments with *B. calyciflorus* (dw = dry weight). Furthermore, Boraas (1983) and Bennett and Boraas (1989b) found K_s to decline with increasing dilution rate.

Growth rate determinations with the "rotating culture" technique have shown that the effects of food concentration, given a defined food, are well described by a modified Monod model, the parameters of which depend on the ingestability and the nutritional quality of the food algae (Rothhaupt 1990a). The food concentrations at which half the maximal growth rate was reached was usually well below 1 $\mu g\,C\,ml^{-1}$ (0.22–0.70 $\mu g\,C\,ml^{-1}$ for *B. rubens* with various food algae; 0.48–0.90 $\mu g\,C\,ml^{-1}$ for *B. calyciflorus*). The only exception was *B. calyciflorus* when fed *Chlorella minutissima*. In this case, a food concentration of 4.3 $\mu g\,C\,ml^{-1}$ was necessary to reach 1/2 r_{max}. This is due to the fact that this very small alga is poorly ingestible by *B. calyciflorus* (Rothhaupt 1990b). Walz (1987b) reported significantly lower K_s of birth rates in rotating cultures (0.07 $\mu g\,C\,ml^{-1}$) than in chemostats (0.78 $\mu g\,C\,ml^{-1}$) for *B. angularis*.

The generally higher food levels in chemostat steady states, manifested by high estimates of K_s, are presumably due to the lower average food quality the animals encounter. The share of unused algae, and hence the food quality, probable increases with dilution rate in chemostat steady states. This would be one explanation for the decreasing K_s found by Boraas (1983) and Bennett and Boraas (1989b). For different foods, K_s has been shown to depend on the nutritional quality (Rothhaupt 1990a).

7.1.3.2 Functional Responses

The functional response of planktonic suspension feeders, i.e., the dependence of food intake on food concentration, can either be described as ingestion rate (IR) or as clearance rate (FR). The clearance rate, and how it changes with food concentration or growth conditions, in some respects is the more interesting variable. Its pattern may be a clue to underlying feeding mechanisms (Rothhaupt 1990c), and it determines how the grazers are influenced by, and in turn influence, their trophic environment. Decreasing clearance rates at low food

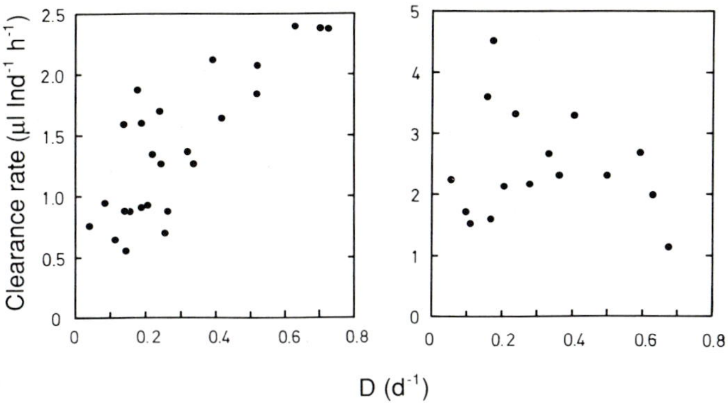

Fig. 7.1.3 Clearance rates, calculated from steady-state data, as a function of dilution rate: *left panel, B. rubens; right panel, B. calyciflorus*

concentrations, for example, tend to stabilize grazer-resource interactions (Steele 1974). Such a feeding pattern was predicted for energetic reasons (Lehman 1976; Lam and Frost 1976) but was not confirmed in short-term feeding experiments with *Brachionus* (Walz and Gschloessl 1987; Rothhaupt 1990c). As a general trend, rotifer clearance rates usually decrease with food concentration, resulting in saturating ingestion rates (Starkweather 1980), but often are constant and maximal up to a certain threshold concentration (Walz and Gschloessl 1988; Rothhaupt 1990c).

Clearance rates deducible from chemostat steady states (Eq. 7.1.3) are quite low, reaching at maximum 2.4 μl individual^{-1} h^{-1} for *B. rubens* and 4.5 μl individual^{-1} h^{-1} for *B. calyciflorus*, and for *B. rubens* appear to increase with dilution rate (Fig. 7.1.3). Similarly increasing clearance rates have been reported by Schlüter et al. (1987) and Scott (1988). Bennett and Boraas (1989b) described a convex pattern with highest clearance rates at intermediate growth rates and decreasing clearance rates at both low and high growth rates. These patterns of increasing clearance rates with growth rate are not easily interpretable. If rotifer steady-state growth rates are limited by food concentration, the growth rate and food concentration should be positively correlated. Hence, according to the above-mentioned general trend, the clearance rate would be expected to decrease with the dilution rate in chemostat steady states or perhaps remain constant over a certain range as long as the food concentration is below the threshold of maximum ingestion.

If in the calculation of clearance rates the reactor algal concentrations measured (C_2) are substituted by the theoretical residual algal concentrations from the "rotating cultures" growth rate determinations (the line in Fig. 7.1.2), the estimated clearance rates are higher and the functional patterns are in accordance with expectations (Fig. 7.1.4). Up to dilution rates of about 0.5 day^{-1}, the estimates appear to scatter around a constant value (11.55 ± 1.77 μl

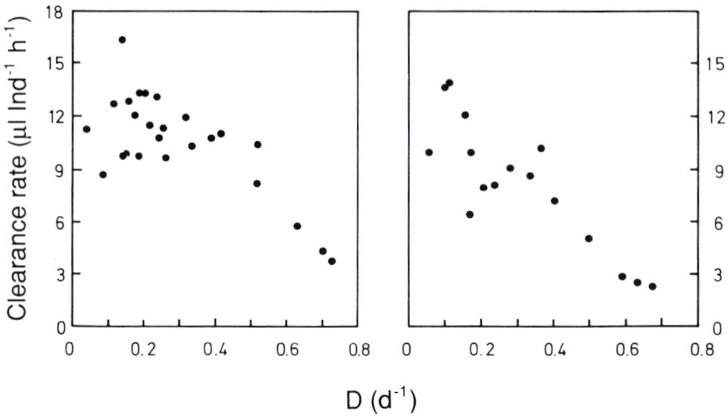

Fig. 7.1.4 Clearance rates as a function of dilution rate, calculated from steady-state data, but C_2 substituted by the expected residual algal concentration (line in Fig. 7.1.2): *left panel, B. rubens; right panel, B. calyciflorus*

individual^{-1} h^{-1} for *B. rubens*; $9.54 \pm 2.43\,\mu$l individual^{-1} h^{-1} for *B. calyciflorus*), and they decrease at higher steady-state dilution rates. These estimates compare favorably to clearance rate determinations in short-term radiotracer experiments with the same rotifers and food alga (maximal clearance rates below the threshold: $10.5 \pm 2.17\,\mu$l individual^{-1} h^{-1} for *B. rubens*; $9.04 \pm 1.68\,\mu$l individual^{-1} h^{-1} for *B. calyciflorus*; Rothhaupt 1990c).

Short-term ingestion measurements with radiotracers, using rotifers (*B. rubens*) that had been preconditioned by growing them to steady state in chemostats at varying dilution rates, also confirmed the view that clearance rates do not decrease with growth rate (Rothhaupt and Lampert, 1992). Maximal per individual clearance rates were not lower for animals preconditioned at very low dilution rates (down to 0.04 day^{-1}) but remained essentially unchanged. Due to a trend of decreasing average body masses with decreasing growth rates, carbon-specific clearance rates even increased for animals from low growth rates.

These considerations make clear that due to the poorly defined nutritional base in rotifer chemostats, it may be misleading to calculate clearance rates from steady states. The magnitudes as well as the patterns may be very different from results of more direct approaches to measuring functional responses. The derivation of clearance rates from chemostat data is very sensitive to the exact determination of the reactor algal concentration (C_2), since this variable appears as a factor in the denominator of the formula:

$$FR = \frac{D(C_1 - C_2)}{C_2}. \tag{7.1.3}$$

This does not equally apply to the derivation of ingestion rates. Here, C_2 appears only as a part of a difference ($C_1 - C_2$), and since usually $C_1 \gg C_2$, the estimation of ingestion rates is comparatively robust to errors in C_2 (Appendix).

7.1.4 Limits of Rotifer Chemostat Experiments

The preceding considerations suggest that some of the assumptions, ideally thought to apply to rotifer chemostats, may have to be revised. There is theoretical as well as empirical evidence that time-invariant populations are impaired by the occurrence of oscillations, and comparisons with results obtained by other techniques suggest that the known input of food algae in a defined physiological state does not result in an equally well-defined trophic environment for the animals. In fact, there is reason to assume that the food quality rather than the food quantity limits population growth in rotifer chemostats. Like the methods to estimate the reactor algal concentration (C_2), the rotifers are probably not able to distinguish between particles of different nutritional quality within their food size range. *Brachionus* has been shown to have distinct food size preferences but to feed unselectively within this size range. The situation in a rotifer chemostat thus appears to resemble that in the field, where the animals also encounter a variety of particles of an ingestible size.

In conclusion, neither the temporal pattern nor the nutritional base in rotifer chemostats appear to be as well-defined as originally thought. This does not preclude the two-stage rotifer chemostat from being a useful tool for ecophysiological research, but it may interfere with some interpretations if the peculiarities are not kept in mind.

7.2 Chemostat Regulation Principles in Natural Plankton Communities

NORBERT WALZ

7.2.1 Introduction

Regulation of zooplankton population densities was shown in rotifer chemostats and in simulation models of chemostats (Chaps. 3.3, 5.1). The next paragraph demonstrates the utility of an experimental chemostat as a model ecosystem. In these models, zooplankton and phytoplankton are interrelated in a regulated cybernetic control loop. Can these models explain the regulation principles in the much more complicated plankton community, with many other species as competitors, predators, and food for rotifer and other zooplankton species (see Chap. 6.3)? This is a problem of scaling up the models and the experiments on the temporal and spatial scales (e.g., Frost et al. 1988). Chemostat work is surely a method at the level of laboratory experiments. It is a good technique to recognize causal relationships but requires an artificial system. These results must be extended to the much more complex lake. Upscaling has to follow hitherto not fully understood principles of similarities of structure (by dimensional analysis) and of function (Uhlmann 1985). This has been done by many aquatic ecologists from laboratory experiments to enclosures in lakes up to manipulations of whole lakes (see review of Lundgren 1985).

The chemostat is very suitable as a model for the functioning of a plankton community because of its self-regulating character. At first, it seems to be too special and, because of its dilution, too strange to be a tool for analyzing natural systems. In the following, it will be made clear why exactly this fact is characteristic for natural systems. Regarding the dilution rate as the turnover of the community brings a substantial improvement of ecological understanding and a possibility to explain plankton community structure. Because of this high turnover, it was not pure chance that basic relationships were revealed first in plankton communities. These principles, however, are to be found in many other communities, as well.

7.2.2 The Chemostat as a Model Ecosystem

Disregarding the much higher complexity in natural ecosystems, the two-stage chemostat and the situation in the open pelagial are comparable in many points (see Preface). Herbivore zooplankton populations nourish on phytoplankton, increasing their growth rate in relation to the phytoplankton concentration, often according to the Monod model. Without doubt, zooplankton grazing activity functions as a negative feedback control for phytoplankton density. Both topics will be discussed later in more detail. But one question appears immediately: What is the relationship of the dilution rate of the chemostat to the pelagial? If we go back to:

$$dX/dt = r' * X = r * X - D * X, \tag{3.3.4}$$

we shall see that D is the dilution rate. At the same time, D is the specific "output" rate of organisms to the next trophic level (Fig. 3.3.5). Considering a population in the pelagial as an open compartment, then the output rate is a forced tribute of individuals to the outside, out of this compartment to the next trophic level. Biomass output is directly proportional to the dilution rate (harvesting rate $= D * X$, Chap. 3.3.5). In general, in the pelagial such a tribute must be equivalent to the pressure of a predator and, therefore, especially to its feeding rate or to the loss of individuals by the lake outflow. A similar model was created by Armstrong (1980) for the herbivory of zooplankton in algal chemostats.

In the chemostat, this inverse "feeding rate" is fixed at constant values. In the steady state, the prey population must produce a surplus for the "predator." The "capital" is not touched by the predator, which is living on the "interest" only. If the dilution rate (i.e., the predation rate of the predator) is changed, the prey enters into a transient state. If the "feeding rate" of the predator increases, i.e., the dilution rate rises above the growth rate of the prey organisms, the net growth rate (r') becomes negative, and the population density decreases. Afterwards it may reach another steady state with lower grazing followed by an increased algae density. Alternatively, if r' becomes too low, the population washes out to extinction.

For these reasons, the chemostat can be imagined as a model of a very simple food chain (Fig. 7.2.1) which is divided into three compartments corresponding to the following three trophic levels:

1. Primary producers (algae), whose density (C_2) is the result of the supply rate ($C_1 * D$; in the pelagial, the net rate of primary production) and the feeding of the herbivores: $C_2 = C_1 -$ grazing (Eq. 3.3.9).
2. Consumers (herbivores), whose density is regulated by the algae supply and by their growth efficiencies: $X = Y\ (C_1 - C_2)$ (Eq. 3.3.6).
3. Predators, which may be identified by the output of the culture ($D * X$).

In the chemostat, the transfer between the three trophic levels is easily studied. In steady states, no time variant processes are to be expected; the state variables

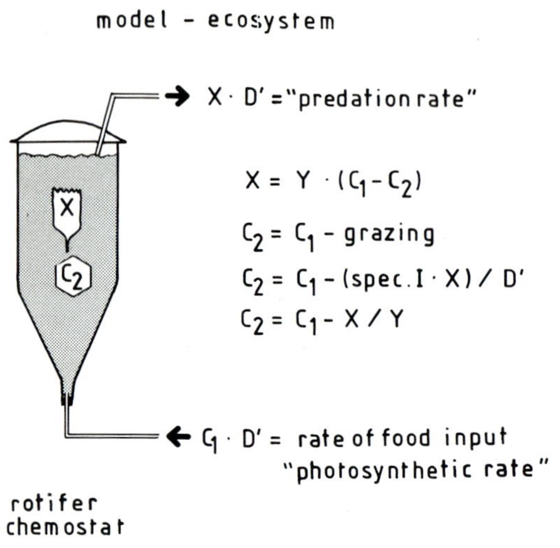

Fig. 7.2.1 The chemostat as a model ecosystem with three trophic levels

and the exchange rates can be calculated directly from the steady-state densities. Studies of transient states additionally show temporal reactions to changes of input food concentrations (C_1) or to the dilution rates. Differences between the chemostat and the pelagial are discussed in detail later in this chapter.

Hitherto, the chemostat has only been used for ecological studies in microbiology with mixed continuous culture studies on nutrient limitation, predator-prey interactions, and competition (Veldkamp and Jannasch 1972; Veldkamp and Kuenen 1973; Jannasch and Mateles 1974; Veldkamp 1976, 1977). Experimentation with plankton ecosystems (phytoplankton, herbivores, bacteria) by semicontinuous or homocontinuous cultivation methods are very rare. One example is a study of the time course and variation of biological constituents, species composition, and energy flow (Uhlmann et al. 1980).

7.2.3 Transferability of Regulation Principles to Pelagial Communities

The application of chemostat regulation principles to the pelagial communities presumes similar regulation mechanisms in the chemostat and in the pelagial. Zooplankton species in the field must possess similar kinetics of bioenergetic rates (growth rate, feeding rate) as dependent on food conditions as in the laboratory. From the energetic relationships, the same food requirements result in both environments (Veldkamp 1977). An evaluation of the transferability was made on hand of extensive data on population dynamics of rotifers (abundance dynamics, birth rates, growth rates) available from studies in Lake Constance in 1977/1978 (Walz, unpublished data). Walz et al. (1987) give the background numbers of rotifers and particulate organic carbon (POC) for phytoplankton

Table 7.2.1 The correlation coefficients between the rotifer birth rates (b), the POC concentration ($\mu g\,C\,ml^{-1}$), and the temperature, T (°C) in Lake Constance

	Pearson correlation				Partial correlation		Multiple correlation
Species	*n*	b − POC	b − T	POC − T	b − POC (T = constant)	b − T (POC = constant)	b − (POC + T)
Keratella quadrata	44	0.593*	0.016	0.064	0.575*	0.067	0.595*
K. cochlearis	30	0.571*	− 0.089	0.027	0.578*	0.090	0.600*
Polyarthra spp.	41	0.562*	− 0.341	− 0.308*	0.510*	− 0.150	0.575*
Kellicottia longispina	17	0.778*	0.327	0.216	0.766*	0.259	0.794*

n, Number of pairs of values; *significant for $P < 0.05$

densities. The general, from year to year mostly regular rotifer development was further confirmed by Pauli (1990).

7.2.3.1 Energetics of Population Dynamics of Lake Rotifers

A correlation matrix of the birth rates as dependent on the POC (of the phytoplankton fraction $< 10\,\mu m$) and on the temperature is shown in Table 7.2.1. The simple Pearson correlation coefficients are virtually identical with Spearman rank correlations for nonlinear relationships (not shown). The birth rates of all rotifer species studied are significantly correlated to POC. The study of birth rates assumes knowledge of the duration of the embryonic development of eggs (Paloheimo 1974), therefore only egg-bearing rotifers are involved in this analysis (for the duration of embryonic developments, see Sect. 6.3.6 and Fig. 6.3.8).

With partial correlations, the birth rates may be related to distinct factors when the other factor is formally held constant. This shows that birth rates are dependent on POC (or to another factor directly coupled with POC), but the influence of temperature is not significant in contrast to the embryonic development rates themselves. Both factors together give a multiple correlation coefficient not very different from the partial coefficient with POC as the unique factor. This underscores that the pelagial food concentration is an important factor for the population dynamics of rotifers.

7.2.3.2 Kinetics of Population Dynamics of Lake Rotifers

Similar kinetics of rotifer birth rates with phytoplankton concentrations (POC) were found in both chemostat experiments and in Lake Constance (Sect. 4.2.3). When the relationship between the birth rate of egg-bearing rotifers and the POC concentration ($< 10\,\mu m$), which was present before in the lake at the time of 50–100% of embryonic development (De), is known, a Monod-type saturation

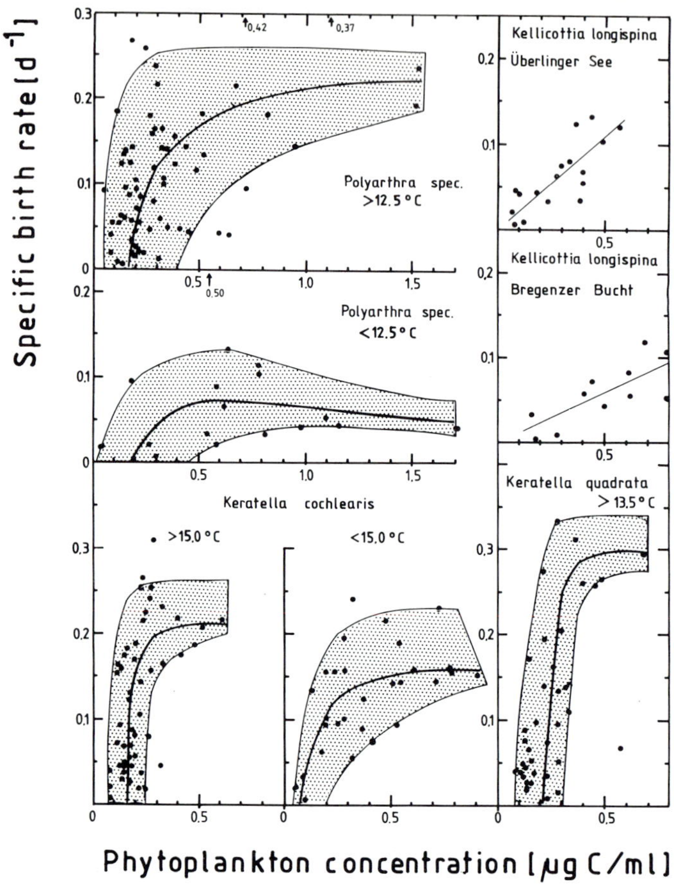

Fig. 7.2.2 Specific birth rates of egg-bearing rotifers in Lake Constance 1977/78 in dependence on the phytoplankton concentration (Particulate Organic Carbon < 10 µm). *Gray areas* include 95% of the points (After Walz 1986a)

kinetics of the relationship can be derived (Fig. 7.2.2). As was expected, the variability is large, and the apparent effective concentrations are lower in the field than in the chemostat experiments. This speaks for the argument of Rothhaupt (Sect. 7.1.3) that the food value of the algae in the chemostat is lowered by feces. These field kinetics are surely biased as no truly planktonic r_{max} strategists (Sect. 6.3.7) are involved since, e.g., *Asplanchna* and *Synchaeta* spp., which attain the highest r_{max} values, are not egg bearing.

Keratella cochlearis birth rates reach a maximum value of 0.3 day^{-1}, not much lower than in the chemostat (Table 4.1.3). Birth rates of *K. cochlearis* and *Polyarthra* spp. differ intraspecifically at low and high temperatures, although this parameter has already been considered in the duration of embryonic development (De) which is a part of calculations of birth rates. By trial and error, two temperature ranges were found which could be related to two populations with different saturation values. Edmondson (1962) showed for *K. cochlearis* similar curves for 10–15 °C and > 15 °C. Perhaps the two populations of *Polyarthra* may be identified as the cold-stenothermal *P. dolichoptera* and the warm-eurythermal *P. vulgaris* (Stenson 1983). Both species have been identified in Lake Constance (Walz et al. 1987), but they have been counted together routinely.

Similar field relationships between clutch size per adult and POC were found by Lampert (1978) for *Daphnia* spp. and by Schober (1980) for *Daphnia* spp. and *Bosmina* spp. in Lake Constance.

7.2.3.3 Feedback of Grazing on Phytoplankton Density

Grazing feedback effects of zooplankton are well-known in many lakes. The effects on phytoplankton have been exclusively ascertained with cladocerans due to their higher filtration rates. In Lake Constance, one of the first examples in which this grazing effect was demonstrated, *Daphnia* spp. are responsible for the early summer clear-water phase (Lampert and Schober 1978). Many other grazing studies and a great number of biomanipulation studies all show a significantly greater water transparency when *Daphnia* spp. are dominant (e.g., Stenson et al. 1978; Neill 1984) than when micrograzers prevail, see below (Shapiro 1980; Lynch and Shapiro 1981).

In contrast, there are only a very few examples in which the effect of the grazing activities of dominant rotifers is clearly observed on phytoplankton densities (Scheda and Cowell 1988; Havens 1991). In Lake Baldeneysee (an impounded river lake near Essen, Germany), rotifers are prevalent, evidently due to the high flushing rate of the lake. *Brachionus* spp. (especially *B. calyciflorus*) densities reach such high abundances that the phytoplankton density is reduced significantly in spite of the lower specific grazing rates of the rotifers (Nusch 1978). In such lakes, cladocerans are nearly completely excluded because of their higher generation time (longer than residence time and, hence, dilution losses).

7.2.3.4 Conceptual Differences Between Chemostat and Field

The situation in chemostats and in the field differs especially in two aspects: In chemostats, the state variables in steady states are time invariant, and the selection of organisms is different in the field.

Steady states are not exclusive conditions in continuous cultures, but the study of such situations is especially fruitful as knowledge of the state variables and the dilution rates enables one to determine the transport rates of the system. Transient states correspond rather better to natural conditions. Nevertheless, the study of steady states is most important. Just like the “equilibrium” state is an "attractor" for nonequilibrium states in thermodynamics (Prigogine 1978), the steady state is an "attractor" for transient states. That is, steady states show the direction of the imminent development of the system.

In steady states, time invariance excludes, e.g., diurnal differences by vertical migrations. But in transient states, the study of such phenomena seems possible, e.g., by the time-dependent addition of different enriched algae. Likewise, abiotic factors such as temperature and light may be changed. In "mixed" (multispecies) chemostats, the time-dependent variability makes the existence of more species possible than correspond to the number of resources (Sommer 1985). Coexistence or extinction is a steady-state "attractor" of competition. In mixed chemostats, it is possible to study the invasion of better adapted species. In this context, "invasion" may be understood as the development of a population from a pool of hitherto "hidden" species. This enables one also to study succession of species in continuous cultures (Sommer 1988a, Sect. 6.15, this Vol.).

As demonstrated by Rothhaupt (Sect. 7.1.3), the residual food concentration in a chemostat reactor (C_2) does not necessarily correspond to the natural phytoplankton density. In this respect, the chemostat is a tool for studying regulation processes to gain a deeper understanding of the interaction of phytoplankton density and the grazing of herbivores, but not for the evaluation of the exact values of the resulting phytoplankton concentration. This question requires other methods.

In contrast to the field, other selection factors are active in chemostats. This seems to be the severest restriction of continuous cultures. In continuous culture, the "predation rate" of a "predator" is not size selective as "the undiscriminative removal of cells (i.e., organisms) by dilution results in simple selection for that species attaining the fastest growth rate under the particular conditions" (Jannasch 1974). In contrast, in the field the prey may be selected by fish according to its size and its visibility (Brooks and Dodson 1965; Kerfoot 1980). Invertebrates, on the other hand, select smaller, unprotected zooplankton species. Moreover, the two-stage chemostat as a model for plankton regulation is more deficient than the situation in the pelagial as there is no feedback from the prey to the "predator" (i.e., to the dilution rate).

7.2.4 Cybernetic Structure of Pelagial Communities

7.2.4.1 Flux of Information in the Ecosystem

The organisms in any ecosystem are structured in an energetic food chain with a flux of energy that maintains its physical structure far from equilibrium according to the constraints of the second law of thermodynamics (Prigogine 1978). This energy "flows" up the trophic levels (Lindemann 1942) from the phytoplankton to the herbivorous and carnivorous zooplankton and to the fish. In parallel, and also in the opposite direction, a flux of information is directed. For example, the herbivorous zooplankton changes the structure of the phytoplankton, its size distribution, and its defense strategy by feeding activity (Berquist et al. 1985). Information is transferred not only by feeding selectivity; even the simple presence of a predator may change the behavior of the prey organisms. To avoid predators, some prey planktivore fish shift their habitat in the presence of piscivores from open waters to vegetated habitats so that the community structure changes substantly (Turner and Mittelbach 1990). Another kind of transfer of information is the secretion of water-soluble organic molecules by which *Asplanchna* triggers the building of lorica spines of *B. calyciflorus* used as a defense mechanism (Gilbert and Waage 1967). These examples demonstrate some exchange of information within the plankton community.

This flux of information is necessary to achieve and to maintain the best adapted structure of the community; for example, through this information the phytoplankton avoids being totally destroyed by the zooplankton. As information is equivalent to negative entropy (Brillouin 1956), the community needs energy to maintain this organizing flux of information (Patten 1959). Hence cybernetic regulation loops with feedback result which are able to damp disturbances (Patten and Odum 1981). Characteristics of these cybernetic processes, therefore, may be recognized firstly after perturbations, for example, during the eutrophication of the waters. During the course of the eutrophication process in Lake Constance, for instance, within a regulation reaction for a long time the higher nutrient input was used up by an increase in the abundance in higher trophic levels (phytoplankton, zooplankton, later fishes). The nutrient input itself and its concentration in the lake were not measurable at that time (Walz et al. 1987; see also Persson et al. 1988). This is the reason why rotifer species show reactions very early during eutrophication and may be used as "indicators" (Karabin 1985; Beaver and Crisman 1990).

Flux of information is a prerequisite for the cybernetic structure of the ecosystem, but the backward direction of this flux, as a feedback generating a loop, is critical (Sect. 5.1.2). The action of the regulator can be increased (positive feedback) or slowed down (negative feedback). In ecological control loops, there can be no "desired values" as in engineering loops, but defined states are stabilized depending on the regulation characteristics of the species present and the input conditions. Surely the species have adapted to their specific role by coevolution. The ecosystem, therefore, is not goal oriented and clearly does not function like an "superorganism" (Patten and Odum 1981). The cybernetic structure of an ecosystem develops by self-organization through the pure existence and, consequently, interaction of its organisms on the same and other trophic levels by natural selection (see Preface). The control loop in a chemostat (Fig. 5.1.1) is only the most simple and primitive starting position for more complicated ecosystems.

This cybernetic view of ecosystems has been heavily criticized (e.g., Engelberg and Boyarsky 1979; Oksanen 1988). According to these authors, the Darwinian struggle for existence is the sole organizing principle, and "no cybernetic and mutualistic controls (ultimately superorganismic qualities) are needed to understand ecosystem structure" (Persson et al. 1988). This is a misunderstanding of the cybernetic structure, which is perhaps caused by a way of thinking based on the Lotka-Volterra model (Sect. 5.1.7, Chap. 5.2). In this model, regulation is implicit and performed by the "carrying capacity" on the same trophic level. Interactions are only described by correcting coefficients in the absence of all causal mechanisms (see Preface).

7.2.4.2 Regulation Model for Rotifers with Different Life History Strategies

Rotifers are a good example to study how different life history strategies are involved in the regulation processes. This was an important topic in models with two rotifers species competing for one food (Chap. 5.1). In Sect. 6.3.7, the different life history strategies of rotifers are demonstrated for two extremes (r_{max} and K_s strategy). This represents a model for the plankton community in which many more organisms with a variety of life history patterns are present at the same time or successively. In this model, both extreme types are in competition on the phytoplankton food resource.

As shown in Sect. 6.3.7 r_{max}-strategy rotifers are more specialized on specific food species than K_s-strategy rotifers, which as suspension feeding generalists feed unselectively on the general pool of "edible" algae up to about 30 μm in diameter, which they have to share with the cladocerans.

A regulation model with a broader food base may be constructed as follows (Fig. 7.2.3). A prerequisite for the specialization on a few algae is the shorter lag phase of the r_{max} strategists, which makes it possible for them to grow up quickly using a frequent but qualitatively restricted food without time delay. Different food particles may be used successively by the same or different species. Population dynamics, therefore, may show diauxie-like phenomena (Kompala et al. 1984). K_s-strategy rotifers, on the other hand, feed on a broader spectrum

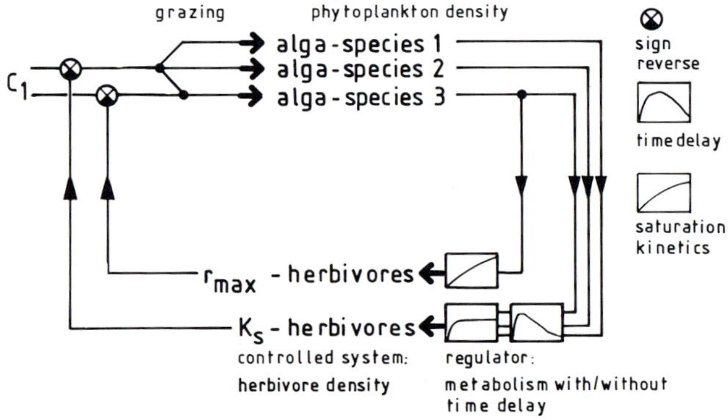

Fig. 7.2.3 Model for regulation loops of herbivores with different life strategies and food selections (After Walz 1986a)

of algae in lower concentrations, but the time delay for their population development is longer. Cladocerans have almost the same food size selection as K_s-strategy rotifers (see Sect. 6.3.7), but their time delay is even greater.

In contrast to that model, Pianka (1972) and others developed the conception on the basis of Lotka-Volterra models that r_{max} strategists should have wider niche overlaps and, therefore, use a larger food range, because they do not happen to meet competitors so often.

This model illustrates that the regulation characteristic of a consumer population depends on its life history strategy and explicitly on its growth kinetics in relation to the food concentrations.

7.2.4.3 Regulation Model for Pelagic Communities

The representation of a biocoenotic regulation model with the different groups of the plankton community in the pelagial of a lake results, even in the most simple case, in the complex interlocking of regulation loops in which a controlled system of one loop functions as the regulator of another loop (Wilbert 1968). A controlled system may also be regulated additively by several regulators. The influence of other regulators then is a command variable of the respective loop (Walz 1986a; Fig. 7.2.4).

The phytoplankton by itself is involved in regulation loops with nutrients and light. The light quantity diminishes with phytoplankton biomass by self-shading. Nutrients will be reduced by high primary production.

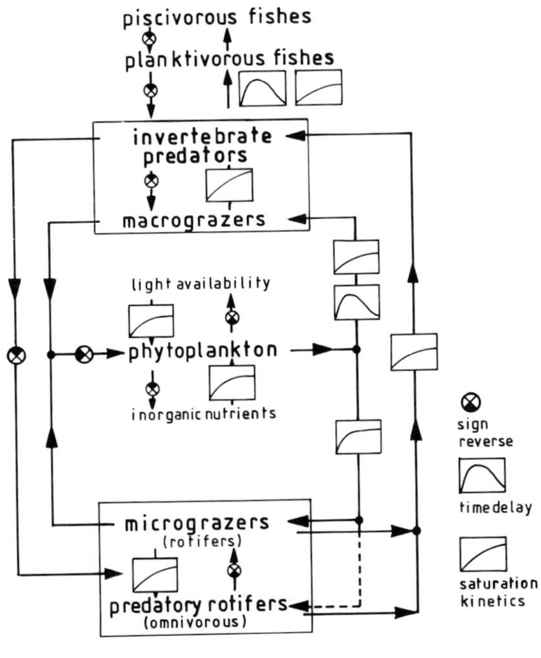

Fig. 7.2.4 Model for regulation loops in the plankton community of a lake (After Walz 1986a)

Simultaneously, the phytoplankton is grazed down by any species of herbivorous zooplankton according to:

$$C_2 = C_1 - \sum_{r=1}^{r_{max}} \left(\frac{X_r}{Y_r}\right) \tag{7.2.1}$$

where r = 1, 2, etc. represents consumer populations.

The sign of the information flux is reversed by this operation. Grazing is negative on the biomass of phytoplankton (for evidence in the field, see above).

Equation (7.2.1) is an extension of Eqs. (3.3.10) (Sect. 3.3.4) and (5.1.2) (Sect. 5.1.6). Similar equations were formulated by Waldon (1975) and Taylor and Williams (1975). The difference $C_1 - C_2$ represents the carrying capacity of the system, which may be identical to the sum of ingestion of all herbivores ($I = X/Y$; Sect. 3.3.4). For that reason, $D*C_1$ has to be designated as the net primary production rate. Sedimentation is another loss process in real lakes removing phytoplankton from the epilimnion.

The response of the herbivorous zooplankton on the phytoplankton is variable according to its concentration. This is based on the different characteristics of the metabolism of the herbivores as the growth rates react in different amounts and with different times to the phytoplankton concentration (Chap. 6.3). This leads to different reaction times since the times for the populations to achieve net growth rates ("age of zero growth" and generation time) are different and more pronounced for K_s-strategists (Sect. 6.3). In general, this time seems to be inversely proportional to Y and increases beginning with protozoans (for which only scarce information is available) over r_{max} rotifers, K_s rotifers to cladocerans and copepods (Table 6.3.1). Therefore, time delays (Chap. 5.1) appear in the regulation loops, which give the rotifers decided time advantages at peak phytoplankton densities. At those times, the rotifers are able to use the higher food supply in higher concentrations.

An even greater time delay is incorporated in the regulation loops by the planktivorous and predatory fishes as their lifetime extends over many years. A feedback control on the zooplankton by enhanced fish growth is to be perceived only after some years (Lampert 1987). In general, for that reason the fish act like a command variable (Sect. 5.1.2) in the regulation loop without an instant feedback. Therefore, their long life span is a stabilizing element in plankton communities (Mills et al. 1987).

With the progressive development of the pelagic community, time advantages are of less importance considering the simultaneously ever more grazed down food basis. Populations with lower K_s and Y values (see Sect. 6.3.4, Table 6.3.1) will be more successful in the same succession as above. With a progressively lower Y, the achieved biomass of the herbivores will also decline.

Under particular conditions, zooplankton are able to promote phytoplankton in a positive feedback reaction. If phytoplankton biomass is removed by extensive grazing actions, its limitation by nutrients is relaxed. In such situations, the photosynthetic rate is stimulated and maintained by the excretion of nutrients by zooplankton (Lehman 1980). Grazing also stimulates algae which are protected against digestion. They take in nutrients during their passage through the gut (Porter 1977). The growth of phytoplankton is enhanced by compounds of low molecular weight as growth factors released by copepods (Wynne and Gophen 1986). In the same way, by a positive feedback higher growth rates of the prey are inforced by the predators.

This model explains why there are in the field at some times positive and at other time negative trade-offs between zooplankton and phytoplankton (e.g., Lewis 1979; Precht 1983; Walz, unpubl. data). If the consumer populations are small, they are accompanied by higher phytoplankton concentrations. After having surpassed a threshold (community grazing > growth rate of phytoplankton), the relationship becomes negative. The consumer density, therefore, determines whether the phytoplankton is a steering factor or involved in a control loop. In contrast to the influence of a predator, which is effective only at high prey densities and, therefore, more likely to be detected, food limitation affects consumers at low densities as well.

A regulation model like that in Fig. 7.2.4 has little to do with the usual energy flow models. This figure shows only the principle of regulation. For this reason, the microbial loop (transfer of energy from phytoplankton via DOM, bacteria, heterotrophic protozoa, zooplankton; e.g., Porter et al. 1988) was not included. Arndt (1992) used chemostats as a particularly useful tool to study microbial webs, i.e., the interaction between algae, protozoans, and zooplankton.

7.2.4.4 Lake Outflow as Dilution Rate of the System

Natural lakes with a hydraulic outflow have a dilution rate ($D = Q/V$) like the chemostat model. This is recognized in lakes only if the flow through (Q) is large enough in comparison with the lake volume (V) to wash out the populations. Only those populations may be held in the lake which present higher growth rates than the dilution rate. For this reason, nanophytoplankton, small algae cells which reproduce quickly, dominate over large, more slowly reproducing forms in a lake with flushes as short as 2.5 days (Dickman 1969).

Gnauck (1974) treated lakes with high dilution rates, for example, impounded river lakes, as completely stirred tank reactors. Uhlmann (1972) compared different lakes with different dilution rates. The output of production of the lake increases with the dilution rate up to a critical D, at which losses are compensated for by high production performances. This is directly comparable to the "utilizable harvest" in a chemostat (see Fig. 3.3.6). The critical level for holding a phytoplankton population is about 2.5 days or $D < 0.4$ (day^{-1}; Uhlmann 1971).

As indicated above, rotifers achieve high selection advantages in lakes with high dilution rates. In an impounded river lake, Lake Postmünster (Upper Bavaria), rotifers reach over a mean residence time of 3.6 days 3–5 times higher densities than the crustacean zooplankton which dominates in periods with lower flushing rates (Krause-Dellin 1983; Pappers 1985; Gruhl 1986). Other lakes with dominating rotifer populations and a high flow-through are, for example, Lake Bielersee (Switzerland; Berner-Fankhauser 1986) and ponds with high flushing rates in Newfoundland (O'Connell and Andrews 1987).

Impounded river lakes often show complicated water currents, so that the theoretical dilution rate is not representative of the actual washout. If the mixing is incomplete, parts of the lake will have higher residence times, allowing the existence of slower growing populations. The impounded river lake Ismaninger Speichersee near Munich (Germany) with a theoretical residence time of 1.8 days (1/D) houses a *Daphnia longispina* population in the still water zone which would not usually be viable at that rate (Mosel 1991). With a lower hydraulic flushing rate, internal feedback processes gain more importance.

7.2.4.5 Predation Rate as Dilution Rate of the System

As mentioned in Sect. 7.2.2, the dilution rate in the chemostat may be considered as the predation rate. As with any other point, the model makes a serious simplification. As mentioned earlier, such simplifications are only justified under the specific purpose of the model (Chpt. 1).

Disregarding the various forms of selection (see above), the influence of a predator is always experienced in the field. The prey species respond with different strategies of avoidance (e.g., loricas, size, helmets, jumping). However, these are only partly successful. The inevitable toll has to be equalized by growth. The escaping capacities of the prey are closely connected with its r_{max} and K_s strategies (Sect. 6.3.7) as r_{max} strategists are more qualified to compensate by growth and K_s strategists, to protect by defense.

Just as the output of production of a lake is higher at higher dilution rates (see above), a predator may enhance the growth rate and the production of the prey population (if the elimination is not too intensive and if not only the ovigerous females are selected). This stimulating effect is known from harvested populations and from fisheries; the fecundity of the surviving adults and birth rate of the populations increase, whereas the natural mortality decreases (for blowflies, *Lucilla cuprina*, Nicholson 1954; for *Daphnia pulicaria*, Slobodkin and Richman 1956; for guppies, Silliman and Gutsell 1958, after Begon and Mortimer 1989). With the same types of reaction (higher fecundities and survivorship of cladocerans which have avoided fish predation), zooplankton populations respond to the pressure of predators (Vanni 1986b, 1987).

Unlike the Lotka-Volterra models for the equilibrium of chemostats, the growth rate is not zero but is positive in the better defined steady state (Eq. 3.3.4). For an observer who does not know of the existence of the dilution or of a predator in that equilibrium, the biomass does not seem to grow. He sees only the "apparent" growth rate:

$$dX/dt = r' = 0. \qquad (7.2.2)$$

This will be calculated according to:

$$r' = \ln X_2 - \ln X_1/t_2 - t_1. \qquad (7.2.3)$$

In the field, it is only possible to determine r′. The development of the abundance of individuals of zooplankton species allows no possibility to realize the actual growth rate of the individuals as the population may be under the pressure of a predator or of a hydraulic lake outflow which destroys a part of its increase. At most, a value of r'_{max} in a season may be considered as r_{max} when one accepts that the true r_{max} will only appear when no predator is present and environmental conditions are excellent (Fig. 6.3.7).

The birth rate (b) shows the potential growth rate which the population should have without natural mortality (d) and predation rate (D).

As $b = r + d$ (3.3.13)

then $r' = b - d - D$

and $r' = r - D$. (3.3.4)

This means with the usual calculation in field ecology, $b - r'$, one obtains the total mortality ($D + d$), which combines mortality due to predation, to hydraulic loss, and to natural mortality. In chemostats, it is possible to distinguish between these mortalities. Figure 4.1.10 shows that natural mortality plays an important role under food limitation.

7.2.5 Regulation Principles in Plankton Communities

7.2.5.1 Competition Between Micro- and Macrograzers

As discussed in Sect. 6.3.8, rotifer generalists and cladocerans have to use generally the same phytoplankton fraction up to about 20 or 30 μm in diameter, the "edible algae." Only rotifer specialists are able to use larger food particles, but these are side branches in the energy flow and are not considered here.

Due to their different filtration capacities and body sizes, competition between generalist rotifers and cladocerans for the same food fraction may be seen as a competition of micro- and macrograzers. Comparing the clearance rates of rotifers and cladocerans, one finds a factor of up to 10–1000 by which cladocerans filter more water than rotifers (Starkweather 1980; Peters 1984).

This two-way canalization into two food chains finds its correspondence in Fig. 7.2.4, in which herbivorous rotifers and ciliates are the micrograzers and herbivorous crustaceans are the macrograzers. In this context, small crustacean filtrators belong to microfiltrators (e.g., *Bosmina* spp. and nauplia; Reich 1988). This distinction between two grazer systems was also made by Neill (1988) in oligotrophic lakes and Mazumder et al. (1990) in enclosures.

This approach has many advantages. Which group dominates depends on the boundary conditions of the system (predation and dilution, primary production as food input), but it has different consequences for the flux of energy and information in the lake. This is especially important for eutrophication,

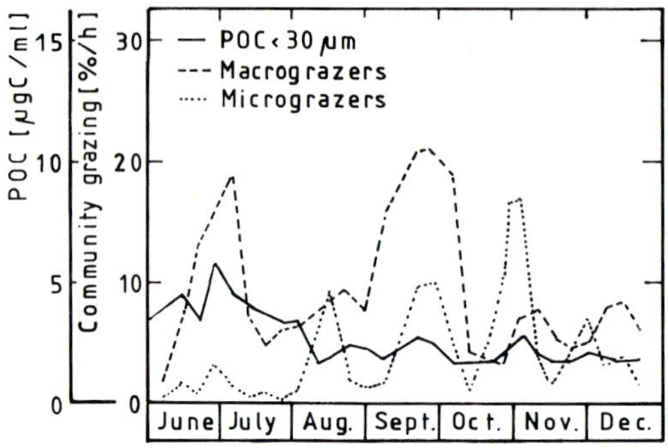

Fig. 7.2.5 Community grazing of macro- and micrograzers and phytoplankton concentration (Particulate Organic Carbon, *POC* < 30 μm) in the Fasaneriesee, a gravel pit lake near Munich (After Platzek 1986 and Reich 1989)

i.e., at an increased primary production compared with the state before. If the boundary conditions favor the macrograzer system, the community grazing rates are higher than those of micrograzers (Fig. 7.2.5; see also Mazumder et al. 1990). More energy is definitely channeled into the food chain than into the sedimentation sinking to the hypolimnion and the bottom of the lake. Carbon, for example, is respired in the epilimnion, which does not negatively affect the oxygen content of the hypolimnion. Sedimentation depends on the phytoplankton concentration, but this is grazed to such low concentrations that the sedimentation process is reduced.

On the contrary, in lakes where the micrograzers are supported, the phytoplankton concentrations remain much higher, and sedimentation is more intensive. Remineralisation of that material takes place more in the hypolimnion, leading to oxygen deficits and to positive feedbacks on the primary production by the remobilisation of phosphorus from the sediment.

7.2.5.2 Producer and Consumer Controlled Regulation of Plankton Communities

Control of the regulation processes in the plankton community may be exercised by the consumers or the producers. For both relationships, there are many examples as the earlier, often controversial debate arose as to whether bottom-up or top-down regulation had occurred (Lampert 1987; Kerfoot and Sih 1987; Carpenter 1988). According to the regulation model of Fig. 7.2.5, these views are not contradictory.

The control of producers has long been recognized, for example, by the "trophic-dynamic aspect" of Lindemann (1942). In studying eutrophication causes and effects, positive statistic regressions between nutrients and phytoplankton densities were found (Dillon and Rigler 1974). The zooplankton density reaches higher levels in lakes of higher trophic status depending on the phytoplankton biomass (McCauley and Klaff 1981). Long-term studies in lakes undergoing eutrophication reveal that a main consequence of the higher nutrient input is an increase in the zooplankton density (Smyly 1978; Karabin 1985; Walz et al. 1987) and fish stock and catchment numbers (Nümann 1973;

Hartmann 1978; Hanson and Leggett 1982). These effects are based on higher amounts of energy becoming available for the higher trophic levels after primary productivity has increased. Productivity of all trophic levels is ultimately based on the nutrient and radiation input into an aquatic system (McQueen et al. 1986). But these statistical trade-offs involve only mean numbers of organisms and do not explain the regulation of the actual abundances in different trophic levels at any time.

Consumer control of the plankton community became manifest in the studies of Hrbaček et al. (1961). Biomanipulation experiments in small lakes and enclosures with removing or stocking up of planktivorous or piscivorous fish elucidated the control of the underlying trophic levels by consumers (Shapiro 1980). This consumer control is propagated as an "cascading trophic interaction" down over several trophic levels (Carpenter et al. 1985; McQueen and Post 1988), finally reaching primary production (Carpenter et al. 1987). The response to fish predation is highest for zooplankton and lower for phytoplankton, and predictability reduces with the trophic distance (Mills et al. 1987). Interactions are more complex as the fish excretion of phosphorus enhances phytoplankton productivity (Vanni and Findlay 1990). These top-down effects are stronger in eutrophic lakes. On the other hand, in eutrophic lakes with cyanobacteria-dominated algal communities, the response to grazing is very weak (Elser and Goldman 1991). Communities in oligotrophic lakes are more limited and controlled by the producers (McQueen et al. 1986).

The consumer and the producer control is active, without doubt, in the plankton community. Mills and Forney (1988) found "that both consumer and producer forces function in concert to structure and regulate biomass." From this, it is only a small step to accept the interaction of both controlling factors in cybernetic control loops (Fig. 7.2.4).

7.2.6 Some Remarks on Regulation Processes in Other Ecosystems

Without doubt, regulation principles and cybernetic structures are also present in other ecosystems, but they are mostly obvious in aquatic and especially pelagic ones. The reason why pelagic systems are over-represented is to be found in the easy evaluation of abundances and in the short times of development, because the participating organisms are very small and have very short generation times (Paine 1980). Regulation processes are relatively easy to follow by comparing numbers of organisms.

More highly evolved organisms, especially birds and mammals, become to some extent emancipated from direct environmental influences, not only from temperature but also from food supply regulation. They often live in territories, and their population densities show no visible relationships to the actual food amounts in those areas. The mechanisms of density regulation are located inside the population ("self-regulation," according to Stugren 1986). The density of many mammals in terrestrial ecosystems is hormonally controlled by an increase of aborts or by direct inhibition of ovulation in the females (Remmert 1986). The population density is also stabilized by the dispersal of recently mature females (Myers and Krebs 1974) and by the "semicontinuous dilution" of unsuccessful – and consequently not reproducing – competitors. However, the required size of the territories is dependent on the amount of food (Remmert 1989). This reveals also in this case the underlying food control (Wynne-Edwards 1965). Territorial animals displace their competitors out of their territory. Thereby, they avoid an excessive overload of the food base. As a consequence, that part of the population which reproduces does not suffer as much from starvation as would be the case if the food had to be shared with all other population members. On the population level, this strategy has the favorable result that the possibility of becoming extinct is minimized and genetic selection is progressed. In spite of this complication, the population densities are regulated by cybernetic processes in these "higher evolved" communities also, although the information flows in other ways.

7.3 Application of Rotifer Continuous Cultures to Ecotoxicology

DIANNE B. SEALE, MARTIN E. BORAAS and
LORIE A. BUCHHOLZ

7.3.1 Introduction

Ecotoxicological assays may exploit key features of open culture systems, with the underlying assumption that toxicity is best understood within the context of an overall examination of the physiological and environmental factors influencing growth. Over the past decade, a number of methods have been developed in our laboratory to apply the theory and practice of microbial continuous culture systems (Monod 1950; Novick and Szilard 1950) to understanding rotifer population growth (Boraas 1980, 1983; Boraas and Bennett 1988; Chaps. 2.1, 2.3, 3.1, 3.2, 3.4, 5.2, 6.1, this Vol.). Recently, our research has focused on the potential power of open culture systems (chemostats, turbidostats, and semicontinuous cultures) to assess the effects of various toxins (NaCl, heavy metals) on algae (Seale et al. 1987; Bennett 1988, 1989, 1990). All such systems provide a regular supply of toxicants along with fresh nutrient, and the concurrent removal of organisms and metabolic wastes. These characteristics minimize the confounding effects of variability of some factors which cannot be held constant in batch culture, such as metabolic products, food, and toxin level. This paper explores the potential utility of these approaches for assessing toxic effects in rotifers.

The open-culture ecotoxicological assay, in principle, is the most sensitive of all growth-based toxicity assays. The method capitalizes on the fact that continuous cultures promote steady-state populations, in which the specific growth rate, r, equals the dilution rate of the culture, D (D = flow rate/culture volume; Boraas 1983; Chaps. 2.1, 2.2, 3.1, this Vol.). Once r is determined for the test species under a given set of environmental conditions, a toxin may be introduced into the growth medium. Unlike other culture methods, the population of organisms must reproduce and grow to remain within the culture vessel; this characteristic may be used to advantage in a toxic assay. Any deleterious effect due to a toxin on the physiology of the organism (e.g., on fecundity, general metabolism, feeding, life span) will be detectable as decreased population densities or as a reduction in r. If the added toxin causes r to become $< D$, the population will simply be diluted out of the culture, which can used to quantify the toxic effect. If r remains $> D$, a new steady state may be achieved, for which population densities and other growth parameters may be evaluated.

Animal population dynamics in open cultures can be modeled using the microbial growth model developed by Monod (1942) and Herbert (1958) as described by Boraas (1983; Chap. 5.2). This model has four biological parameters, r_{max} (the maximum specific growth rate), K_s (the half-saturation constant of growth), Y_g (the maximum food conversion efficiency), and μ_c (the maintenance respiration rate). The probable mechanism for toxic action may be deduced by comparing the model output (where the toxin affects one or more of these parameters) to observed population growth dynamics, for steady-state or transient population growth.

Halbach (1984) and his colleagues (Beuter et al. 1981; Halbach et al. 1981, 1983) designed a growth-based toxicity assay using rotifers. These workers were able to resolve minute sublethal effects, e.g., reductions in the life span and fertility of individuals, and to observe both adaptations to pollutants and chronic sublethal effects that developed over many generations, especially for chlorinated hydrocarbons. They added algal food to rotifer cultures and removed an equal volume of medium but not animals. Therefore, the major population loss was individual mortality. Such populations tend to oscillate, with a frequency and period that are dependent on the temperature and toxin concentration (Halbach 1984). The effects of toxins were detected with a variety of statistical and time-series analyses, e.g., autocorrelation, Fourier transforms, etc. (Beuter et al. 1981) and with simulation modeling. While very sensitive, their techniques were extremely labor intensive and required culture durations that extended for months.

This paper describes a straightforward application of continuous culture to ecotoxicological studies in rotifers. In these studies, rotifers (*B. calyciflorus*) were grown using established methods for semicontinuous culture (SC; Chaps. 2.1, 3.1, this Vol.), with *C. pyrenoidosa* as the feed algae (from a steady-state chemostat). Each culture served as its own control (i.e., pre- and postexposure conditions). After exposure, the toxin was removed, and the population dynamics were followed to examine lags and the reversibility of effects.

7.3.2 Special Methods

We first established steady-state algal cultures using methods previously described (see Boraas 1983) to provide food for the rotifers in second-stage, SC (Chaps. 2.1, 3.1). After obtaining steady states in each of the rotifer cultures, ecotoxicology assays were initiated. First, a range of concentrations of PbCl was introduced as pulse exposures, beginning with very low levels (0.02 mM or 20 μM). The concentrations were increased stepwise until a threshold and lethal values, i.e., lower and upper concentration bounds for toxic action, were determined empirically. Secondly, the system was flushed free of toxicants, allowing the rotifer populations to recover, if possible. Then, the toxin was pumped in at sublethal levels, and time-course data were obtained for rotifers and residual algae. Parameter estimates for the models were obtained empirically from the time-course and steady-state data.

Algae (*Chlorella pyrenoidosa* Chick, UTEX No. 26) from a chemostat provided the sole source of food for rotifers in replicated two-stage SC (Chap. 2.1). The algal chemostat was maintained under constant light, temperature, and dilution rate conditions with limiting nitrate at 0.5 mM, using the medium given in Boraas (1983). The 485-ml first-stage algal chemostat was held at 25 °C under constant illumination (about $15 \times 10^3\ \mu E\ cm^{-2}\ s^{-1}$) with cool-white fluorescent lamps and air (0.2-μm filtered) at a flow rate of $800\ ml\ min^{-1}$. The resulting algal cultures (first stage) produced *C. pyrenoidosa* at a constant input concentration, C_1, which was sufficiently high to allow nutrient-saturated growth of the rotifers (see Chaps. 2.1, 6.1; Boraas et al. 1990). The rotifers used (*Brachionus calyciflorus*) initially were isolated from Milwaukee Harbor, Wisconsin, USA. During periods of steady-state growth, outflow volume from the second stage was measured daily to 0.1 ml with a top-loading balance (assuming $1.0\ g = 1.0 \pm 0.1$ ml at room temperature).

The SC were operated simply by collecting outflow from the algal chemostat and using this to replace a volume removed from the rotifer culture (see Chap. 2.1). The rotifer cultures were replicated, foil-covered, 250-ml, polycarbonate Erlenmeyer flasks filled to 200 ml, at 25 ± 1 °C. The cultures were mixed gently each day; a predetermined fraction of the culture volume was poured off for quantification, which was immediately replaced with an equal volume of fresh algal suspension. The cultures and replenished algae were poured into a second, clean Erlenmeyer flask, to minimize biofilm accumulation. The second stage was operated in near darkness to inhibit growth of the algal prey. Steady-state rotifer cultures were established at a dilution rate of $0.29\ day^{-1}$, where 25% of the culture volume was replaced daily. This D (= growth rate) is far below the r_{max} previously determined for these *B. calyciflorus* (Boraas 1983; Boraas et al. 1990; Chap. 6.1).

The *Chlorella* populations (numbers and biovolume) in the first stage (C_1) and their residual densities in the second stages (C_2) were measured with a Particle Data Celloscope, using a 60-μm aperture (Boraas 1983; Boraas and Bennett 1988; Chap. 2.1). Rotifers (X) were counted manually with a dissection microscope. Two independent aliquots (each 10–50 ml) were counted.

7.3.3 Experiments Examining Step Perturbations and Pulse Additions of PbCl

The first phase of the toxicity study (Fig. 7.3.1) determined the critical concentrations for threshold, sublethal, and lethal effects of Pb in parallel duplicate cultures from time-course data for rotifer concentrations (X). Initially,

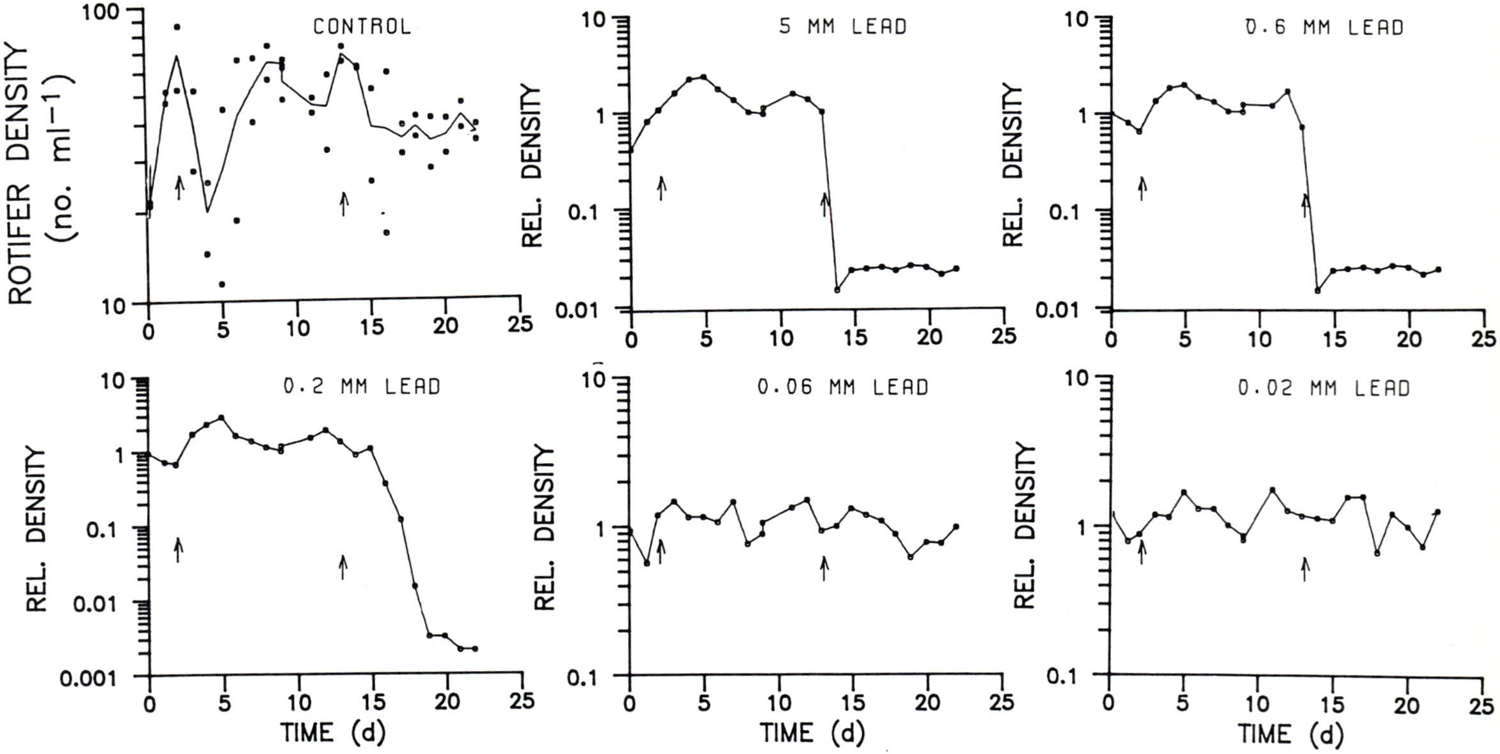

ROTIFER DENSITY (no. ml^{-1})
CONTROL
5 MM LEAD
0.6 MM LEAD
0.2 MM LEAD
0.06 MM LEAD
0.02 MM LEAD
REL. DENSITY
TIME (d)

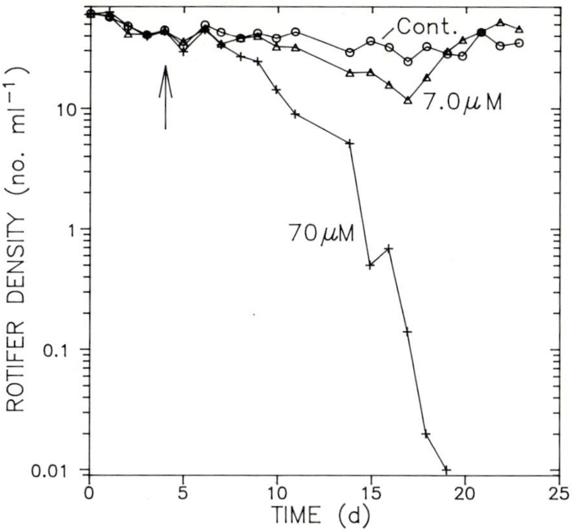

Fig. 7.3.2 Time course of the effects of 70 μM Pb and 7 μM Pb to steady-state SC rotifer cultures. Each line is the average of 3 (70 μM, *crosses*; 7 μM, *triangles*) or 4 (control, *circles*) simultaneous, independent cultures. Wash out occurred at 70 μM but not at 7 μM. From Figs. 7.3.1 and 7.3.2, lead becomes abruptly toxic between 60 and 70 μM when rotifers are growing at a dilution rate of 0.29 day^{-1}

the rotifer cultures were grown without replacement, i.e., as conventional batch cultures. At 2.1 days (leftmost arrows in panels of Fig. 7.3.1), each culture was converted to an open SC system, at $D = 0.29\,day^{-1}$. At each rightmost arrow (Fig. 7.3.1), PbCl was added as a step change to the indicated concentration. Lethal effects (manifested as immediate wash out) were seen at 0.6 and 5 mM Pb. A sublethal effect was observed at 0.2 mM (rate of washout indicative of a $r \leqslant D$), and no effects were noticeable at or below 0.06 mM Pb (see description in legend to Fig. 7.3.1). In a second step-change experiment (Fig. 7.3.2), the SCs were exposed to 70 μM (0.07 mM) and 7 μM (0.007 mM) Pb, for comparison with parallel control cultures. Wash out occurred at 70 μM, but not at 7 μM.

←

Fig. 7.3.1 Growth of *Brachionus calyciflorus* in semicontinuous culture on *Chlorella pyrenoidosa* before and after step-change additions of Pb (as lead chloride). *Arrows* at 2.1 days indicate the onset of daily dilution of 25% of the culture with fresh media ($D = 0.29\,day^{-1}$); cultures were operated in batch mode from 0 to 2.1 days. *Arrow* at 12.5 days indicates addition of the toxin to the steady-state rotifer cultures. The lead chloride was introduced with the algal input medium as a step change at the concentrations shown and maintained with subsequent additions. Each panel shows the time course for daily counts of live *B. calyciflorus*. Data in the control panel (*upper left*) are from two replicate cultures; the *solid line* is the average of the two. The apparent variability in the rotifer population density is largely due to variations in the density of the input algae. For the remaining panels, the rotifer density was divided by the mean control rotifer density to eliminate variability in input algae. The normalized values are, in general, less variable than the absolute values. Lead exposures were step changes in PbCl ranging from 0.02 to 5 mM Pb. Lead was acutely toxic at 0.6 and 5 mM; the rate of rotifer loss was greater than the dilution rate, indicating rotifer mortality. At 0.2 mM Pb, rotifers were washed out of the culture at a rate nearly equal to the dilution rate, indicating cessation of growth, but no excess mortality. There were no detectable effects at 0.06 mM (60 μM) and 0.02 mM (20 μM) Pb

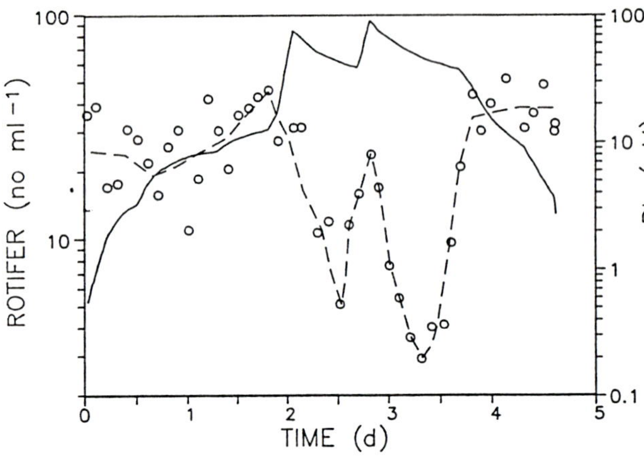

Fig. 7.3.3 Pulse of Pb into a steady-state rotifer culture. Lead chloride (*solid line*) was first added gradually into a rotifer chemostat to a concentration of about 10 μM (daily live rotifer density is indicated by *open circles*). About 1.7 days after initiation of the culture, Pb was pulsed in to a concentration of 70 μM. There was an immediate decrease in rotifer density following the pulse. The Pb washed out (the daily 10 μM Pb addition was maintained). When the Pb concentration reached 40 μM, the rotifer population resumed exponential growth. A second pulse replicated results from the first pulse. When Pb was removed, the rotifer population reentered a steady state

A separate experiment (Fig. 7.3.3) examined the transient effects of Pb in a nonsteady-state rotifer culture. First (at about 1.7 days), PbCl was pulsed in at a concentration of 70 μM from a background concentration of 10 μM. After the Pb pulse, both the rotifers and the lead began to wash out of the culture, due to system dilution. However, the rotifer population first began to recover when the Pb level reached about 40 μM. A second pulse (at about 2.5 days) resulted in a repeat of the first observation. Removal of Pb from the input medium caused the Pb to begin to be diluted out of the system (at around 3.3 days). This removal of Pb by dilution allowed the rotifer population to recover to its previous concentration. Thus, the culture served as its own control. The effect of the lead toxicity was clearly nonlethal and reversible.

7.3.4 Modeling

The effect of the continuous addition of lead to a rotifer chemostat was simulated with a mathematical model (Fig. 7.3.4), using empirically determined parameters. The simulation was based on the rotifer growth model described in Boraas (1983), in which the rotifer instantaneous growth rate is a hyperbolic function of the algal concentration in a chemostat. In the simulation, Pb is added from

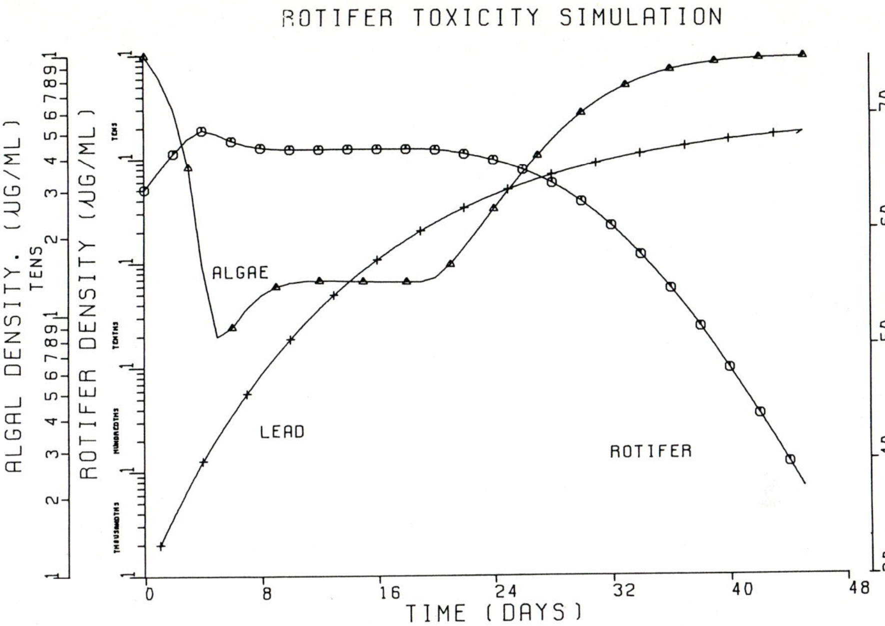

Fig. 7.3.4 Simulation of effect of increasing toxicant concentration on rotifers in chemostat culture. The rotifer growth model of Boraas (1983) assumes that the rotifer instantaneous growth rate is a hyperbolic function of the algal concentration in a chemostat. The toxin concentration is assumed to vary according to the equation: $dC/dt = k(Tf-T)$, where T is the instantaneous toxin concentration, Tf is the final concentration, and k is the rate parameter (here assumed equal to the dilution rate of the chemostat)

a reservoir containing $70\,\mu M$ Pb and $100\,\mu g\,ml^{-1}$ algae. The model assumes that the rotifer r_{max} declines linearly with increasing toxin concentration, once the toxin exceeds the threshold ($60\,\mu M$; after Fig. 7.3.1). When a level of $70\,\mu M$ Pb is reached within the culture vessel, r_{max} is assumed to become zero (Fig. 7.3.2). The model assumes that the reduction in r_{max} is the only effect of Pb.

The simulated rotifer and residual algal populations enter and maintain steady states within a few days (Fig. 7.3.4). Rotifer growth is unaffected by Pb until the critical concentration of $60\,\mu M$ is reached. At this point, the rotifer population begins to decline. As the Pb increases within the culture, r_{max} decreases even further, resulting in an accelerating decline in rotifer density. The residual algal population (C_2) rises rapidly, approaching its maximum value of $100\,\mu g\,ml^{-1}$ (C_1) as the rotifer population approaches zero. The algal population is a nonlinear inverse function of rotifer density. In this model, the rate of change of the toxin concentration is set to be gradual, and the rate of change of the rotifer population is also gradual, here extending over several weeks of simulated time.

7.3.5 Application of Open Culture Systems to Toxic Assay: General Comments

The characteristics of open culture systems were exploited to improve on the accuracy and sensitivity of assessments of the effects of selected toxicants on rotifer population growth parameters. We used a combination of steady-state and transient approaches in our experiments (Figs. 7.3.1, 7.3.2, 7.3.3) to assess critical concentrations and incubation conditions for toxicity. Steady-state bioassays are inherently more accurate than bioassays relying on transient effects. However, transient or nonsteady-state bioassays produce results over a wider range of toxin concentrations much more rapidly and with fewer cultures.

7.3.6 Critical Concentrations of PbCl for *B. calyciflorus* Toxicity

We were able to determine thresholds and ranges of concentration for both sublethal and lethal effects. In addition, we were able to assess lags and the reversibility of effects, which could not be monitored with batch assays. The onset of the toxic effect was abrupt (Figs. 7.3.1, 7.3.2, 7.3.3). There was no detectable toxic effect until the apparent threshold (60 μM Pb) was reached, and the rotifers began to be washed out of the culture vessel (Fig. 7.3.1). In the step perturbation experiments, there was a rapid onset of toxicity between 60 and 70 μM Pb (Figs. 7.3.1, 7.3.2). In transient, recovery studies (Fig. 7.3.3), Pb was toxic at a concentration just below 70 μM. However, after the transient pulse exposure at 70 μM, growth did not resume until [Pb] was reduced to 40 μM (Fig. 7.3.3). This apparent discrepancy may be due to the retention of Pb within rotifer or algal tissue, inducing a lag in the resumption of growth after removal of Pb from the culture medium.

7.3.7 Comparisons of Model Output to Results for Assessing Mechanisms for Toxic Action

By combining modeling with continuous cultures in evaluating toxic effects, one can gain insight into the mechanisms for toxic action. Classical population growth models may be modified to incorporate the effect of a toxin on important growth parameters. Hence, mathematical models can be used as tools to determine which growth parameters are affected by the toxin, giving clues as to the physiological mechanism of the toxin (see Bennett 1988). In the steady state, virtually all of the assumptions of classical population growth models are met: a constant specific growth rate ($dN/Ndt = r$), a constant and stable age distribution, and constant rates of material transfer among components (e.g., feeding, respiration, excretion). If the addition of a toxic substance has any significant sublethal effect on the population growth dynamics, one or more of these fundamental physiological or ecological variables will be altered. The variable affected can be readily determined by straightforward sampling procedures and by application of appropriate mathematical models (Boraas 1983), minimizing the need for more time-consuming measurements (e.g., oxygen consumption, radiotracer studies).

In this study, we used a simple growth model to describe transient time-course patterns (Fig. 7.3.4). Based on empirical data, the mechanism for action of the toxin was assumed to be a concentration-dependent reduction in the maximum growth rate. The resulting simulation mimics one of the more counterintuitive results observed (Figs. 7.3.1, 7.3.2, 7.3.3): the abrupt (not gradual) onset of toxic effects. We did not uncover any effect on rotifer physiology except for a probable reduction in r_{max}. These analyses do not reveal any detectable toxic effects on respiration, food conversion efficiency, or feeding ability. However, a similar approach could be used to evaluate potential toxic effects on these growth parameters.

7.3.8 Rotifers and Open Culture Systems in Ecotoxicology: Advantages and Future Studies

Rotifers have advantages over other zooplankton species in toxic bioassays: (1) a rapid growth rate (doubling time of 0.5 day or less), (2) a small body size ($< 200\,\mu m$), (3) amenability to automatic particle counting (Boraas 1980, 1983), and (4) a cosmopolitan distribution. In addition, they are readily cloned, and resting eggs can be preserved for long periods, ensuring constancy of the genetic makeup of test organisms.

The technical advantages of open rotifer culture systems can reduce the time needed to characterize an adverse effect by a factor of 5–10 over that needed for traditional studies. Most toxicological studies use static (e.g., LD-50 and LC-50 assays), flow-through (flow of toxin-laden water through a "cage"), or life-table analyses (e.g., USEPA 1979). Life tables provide a sensitive measurement of toxicity (Halbach 1984; Rao and Sarma 1986), but data collection for this approach is very labor intensive and requires skilled personnel. Accurate life tables should extend for the life span of the organism (Halbach et al. 1983) which, for many species, may be a prohibitively long period. Life tables are, therefore, inappropriate as a survey procedure. Other approaches, e.g., physiological, behavioral, and genetic alterations, are most appropriate for measuring the effects of toxins on species of particular interest, e.g., *Daphnia* (Goodrich and Lech 1990) or fish (Weber et al. 1991).

Assays that measure r_{max} directly may be more useful for toxicity measurement than chemostats or SCs. Chemostats or SCs are useful for studying growth at low to medium rates but become unstable as the dilution rate approximates r_{max}. The turbidostat is particularly suited for studying growth near r_{max} (see Bennett and Boraas 1988, 1989a, b; Boraas and Bennett 1988). Since the turbidostat operates near the maximum specific growth rate of the organism, it produces population results as rapidly as possible, if a toxin is added to the culture, any effect on r_{max} will be seen immediately as a reduction in the system dilution rate.

Our rotifer chemostat culture device may be used as a turbidostat (Chaps. 2.3, 3.4), which may greatly improve the accuracy and sensitivity of the approach for future ecotoxicological studies with rotifers. In our computer-regulated rotifer turbidostat, the turbidity falls because of the consumption of food cells and rises when fresh food-cell suspension is metered in. We have used this system to select rotifers with the highest maximum growth rates yet reported for any metazoan (Bennett and Boraas 1988, 1988a, b; Chap. 3.4, this Vol.). In our laboratory, algal turbidostats have been used successfully to provide a superior approach for assessing toxic effects in algal cultures (Bennett 1988, 1990; Bennett and Brooks 1989). The next logical extension of this research is to use turbidostat culture to examine rotifer toxic effects.

We believe that open culture systems may revolutionize ecotoxicological assays for growth inhibition in animals. Measurements of the degree of inhibition, if any, of lower trophic levels due to toxic substances are essential for calculating food-chain transfer rates, sedimentation rates, etc. (Bennett et al. 1986). The effects of toxins on zooplankton growth and survival may be examined in steady-state SC, continuous, and turbidostat cultures, similar to those developed for toxicological studies in phytoplankton to minimize the shortcomings of the batch culture system (see Lederman and Rhee 1981; Gotham and Rhee 1982; Seale et al. 1987; Bennett 1988, 1990). Models for predicting the impact of toxins on zooplankton growth are needed to assess the effects of the toxicants on secondary production in natural ecosystems (Lassiter 1979). A steady-state ecotoxicology assay can provide a sensitive and powerful method for assessing a water-soluble toxicant in several respects. (1) Partitioning of the toxicant can be determined in the water column vs. the biomass of the biota at the steady state. (2) Rates of accumulation of the toxin within the biota and efficiencies of transfer of the toxin across trophic boundaries can be determined in steady-state culture (a steady state is required for accurate estimates of the rates of transfer). These first two advantages allow accurate estimates of bioaccumulation factors. (3) Rates of production, yields, maximum growth rates, and size-age distributions for steady-state populations can be compared in the presence and absence of the toxicant.

Acknowledgement. This research was supported by Environmental Protection Agency Grant R-810871-01-0 to M.E. Boraas and D.B. Seale at the University of Wisconsin, Milwaukee.

This is contribution no. 373 from the Center for Great Lakes Studies.

7.4 Use of Rotifer Chemostats in Aquaculture

CHARLES M. JAMES

7.4.1 Introduction

The ubiquitous rotifer *Brachionus plicatilis* is one of a few planktonic species that inhabits regions of varying salinity. Because of its adaptability, it is used as a primary food not only for marine, but also for freshwater fish and crustacean larvae (Hirata 1980; Watanabe et al. 1983a; Cruz and James 1989). It is essential that rotifers be cultured at relative high densities to provide an adequate feed supply for hatcheries producing a large number of fingerlings (Kafuku and Ikenoue 1983; James et al. 1983a, b, 1987). The success of a marine fish hatchery, especially during the early larval rearing stages, exclusively depends on the high quantity and quality of live microorganisms such as rotifers (Scott and Middleton 1979; Hirata 1980; Gatesoupe and Luquet 1981; Watanabe et al. 1983; James et al. 1983b, 1987; Rezeq and James 1987; James and Rezeq 1989a).

In recent years, many attempts have been made to improve rotifer culture methods and the nutritional quality of the rotifer produced for aquaculture (Watanabe et al. 1983; James et al. 1983b, 1987; James and Abu-Rezeq 1988). Also, different strains of rotifers have been mass produced to meet the prey size requirements of marine fish larvae (James and Abu-Rezeq 1989a). It is generally known that mixed diets of algae and yeast produce higher population growth rates in the rotifer cultures than a diet of algae or yeast alone in the culture system (Gatesoupe and Robin 1981; Hirayama and Funamoto 1983; James et al. 1983a, 1987; Abu-Rezeq and James 1989). However, the essential fatty acids required for the growth and survival of marine fish larvae are mostly provided by the microalgae in the culture system (Watanabe et al. 1983; James et al. 1983b, 1986, 1987).

The usual technique for the production of microalgae and rotifers consists of a multistep backup batch culture system involving several tanks that occupies a great area of hatchery space and is labor intensive (Fulks and Main 1991). In recent years, several methods for maintaining zooplankton in steady-state growth in continuous cultures have been explored (Droop 1970, 1976; Scott 1977, 1980; Droop and Scott 1978, 1982; Boraas 1980, 1983; Trotta 1980; Walz 1983b; Rothhaupt 1985). Taub (1980) concluded that continuous cultures would not only provide constant yield, but also stabilize the nutritional quality of the cultured organisms. Boraas and Bennett (1988), while discussing the steady-state rotifer growth in a computer-controlled turbidostat culture system, observed that such systems provide a means to explore the population dynamics of a metazoan maintained near its maximum specific growth rate (r_{max}) under experimental conditions. The above investigations, however, were restricted to small-scale experimental conditions.

Considering the high efficiency of chemostats in small-scale experiments, James et al. (1988) first developed an algal chemostat for aquacultural purposes and later (James and Abu-Rezeq 1989b) applied the general principles to develop a rotifer chemostat. Microalgae and rotifer chemostats have been practically utilized for aquacultures in Kuwait (James and Abu-Rezeq 1989b, c; James and Al-Khars 1990).

7.4.2 Rotifer Chemostat Design

The rotifer chemostat continuous culture system for aquaculture consists of a stage-1 algal chemostat and a stage-2 rotifer chemostat as shown in the schematic diagram (Fig. 7.4.1). For the stage-1 algae chemostat, James and

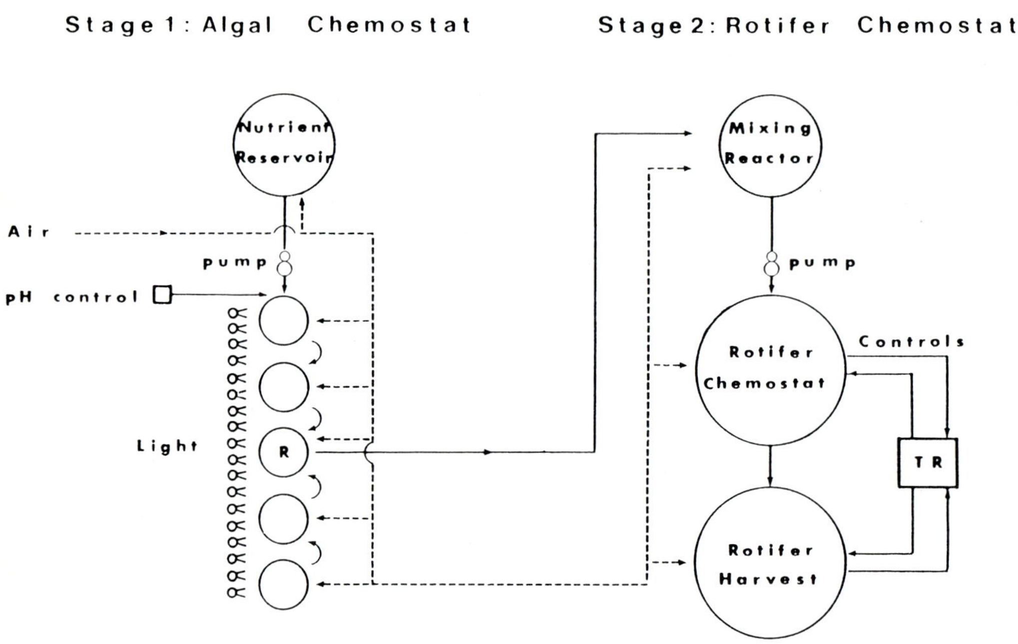

Fig. 7.4.1 Schematic diagram of the two-stage rotifer chemostat system; *R*, algal reservoir; *TR*, temperature monitor and regulator (After James and Abu-Rezeq 1989b)

Al-Khars (1990) used 200 l capacity, translucent, vertical tubes (30 cm in diameter) made of corrosion-free, fiberglass-reinforced, polymer sheets (Solar Components Corporation, USA) and equipped with an inlet for nutrients and outflow for algal suspension. Aeration is provided at the bottom of the tubes through polyethylene tubing, which is also used as a sampling port. Each continuous culture unit consists of five such vertical tubes (photobioreactors) with the nutrient and algal flows as indicated in Fig. 7.4.1. Tubes 1–2 and 3–4 are operated as continuous photobioreactors, whereas the fifth tube (in the middle) is conducted as a reservoir. The base medium is a combination of commercial fertilizers and other micronutrients in diluted seawater of 30‰ salinity (James et al. 1988), which is delivered to the algae reactors from a nutrient reservoir by a multichannel metering pump.

Photon energy is provided by a vertical bank of daylight cool fluorescent lights at about 330 $\mu E\, m^{-2} s^{-1}$ quantum scalar irradiance for 24 h per day to achieve optimum algal productivity (James and Al-Khars 1990). The pH of the algae culture is monitored and controlled at 6.5 for *Chlorella* spp. and at 7 for *Nannochloropsis* spp. The dilution rate is adjusted every day in relation to the instantaneous growth rate of the algae in the culture system. The algal cell density in the chemostats reaches more than 150×10^6 cells ml^{-1} for *Chlorella* spp. and 80×10^6 cells ml^{-1} for *Nannochloropsis* spp. Therefore, the high density of microalgae is diluted to 20×10^6 cells ml^{-1} in the mixing reactor before being pumped to the rotifer chemostats.

The stage-2 rotifer culture system consists of several 1 m^3 capacity chemostat units fed from a 500-l capacity mixing reactor. The desired food level in the rotifer chemostat is synchronized by a piston metering pump from the mixing reactor to the rotifer chemostat. The output (rotifer produced) from the rotifer chemostat is collected in a rotifer harvesting tank. The temperature is controlled at 25 °C. The diet of *Nannochloropsis* is provided at 20×10^6 cells ml^{-1} together with a daily addition of 0.3–0.4 g baker's yeast per 10^6 rotifers. For further details, see James and Abu-Rezeq (1989b, c).

7.4.3 Productivity of Different Strains of *Brachionus plicatilis*

Studies of rotifer chemostats for aquacultures show (James and Abu-Rezeq 1989b, c) that the productivity of rotifers in the continuous culture system depends on the strain of rotifers utilized. In 1 m^3 capacity rotifer chemostats, James and Abu-Rezeq (1989c) observed (Fig. 7.4.2) a mean rotifer output of 308.75×10^6 individuals $m^{-3}\, day^{-1}$ of *B. plicatilis rotundiformis* (S-type, size 95–160 μm) and 186.71×10^6 individuals m^{-3} day^{-1} of *B. plicatilis typicus* (L-type, size 220–300 μm) from the culture system. With a dilution rate of 0.5–0.6 day^{-1} it is possible to achieve a dry weight rotifer biomass of up to 0.0965 g l^{-1} day^{-1} with a mean of 0.0659 g l^{-1} day^{-1} for the S-type and up to 0.0789 g l^{-1} day^{-1} with a mean of 0.0536 g l^{-1} day^{-1} for the L-type rotifers.

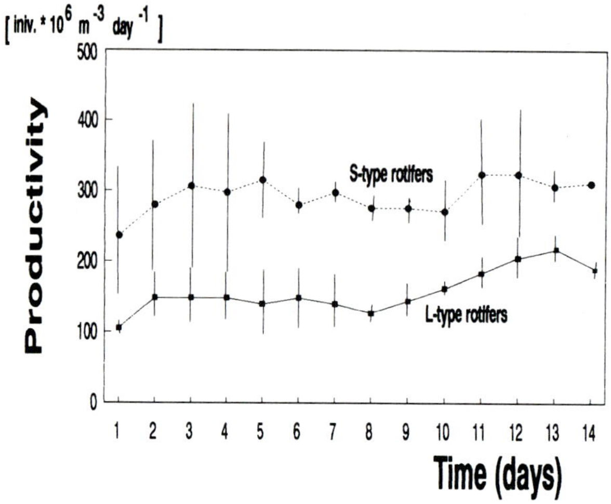

Fig. 7.4.2 Productivity (daily output) of S- and L-type rotifers (*B. plicatilis*) in the continuous culture system; *vertical bars* indicate the range (After James and Abu-Rezeq 1989c)

The food conversion ratio of S-type rotifers (mean 4.43 ± 1.02) is also significantly higher ($P < 0.001$) than that of the L-type rotifers (mean 7.92 ± 3.02) in rotifer chemostats (James and Abu-Rezeq 1989c). This is due to the fact that food consumption and utilization depend on the strain of rotifers employed in a culture system (Yúfera and Pascual 1985). Furthermore, the high productivity of the S-type rotifers is due to their shorter embryonic development time than in other rotifer strains (Yúfera and Pascual 1985).

The output of both S- and L-type rotifers from the chemostat culture system is considerably higher than from any conventional rotifer production system reported to date for aquaculture. The high output of S-type rotifers compared with L-type rotifers observed in our rotifer chemostats is in accordance with the observations of Yúfera et al. (1983) and James and Abu-Rezeq (1989a), who obtained a higher yield of small-sized rotifers (strain-Bs) compared with the larger-sized ones (strain-S1) under the same feeding regime in batch cultures. In addition to the high productivity of rotifers achieved in the chemostats, it is possible to maintain the cultures without any decline of rotifer productivity for more than 6 months (pers. observ.). This reflects the consistency and reliability of rotifer production in our aquaculture chemostats.

7.4.4 Productivity of Rotifers in Relation to Salinity

Although *B. plicatilis* tolerates a wide range of salinities from 1 to 97‰ (Walker 1981), its reproductive rates are strongly influenced by the salinity of the culture medium (Ito 1960; Ruttner-Kolisko 1972b; Pascual and Yúfera 1983; Lubzens et al. 1985). Investigations in aquaculture rotifer chemostats show an inverse relation of rotifer output to increasing salinities (James and Abu-Rezeq 1990).

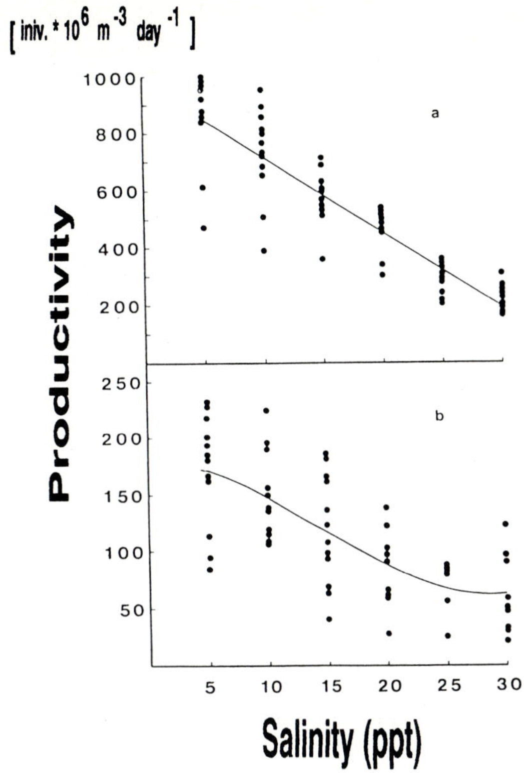

Fig. 7.4.3 a, b Productivity (daily output) of the S-type (**a**) and L-type (**b**) rotifers at different salinities in rotifer chemostats (After James and Abu-Rezeq 1990)

Considerably higher productivity is achieved for both S- and L-type *B. plicatilis* at 5‰ salinity than at higher salinity regimes (Fig. 7.4.3).

In 100-l capacity rotifer chemostats it was possible to achieve a mean output of 856×10^6 S-type rotifers m^{-3} day^{-1} at 5‰ salinity, which is considerably higher than in any of the conventional rotifer production systems to date in aquaculture. The investigations also show that scaling up of the rotifer chemostat culture system from 100 l to 1 m^3 capacity does not affect the rotifer productivity (James and Abu-Rezeq 1989b, c). Thus, the rotifer productivity achieved at 30‰ salinity in 100-l rotifer chemostats is comparable with that in 1-m^3 chemostat units.

Ito (1969) reported optimal reproduction rates in salinities ranging from 3.5 to 11‰ Cl (equivalent to seawater salinities ranging from 6.3 to 20‰). Pascual and Yúfera (1983) observed maximal instantaneous growth rates in seawater of 2–35‰ salinity, whereas Ruttner-Kolisko (1972b) noticed maximal reproduction rates in artificial seawater containing 17 g l^{-1} of salts. In contrast, our investigation showed higher growth rates in lower salinities for both the S- and L-type rotifers with a low salinity of 5–10‰ to be most conducive for the productivity in the chemostat culture system. Differences of growth and productivity compared with previous reports also reflect the fact that the S- and

L-type rotifers in our investigations were preadapted to the tested salinities, resulting in a consistent productivity of rotifers in the respective salinity regimes (Pascual and Yúfera 1983). Furthermore, filtration rates are reduced at high salinities (Hirayama and Ogawa 1972), which may also account for the lower reproductive rates.

7.4.5 Manipulation of Rotifer Size in Chemostats

It is mandatory to produce adequately sized food organisms for feeding fish larvae in aquaculture systems. For example, newly hatched larvae of the marine gilthead sea bream (*Sparus aurata*) prefer small rotifers (191 mol × 138 μm), whereas larvae more than 6 days old select larger (258 mol × 171 μm) rotifers (Helps 1982). Hence, small rotifers must be supplied in greater numbers than larger ones to meet the metabolic demands of developing, newly hatched larvae (Hunter 1980). The size of *B. plicatilis* (Müller) under conventional culture methods varies between 150 and 280 μm (Theilacker and McMaster 1971; Howell 1973; Christiansen and Yang 1976; Girin 1978). However, the body size and shape of *B. plicatilis* are strongly affected by genotype (Serra and Miracle 1985).

Investigations carried out in aquaculture rotifer chemostats (James and Abu-Rezeq 1990; Fig. 7.4.4) show an inverse relation of rotifer size to increasing salinities in the culture system. These results point out that the rotifer size can

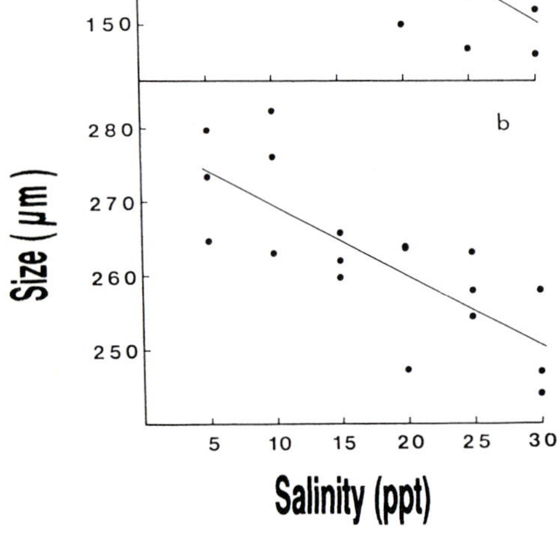

Fig. 7.4.4 a, b Size variations in rotifers in relation to different salinities in the rotifer chemostats: **a** S-type rotifers (size 193.03–1.41x, $r^2 = 0.64$, $n = 18$); **b** L-type rotifers (size 279.00–0.94x, $r^2 = 0.60$, $n = 18$) (After James and Abu-Rezeq 1990)

be manipulated according to larval feed size requirements by adjusting the culture salinity in the chemostat. The lorica length for both S- and L-type rotifers is significantly greater ($P < 0.001$) at 5‰ than at 30‰ salinity. Since the rotifer size is inversely proportional to increasing salinity in the culture system, small rotifers could be produced at higher salinities and vice versa for the specific feed demands of fish larvae.

Although a change in salinity is one of the factors that induces sexual reproduction in rotifer cutures (Hirayama and Ogawa 1972; Hagiwara et al. 1988; Hino and Hirano 1988), in our chemostat culture system mictic reproducon was absent in both the S- and L-type rotifers under all salinity regimes tested (5–30‰), since the rotifers are from an asexually reproducing clone maintained for several years in the rotifer production tanks in the hatchery. Therefore, the populations have lost the ability to produce males, and no males were detected under any salinity regimes tested in the chemostats.

7.4.6 Fatty Acid Composition of Rotifers

The nutritional quality of rotifers for aquacultural purposes is evaluated based on the content of the long chain omega 3 highly unsaturated fatty acids ($\omega3$ HUFA). Among these, the most essential fatty acids (EFA) in rotifers for the growth and survival of marine fish larvae (Watanabe et al. 1983; James et al. 1983b) are eicosapentaenoic acid (EPA 20:5ω3) and docosahexaenoic acid (22:6ω3). The rotifers synthesize these fatty acids from the feed provided in the culture system, especially from microalgae. However, the fatty acid profiles of microalgae, especially the $\omega3$ HUFA, vary among the different species and also depend on the temperature of the culture system (James et al. 1988, 1989). Furthermore, the $\omega3$ HUFA content in rotifers depends on the algal species, algal density, and rotifer strain used in the culture system (Watanabe et al. 1983; Rezeq and James 1987; James and Abu-Rezeq 1988, 1989a–c). The most commonly used species of microalgae in commercial hatcheries is *Nannochloropsis oculata* because of its high $\omega3$ HUFA content (James et al. 1989).

The $\omega3$ HUFA composition of rotifers produced in chemostats for aquaculture shows that the nutritional quality of rotifers depends on the algal species, algal cell density, and flow rate utilized in the chemostat culture system (James and Rezeq 1989b, c). Furthermore, our investigations (James and Abu-Rezeq 1990) in aquaculture chemostats with different salinities demonstrate that 15–20‰ salinity will be most conducive to produce nutritionally enriched S-type rotifers (Fig. 7.4.5). On the contrary, lipid synthesis is most promoted at 30‰ salinity, producing the L-type rotifers. The difference in the content of EPA 20:5ω3 and the long chain 22:6ω3 observed at different salinity regimes between the two rotifer types is strain dependent (James and Abu-Rezeq 1989c).

The presence of $\omega3$ HUFA in rotifers fed with *Nannochloropsis* spp. in 1 m^3 capacity rotifer chemostats indicates (Table 7.4.1) that both rotifer strains have adequate quantities of these EFA (James and Abu-Rezeq 1989b, c). Additionally, no significant differences ($P > 0.05$) have been observed in the

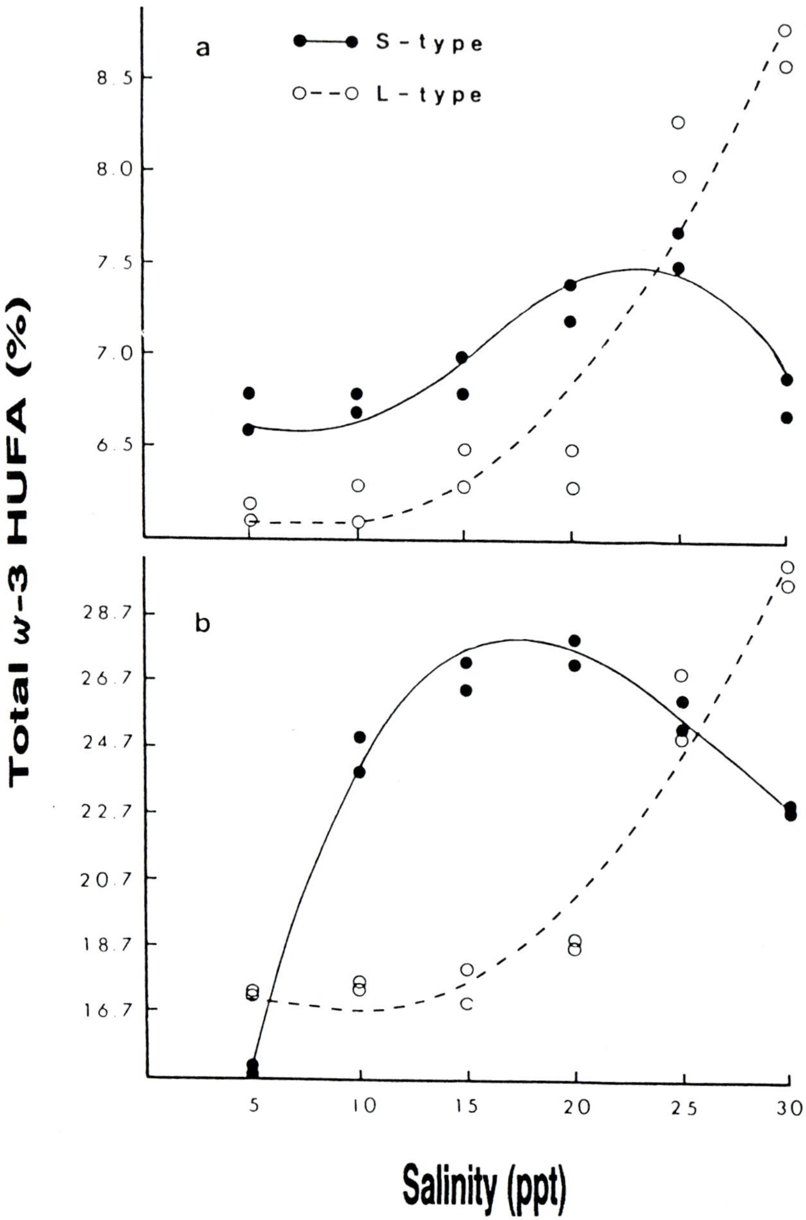

Fig. 7.4.5 a, b Total omega 3 highly unsaturated fatty acids (ω3 HUFA) in the S- and L-type rotifers in relation to different salinities: **a** ω3 HUFA content in rotifers fed with *Chlorella*; **b** ω 3 HUFA content in rotifers fed with *Nannochloropsis* (After James and Abu-Rezeq 1990)

Table 7.4.1 Fatty acid composition (area %) in *Nannochloropsis* strain MFD-2 and in the S- and L-type rotifers produced in chemostats (mean ± SD)

Fatty acid	*Nannochloropsis*	S-type rotifers	L-type rotifers
14:0	7.5 ± 0.15	2.58 ± 0.15	3.20 ± 0.24
14:1 ω9	–	–	–
15:0	0.38 ± 0.08	0.78 ± 0.08	0.55 ± 0.09
16:0	22.63 ± 4.54	15.63 ± 1.48	13.45 ± 2.45
16:1 ω9	18.75 ± 0.96	14.90 ± 0.29	17.03 ± 3.59
17:0	0.33 ± 0.19	0.68 ± 0.13	0.35 ± 0.09
18:0	[a]	[a]	4.88 + 0.42
18:1 ω9	3.00 ± 0.86	23.08 ± 2.19	14.13 ± 3.02
18:2 ω6	3.73 ± 1.08	2.35 ± 0.05	6.25 ± 2.97
18:3 ω6	–	–	0.23 ± 0.13
18:3 ω3	0.33 ± 0.24	–	*1.10* ± 0.87
19:0	0.90 ± 0.19	–	–
20:0	–	–	0.38 + 0.22
20:1 ω9	–	2.05 + 0.55	1.55 + 0.21
20:2 ω6	–	–	–
20:3 ω3	4.23 ± 0.23	5.10 ± 0.07	4.30 ± 0.07
20:4 ω3	–	–	0.23 ± 0.13
20:5 ω3	31.18 ± 4.01	13.88 ± 0.79	18.68 ± 2.24
22:1 ω5	–	0.68 ± 0.13	0.50 ± 0.17
22:3 ω3	–	1.13 ± 0.19	0.53 ± 0.22
22:4 ω3	–	–	–
22:5 ω3	–	–	–
22:6 ω3	–	8.58 ± 0.38	4.20 ± 0.46
24:0	–	0.25 ± 0.12	–
24:1 ω6	–	–	0.23 ± 0.13
Σω 3-HUFA	36.00 ± 3.72	28.68 ± 1.00	29.03 ± 1.75
Total lipid (%)	18.38 ± 2.77	23.90 ± 1.87	23.35 ± 0.45

$n = 4$
[a] 18:0 and 18:1 ω9 are counted together;
–, not detected.

total ω3 HUFA content of S- and L-type rotifers. However, as indicated above, under different salinity regimes, significantly different contents of EPA 20:5ω3 and of the long chain 22:6ω3 were found among the two types of rotifers ($P < 0.05$). The presence of high quantities of EFA 22:6ω3 in both strains of rotifers and its absence in *Nannochloropsis* reveals the rotifers could biosynthesize the long chain EFA such as 22:6ω3 when the feed contains EPA 20:5ω3. This is mandatory for the growth and survival of marine fish larvae (Scott and Middleton 1979; Watanabe et al. 1983; James et al. 1983b). A concentration of 12% EPA 20:5ω3 in rotifers is normally used for feeding marine fish larvae (Watanabe et al. 1983), although the fatty acid 22:6ω3 is highly desired for marine fish larvae rearing. The rotifers produced in chemostats contain about 13% EPA 20:5ω3 in S-type rotifers and about 18% EPA 20:5ω3 in L-type rotifers when fed with *Nannochloropsis* spp. Furthermore, from the high concentrations of EFA 22:6ω3 in these rotifers it can be supposed that adequate quantities of $\omega3$ HUFA are present in them. This is a prerequisite for feeding marine fish larvae. A further rotifer enrichment to enhance their nutritional quality as generally practiced in aquaculture hatcheries (James et al. 1983a, b; Watanabe et al. 1983; Fulks and Main 1991) is not required.

Considering all the factors determining the applicability of rotifers as food for marine fish larvae (rotifer productivity, nutritional quality with respect to fatty acid content, and rotifer size), a salinity of 20‰ is recommended for producing S-type rotifers and 30‰ for the L-type rotifers in the chemostat culture system for aquaculture.

7.4.7 Rotifer Production Levels Required for Aquaculture

Except in Kuwait where rotifer chemostats are used to produce rotifers to carry out research and development programs in aquaculture, all other aquaculture hatcheries around the world produce rotifers using batch and semicontinuous methods. Commercial marine fish hatcheries in Japan must produce about 20×10^9 rotifers per day to feed their fish larvae ((Fushimi 1989). A recent USA-Asia workshop convened at the Oceanic Institute Hawaii on rotifer and microalgal culture systems reports (Fulks and Main 1991) a production of 6×10^9 rotifers per day by Ehime Prefectural Fish Farming Center in Japan and 4.8×10^9 rotifers per day by the Sung Ji Industry in Korea, which are considered the largest producers of rotifers for aquacultural purposes (Table 7.4.2). The next average production levels, by the government-operated finfish hatchery in Singapore and by the chemostat in Kuwait, were about the same, 1.9 and 1.12×10^9 rotifers per day. The Oceanic Institute, USA, Hubbs-Sea World Marine Laboratory, USA, and Thailand have average daily outputs of 400, 305, and 166×10^6 rotifers per day, respectively.

The above figures suggest the enormous quantities of rotifers needed to be produced daily for larval feeding in a hatchery. Since early larval feeds, especially in a marine fish hatchery, depend on the availability of adequate high quality rotifers, a decline in rotifer productivity in a marine fish hatchery results in a high mortality of fish larvae, which is detrimental to commercial ventures. The application of the rotifer chemostat we developed for aquaculture ensures a reliable production of high quality rotifers, employing less space and manpower compared with other conventional methods.

7.4.8 Efficiency of Chemostats vs. Conventional Methods

Conventional rotifer production systems in aquaculture (Fulks and Main 1991) use batch and semicontinuous methods, which are intensive culture systems, in which a culture is inoculated and allowed to grow a given period before the

Table 7.4.2 Rotifer production levels and requirement for aquaculture in the USA and Asia

	USA (a)	(b)	Japan (c)	Korea (d)	S. China (e)	Singapore (f)	Thailand	Taiwan	Kuwait (g)
Average output (10^6 rotifers day^{-1})	400	305	6000	4800	Unknown	1900	166	1000	1222
Average productivity (rotifers ml^{-1} day^{-1})	75	25.7	30	150 (70 in winter)	10	35	30	12	309S 187L
Output required (10^6 rotifers day^{-1})	1000	400	6000–8000	Unknown	Unknown	1700	Unknown	100	1000
Space required (ha)	0.5	< 0.2	0.2	0.9	Depends	0.1	0.1	0.004	0.005
Capital cost (US$)	Unknown	Unknown	281000	10 million	Unknown	1300	2000	40000	10000
Labor (h week^{-1})	196	40	84	36–40	210	20	56	35	7
Algae for 10^9 rotifers (m^3 day^{-1})	105	85	50	45–60	40	10	100	140	5
Tank volume (m^3) for 10^9 rotifers day^{-1}	14.4	42.5	400	15	290	30	50	80	6

Data compiled from Fulks and Main (1991)
(a) The Oceanic Institute, Hawaii; (b) Hubbs-Sea World Marine Lab., California; (c) Ehime Prefectural Fish Farming Center; (d) Sung Ji Industry; (e) Zhanjiang Experimental Station; (f) Marine Aquaculture Section; (g) Chemostat system, C. James

entire volume or a portion thereof is harvested. This approach has been followed because of its technical simplicity, but it is the least efficient method, requiring much hatchery space and manpower (Table 7.4.2). Fukusho (1989), while discussing the current methodology for large scale production of rotifers in Japan, describes a typical Japanese hatchery with 100-m^3 tanks for rotifer culture and a diet of microalgae, usually *Nannochloropsis*, and baker's yeast. After a growth period of a few weeks, rotifer densities of about 100 ml^{-1} of the L-type *B. plicatilis* are reached, and the mass culture is then harvested for several days. Six 100-m^3 tanks hold a standing crop of rotifers from which $1–2 \times 10^9$ rotifers can be harvested daily. In this system 400–600 m^3 water must be managed to produce about 1×10^9 rotifers per day.

Continuous cultures with rotifer chemostats are the most intensive and yield the highest rotifer biomass production per unit of effort reported to date for aquaculture. The productivity of the S-type as well as the L-type rotifers achieved in our continuous culture system is considerably higher than that of the conventional rotifer production systems in aquaculture. In addition to achieving the highest productivity in the chemostat culture system, it is possible to maintain the culture without any decline of rotifer productivity for a long period. Thus, we were able to operate rotifer chemostats to produce nutritionally enriched rotifers for marine fish larvae rearing for more than 6 months in Kuwait, until the system was destroyed by Iraq.

Table 7.4.2 shows the production requirement for various systems in different countries. The labour requirement for rotifer culture, exclusive of microalgal production, is up to 196 h $week^{-1}$ at the Oceanic Institute, USA, which produces 400×10^6 rotifers per day, whereas that with the automated rotifer chemostats in Kuwait, which also includes time for microalgal production in the system for the hatchery, is low. Production from a few 1-m^3 rotifer chemostats would be sufficient to meet the rotifers' needs in most small to medium sized hatcheries. For example, 6–7 m^3 of culture volume is sufficient to produce 1×10^9 rotifers day^{-1}, which is 100-fold less than is required by the batch culture method (Fukusho 1989). Furthermore, the nutritional quality of the rotifers produced in chemostats is far superior to that of rotifers produced in other culture systems (James and Abu-Rezeq 1989b, c). Intensively managing much smaller volumes in rotifer chemostats also results in substantial savings of hatchery space and labor.

James and Abu-Rezeq (1989b, c, 1990) reported no differences in rotifer production when upscaling 100-l chemostats to 1-m^3 chemostat cultures. This means, to meet the requirement of very large commercial hatcheries producing about 20×10^9 rotifers per day, 12–14 units of 10 m^3 capacity rotifer chemostats are sufficient, which would save considerable capital and operating costs in a commercial venture. For these reasons, the application of rotifer chemostats offers the best hope for producing nutritionally enriched high quality rotifers consistently for larviculture.

8 Epilogue

Norbert Walz

The different continuous culture techniques described in this volume offer – in various steps of complexity and technical investments – a world of new aspects both for experimental and on theoretical ecology. This concerns such broad fields as population and community ecology, ecosystem theory, ecotoxicology, and aquaculture. For the application of continuous cultures to these fields, many preparations and prerequisites have been necessary. For example, the mathematically well-formulated chemostat theory worked out for microorganisms had to be modified for metazoans (Chap. 3.3).

Semicontinuous cultures have been developed (Chap. 2.1) as a method well suited to give excellent support to work on questions of the growth characteristics of populations (Chap. 3.1). Another possibility involves studying competition between populations within the community by means of resource partitioning (Sect. 6.2.5) or by "shift-up" and "shift-down" experiments (Chap. 6.1). Another application is in ecotoxicology (Chap. 7.3). This method is in the best sense very simple and at a low technical level, so it may be applied in many laboratories. The characteristic dilution of the system is done mechanically by a regular discard of a fixed part of the culture volume supplemented with a new feed suspension. The manpower requirement, on the other hand, seems to be fairly high. In particular, the working hours needed could restrain this method. Fed-batch cultures (Sect. 2.1.2 may be particularly adapted to experiments on zooplankton population dynamics. The first method has been utilized with rotifers, but nothing speaks against the successful application of both lines to cladocerans and protozoans.

In contrast, the turbidostat is the technically highest developed continuous culture device (Chap. 2.3). Its operation field includes studying the maximum performances of physiological and population dynamic parameters and selecting the fittest populations (Chap. 3.4). Maximum performances (e.g., r_{max}) are affected the most by toxins. This technique, therefore, should be well suited for the assessment of ecotoxicity (Sect. 7.3.8) as a system response. Hitherto, this kind of turbidostat has not been used in aquaculture, for which it should be a very interesting method.

The effort to build up and operate chemostats is shared between both parts. This technique is nearly as time consuming as semicontinuous cultures, while the methodical investment may be a little lower than that for turbidostats (Chap. 2.2). Because of its more constant dilution rate, it supplies more evident results. Because of its substrate limitation, it imitates best the situation in the community and in the ecosystem. The chemostat, therefore, is mostly adapted as a model ecosystem both in its single and two-stage version. This system has been successfully used to study the carbon metabolism and energy balance of different rotifer species (Part 4). This method was applied (among others) to approach experimentally questions of numerical and functional responses of populations, of food quality effects, of competition, of the role of the "microbial loop" (Chap. 6.2), and of community regulation mechanisms (Chaps. 3.2–3.4). A very promising employment is in aquaculture (Chap. 7.4).

The different continuous culture methods have their own benefits and special employments. Above these mostly technically defined applications of continuous cultures, there are more theoretically founded arguments. These refer to two different points which should not be separated:

1. Dilution principle in the model ecosystem
2. Experimental testability of hypotheses

Continuous cultures are model ecosystems or mesocosms most likely adapted to the system of the plankton community because the structure of the plankton community is reflected truly in continuous cultures. Since there is an energy flow along the food chain and simultaneously a real information feedback in continuous cultures from predators to prey [e.g., from the consumer to its food and from the dilution rate, which imitates the action of a predator (Fig. 7.2.1), to the consumer], this system allows us to study the interaction between trophic levels and, therefore, the "emergent properties" of the system (Preface, Chap. 7.2).

By these methods, therefore, many problems of the plankton community may be assessed, especially problems of competition for food resources or of competition under simultaneous predation. With continuous cultures, the principles, of many predator-prey problems may be studied experimentally. Competition can be studied under different regimes of dilution, i.e., of the feeding rate of the "predator," and at different levels of food input. This is a fundamental difference to other mesocosms (e.g., Kersting 1985).

So far, these model ecosystems are very simple and are based on the ecological knowledge of a few species only. In the future, more models with increasing complexity have to used. The energetic and kinetic relationships of more and especially of "key" species, which are important in the food web of the community, have to be studied. The dilution rate as a model "predator" selects for individuals with the highest growth rates. This is a natural situation in highly flooded lakes and rivers only. Indeed, in future experiments the hydraulic dilution must be supplemented by the feeding action of planktivorous fish selecting according to other rules. On the other hand, experiments with hydraulic dilution may reveal the most fundamental – but perhaps too abstract – action of a predator and of the reaction of the community.

In all continuous cultures, the dilution rate is fixed for each stage. But which dilution rate do natural communities have? And which dilution rate do more complicated model systems have, especially if more trophic levels are usually present, e.g., if additional fish are introduced? How are the "dilution rates" of the different trophic levels coupled? Is it right that if there is a top-down control, then the feeding rate of the organisms of the top level must give the rhythm for the turnover of the whole community like the dilution rate in simpler models? But how do the turnover rates of the other trophic levels adapt to this overall turnover? This question seems to be one of the main problems of the upscaling of the conceptions arising from simple continuous culture models to natural communities (Chap. 7.2).

The regulation model which is developed on the basis of continuous cultures and is in accordance with observations must not have any a priori expectations of equilibrium (see Introduction). To be certain, the model is based on continuous cultures in which the system finally achieves a steady state or equilibrium. But with the same structure and with those parameters worked out for steady states, this model can also describe transient states (Chap. 5.1). Equilibria are the result of regulation processes. Models with time lags may not show such steady states.

Regulation models include the food resources explicitly. The population development and its density depend dynamically on these resources, which are controlled by feedback reactions. Consistently, the regulation model dominates over the widespread logistic and Lotka-Volterra models. Those models are not adapted to regulation processes because they do not explicitly include the resources, which are regulated, too. As a consequence, any discussion of the density dependence of populations should include the decisive resources. In future, the resource conception should replace the formulation of density-dependent models. Additionally, for this reason it makes no sense to speak of "density-independent" factors. Of course, further factors are effective which are not regulated if they come from the outside of the system, for example, the temperature.

Having accepted that Lotka-Volterra models are no longer useful for predictions in population dynamics, there are a lot of theoretical consequences for all the fields of ecology concerned, because many fundamental conceptions are based on these models, especially competition models, predator-prey models, and models for community dynamics.

Another topic which contradicts the regulation principle is the identification of a carrying capacity, an important term in the logistic and in Lotka-Volterra models. In the regulation model, the carrying capacity is not a predefined parameter but the result of nutrient or food input, i.e., dependent on steering factors and on boundary limits (see, for example, Chap. 3.3). A second step is the consumption of the supplied resources by the community, which is a regulated process; for example, the growth efficiency is not constant. In contrast, in the logistic and Lotka-Volterra models the carrying capacity has to be explicitly declared as a specific value.

This is also true for the so-called competition coefficients in the Lotka-Volterra competition models which are known only a posteriori and have no predictable value. This consequently invalidates the well-known r/K concept.

With the r_{max}/K_s life history strategy model (Fig. 5.1.6; Chap. 6.3) or with the Tilman model (Chap. 6.2), however, it seems possible to predict the outcome of a competitive situation based on the energetic and kinetic data of the competing species. These models are an extension of the regulation principle.Coexistence is possible by food partitioning (Sect. 6.2.5) and by time lags in the population dynamics (Sect. 5.1.6).

Continuous cultures refer obviously to the pelagial, in which the plankton density is relatively simple to measure. In this community, the food resources may easily be assessed, too. Other populations and communities are thought to be food limited, too, although the exact proof is not easily furnished. Nevertheless, the conclusions drawn from the regulation model –based on continuous cultures as model ecosystems – should indeed be transferred to other communities. As a consequence, they should replace orthodox density-dependent concepts which do not explicitly include the resources. Some of these ideas have been developed in Sect. 7.2.6.

In the past few years this regulation principle, the feedback reaction between the different trophic levels, has been used practically to reduce the eutrophication effects in water management. In many cases, the structure of the plankton community of small lakes could be successfully changed in this way so that the community could fulfill a prominent function in a sanitation concept. This can be done by so-called biomanipulation, food chain manipulation, or top-down control (e.g., Shapiro 1980; Benndorf 1987). The structure and the dynamics of the community are the result of boundary limits. In order to reach a maximum transparency (Secchi depth) of a lake, the large cladocerans, the most effective filtrators, have to be promoted by a reduction of zooplanktivorous fish. This is managed by intensive fish catches or by the introduction of piscivorous fish.

The questions are, how sufficient is this top-down effect and how stable is this community, as otherwise these measures have to be repeated regularly. For the system analysis, the deciding point is the permanent change of the steering factors. Only with an ongoing reduction of nutrients, one external factor having mostly dictated the model ecosystem and the regulation model, could a continuing stability of the community within the desired limits be found (Benndorf 1987). From this standpoint, the action of top predators seems to be an internal factor. It depends on the model itself whether this factor is to be understood as external or internal. As a consequence, without nutrient reduction this community regulates itself to a stable point outside the desired limits. Under these conditions, manipulation has to be considered as "work" to be done against this regulation. It follows that a fundamental knowledge of the regulation characteristics has to be involved to apply the manipulation measures successfully.

A very interesting question deciding the usefulness of any predictive model is the occurrence of a chaotic behavior in it. As for any deterministic nonlinear model, it is known from Lotka-Volterra models that chaotic development depends on the values of some specific parameters (May 1981, Sect. 7.1.2). Even the simulation of the regulation model in Chap. 5.1 demonstrates a chaotic course at higher time lags, depending on the boundary limits of the system. The effect of these conditions, however, is very different in both models. Lotka-Volterra models tend to show cyclic or "strange" attractors (chaotic behavior) at higher algal growth rates and this only if the model has three interacting levels (Scheffer 1991). On the other hand, the regulation model shows a better resistance to cyclic oscillations and chaotic behavior in simulations with higher food inputs and higher dilution rates (Sect. 5.1.5). The stability of the models is lost under exactly opposite conditions.

This means that the conditions necessary to create chaos are dependent on the structure of the model. For this reason, it is too early to draw conclusions between model ecosystems and the real world. Scheffer (1991) found chaotic behavior in the mesocosms of Kersting (1985), but in this model system, no predators are present. In continuous cultures, a stabilization by higher dilution rates, i.e., by predators, seems to be possible. In the above-mentioned sense, top level organisms are able to structure the whole community. A very forward conclusion from the regulation model founded on continuous cultures is that communities generally tend to become chaotic but are stabilized by predators or, in the case of highly flooded lakes, by the dilution rate. The existence of chaos observed in computer simulation models for model ecosystems and in the field remains controversial and should be investigated further in the real world. For this purpose, continuous cultures seem to be ideal test systems.

Appendix

Norbert Walz

1 Terminology and Dimensions

1.1 Masses, Concentrations (dry weight, carbon ml^{-1}*)*

C_0	threshold food concentration (for $r = 0$)
C_1	input-algae concentration
C_2	reactor algae concentration, residual algae concentration
E	egg density (eggs ml^{-1})
I	ingestion
ILL	incipient limiting level = food concentration when reaching the plateau
K_s	Monod saturation "constant," half-maximum food concentration
N	rotifer density (number ml^{-1})
P	production
W	rotifer individual weight (dry weight, carbon)
X	rotifer biomass ($= N*W$)
X_1	rotifer biomass at $t = 1$
X_2	rotifer biomass at $t = 2$
X_e	eliminated rotifer biomass

1.2 Rates (mass × time^{-1} $= \mu g\,C\,d^{-1}$*)*

FR	clearance rate = theoretically particle-free filtered water ($\mu l *$individual$^{-1} *$ day^{-1})
IR	ingestion rate $= FR*C_2$ ($\mu g\,C *$individual$^{-1} *$day^{-1})
IR_0	ingestion rate for maintenance
PR	production rate

1.3 Specific Rates (mass∗unit mass^{-1}*∗time*$^{-1}$ *= day*$^{-1}$*)*

b	specific production rate = specific birth rate
d	specific natural mortality rate
D	(hydraulic) dilution rate (daily rate of water flow/reactor volume)
i	specific ingeston rate

i_0	threshold food ration (for $r = 0$) = specific substrate consumption for maintenance (Pirt 1965)
L	total loss rate
μ_c	specific maintenance rate, minimum metabolic loss rate (Marr et al. 1963)
r	specific growth rate
r′	specific rate of change of population density = "apparent specific growth rate"
spec. A	specific assimilation rate
spec. R	specific respiration rate

1.4 Efficiencies, Quotients, Coefficients (Dimensionless) and General Abbreviations

CI 95%	±95% confidence interval of mean value
EFA	essential fatty acids
EPA	eicosapentaenoic acid
E/N	egg ratio (eggs/females)
F	fraction of daily removed volume in semicontinuous culture
FBC	fed-batch culture
HUFA	highly unsaturated fatty acids
n	number of samples
P	probability of error
PE	production efficiency (b/i)
R	Pearson correlation coefficient
RCM	resource competition model
SC	semicontinuous culture
SD	standard deviation
SE	standard error of the mean
TGF	trace growth factor
Y, K_1	growth yield, gross growth efficiency (r/i)
Y_g, Y_{max}	maximum yield, "true growth yield"

2 Calculations and Derivations

2.1 Derivation: $K_1 = Y$

(see Sects. 3.3.6, 4.1.7, 4.2.6, 6.3.4)

$$Y = \frac{\text{rotifer biomass developed}}{\text{food consumed}} = \frac{X}{C_1 - C_2} \,|\, * D \tag{3.3.11}$$

$$Y = \frac{X * D}{D * (C_2 - C_2)} \qquad\qquad \frac{X}{D * (C_1 - C_2)} = i$$

$$Y = \frac{D}{i} = \frac{r}{i} \qquad | \; r = D, \text{ steady state}$$

2.2 Derivation: Food Concentration $C_2 = C_1 - X/Y$

(see Sect. 3.3.1)

change in algae concentration = input − output − consumption

$$\frac{dC_2}{dt} = DC_1 - DC_2 - IR \qquad \Big| \; \frac{dC_2}{dt} = 0, \text{ steady state}$$

$$DC_2 = DC_1 - IR \qquad | \; \div D$$

$$C_2 = C_1 - IR/D \qquad | \; IR = i * X$$

$$C_2 = C_1 - i * X/D \qquad | \; r = D$$

$$C_2 = C_1 - 1/Y * X$$

(Powell 1958)
(see Eqs. 3.3.6, 3.3.9, 3.3.10)

2.3 Derivation: specific ingestion rate (i)

(see Sects. 3.3.3, 4.1.3)
change in algae concentration = input − output − consumption

$$\frac{dC_2}{dt} = DC_1 - DC_2 - IR \qquad | \; \frac{dC_2}{dt} = 0, \text{ steady state}$$

$$IR = DC_1 - DC_2 \qquad | \; \div X$$

$$i = \frac{D(C_1 - C_2)}{X} \qquad | \; i = IR/X$$

2.4 Derivation: b = Specific Production Rate

(see Sects. 3.3.1, 4.1.5, 4.2.3)

$$P = (X_2 - X_1) + X_e$$

(Boysen-Jensen 1919)

steady state: $X_1 = X_2$, $P = X_e$
production rate PR:

$$PR = \frac{dP}{dt} = \frac{dX_e}{dt} = X * D + X * d \text{ (output-rate + natural mortality rate)}$$

$$\frac{dX_e}{Xdt} = r + d \qquad | \; \div X, \quad r = D, \text{ steady state}$$

$$b = r + d$$

(Lotka 1922; Eq. 3.3.13)

Romanovsky and Polishchuck (1982) also give a derivation for b = specific production rate of rotifers for "nonsteady state conditions" under the assumption that elimination is not selective. This assumption is also valid for the above steady-state derivation.

2.5 *Derivation of Maintenance Rate and Consumption*

Y is the growth yield or gross growth efficiency, a coefficient of substrate consumption used for rotifer biomass building. But in addition to growth, populations have to consume energy to overcome losses for feces, respiration, and mortality. Pirt (1965) introduced Y_g, the true growth yield (see Sects. 3.1.4, 3.3.6, 4.1.4).

If there are no other losses of food concentration (losses to dissolved fractions?):

$$\frac{dC_2}{dt} = -\frac{dI}{dt} = -IR = -(f' + m' + d') - \frac{r * X}{Y_g}$$

$(r * X/Y_g)$ = ingestion rate for growth

$IR_0 = (f' + m' + d')$ = consumption rate to compensate losses of feces, respiration, and mortality

$$\frac{dC_2}{dt} = -IR_0 - r * X/Y_g \qquad \text{(see Eqs. 3.3.2, 3.3.5)}$$

$$-\frac{rX}{Y} = -IR_0 - \frac{rX}{Y_g} \qquad \Big| \quad \frac{dC_2}{dt} = -\frac{r * X}{Y}$$

(see Eqs. 3.3.2, 3.3.5)

$$\frac{1}{Y} = \frac{IR_0}{r * X} + \frac{1}{Y_g} \qquad \Big| \quad \div(-r * X)$$

$$\frac{1}{Y} = \frac{i_0}{D} + \frac{1}{Y_g} \qquad \Big| \quad i_0 = IR_0/X, \text{ Specific consumption rate}$$

| $D = r$, steady state

$$\frac{1}{Y} = \frac{1}{Y_g} + \frac{\mu_c}{Y_g} * \frac{1}{D} \qquad \Big| \quad \mu_c = i_0/Y_g \text{ (Schulze and Lipe 1964)}$$

= specific maintenance rate (day^{-1}; Marr et al. 1963)

If $Y_g = Y_{max}$, $\mu_c = D$ for $\frac{1}{2}$ Y_{max}.
Upon rearrangement, it follows:

$$Y = Y_g * \frac{D}{\mu_c + D} \qquad \text{(see Eq. 3.3.12)}$$

References

Abu-Rezeq TA, James CM (1989) Evaluation of microbial SCP, microencapsulated diets and microalgae (*Nannochloropsis*) for producing the rotifer *Brachionus plicatilis* (L-type) for aquaculture. J Aqua Trop 4: 97–109

Ahlgren G, Lundstedt L, Brett M, Forsberg C (1990) Lipid composition and food quality for some freshwater phytoplankton for cladoceran zooplankters. J Plankton Res 12: 809–818

Allan JD (1976) Life history patterns in zooplankton. Am Nat 110: 165–180

Allen TFH, Starr TB (1982) Hierarchy: perspectives for ecological complexity. University of Chicago Press, Chicago 310 pp

Aloia RC, Moretti RL (1973) Sterile culture techniques for species of the rotifer *Asplanchna*. Trans Am Microsc Soc 92: 364–371

Amren H (1964) Ecological studies of zooplankton population in some ponds in Spitzbergen. Zool Bidr Upps 36: 161–191

Anagnostropoulos GD (1971) Unbalanced growth in a semi-continuous culture system designed for the synchronization of cell division. J Gen Microbiol 65: 23–33

Aris R, Humphrey AE (1977) Dynamics of a chemostat in which two organisms compete for a common substrate. Biotechnol Bioeng 19: 1375–1386

Armstrong RA (1980) Zooplankton as algal chemostats: a theoretical perspective. In: Kerfoot WC (ed) Evolution and ecology of zooplankton communities. University Press of New England, Hanover, pp 305–311

Armstrong RA, McGehee R (1980) Competitive exclusion. Am Nat 115: 151–170

Arndt H (1992) Rotifers as predators on components of the microbial web (bacteria, heterotropic flagellates, ciliates). Hydrobiologia (in press)

Arndt H, Güde H, Macek M, Rothhaupt KO (1992) Chemostats used to model the microbial food web: evidence for the feedback effect of herbivorous metazoans. Arch Hydrobiol Beih (Ergebn Limnol) 37: 187–194

Arnold DE (1971) Ingestion, assimilation, survival, and reproduction by *Daphnia pulex* fed seven species of blue-green algae. Limnol Oceanogr 16: 906–920

Bader FG, Tsuchiya HM, Fredrickson AG (1976a) Grazing of ciliates on blue-green algae: effects of ciliate encystment and related phenomena. Biotechnol Bioeng 18: 311–331

Bader FG, Fredrickson AG, Tsuchyia HM (1976b) Dynamics of an algal-protozoan grazing interaction. In: Canale RP (ed) Modeling biochemical processes in aquatic ecosystems. Ann Arbor Science Publ 389 pp

Bailey JE, Ollis DF (1986) Biochemical engineering fundamentals, 2nd edn. McGraw-Hill, New York, 984 pp

Baker K (1968) Low cost continuous culture apparatus. Lab Pract 17: 817–824

Baltzis BC, Fredrickson AG (1984) Competition of two suspension-feeding protozoan populations for growing bacterial populations in continuous culture. Microb Ecol 10: 61–68

Banse KS (1980) Adult body mass and annual production/biomass relationship of field populations. Ecol Monogr 50: 355–379

Banse K (1982a) Cell volumes, maximal growth rates of unicellular algae and ciliates, and the role of ciliates in the marine pelagial. Limnol Oceanogr 27: 1059–1071

Banse K (1982b) Mass-scaled rates of respiration and intrinsic growth in very small invertebrates. Mar Ecol Prog Ser 9: 281–297

Banse K, Mosher S (1980) Adult body mass and annual production/biomass relationship of field populations. Ecol Monogr 50: 355–379

Barclay H, Van den Driessche P (1975) Time delays in ecological systems. J Theor Biol 51: 347–356

Barford JP, Pamment NB, Hall RJ (1982) Lag phases and transients. In: Bazin MJ (ed) Microbial population dynamics. CRC Press, Boca Raton, pp 55–89

Beaver JR, Crisman TL (1990) Use of microzooplankton as an early indicator of advancing cultural eutrophication. Verh Int Ver Limnol 24: 532–537

Begon M, Mortimer M (1989) Population ecology. A unified study of animals and plants, 2nd edn. Blackwell, Oxford, 220 pp

Bell G (1984) Measuring the cost of reproduction. I. The correlation structure of the life table of a plankton rotifer. Evolution 38: 300–313

Benker U (1992) Einfluß der Futterkonzentration auf strukturelle und populationsdynamische Merkmale von *Brachionus urceolaris* var. rubens Ehrenberg und *Brachionus calyciflorus* Pallas (Rotatoria). Diploma Thesis Fac Biology, University of Munich, 111 pp

Benndorf J (1980) The applicability of the Monod-Model to the calculation of the phytoplankton growth rate in aquatic ecosystems. Contin Cultiv Microorganisms, Proc 7th Symp Prague, July 10–14, pp 305–312

Benndorf J (1987) Food web manipulation without nutrient control: a useful strategy in lake restoration? Schweiz Z Hydrol 49: 237–248

Bennett WN (1988) Assessment of selenium toxicity in algae using turbidostat culture. Water Res 22: 939–942

Bennett WN (1990) Measurement of manganese amelioration of cadmium toxicity in *Chlorella pyrenoidosa* using turbidostat culture. Arch Environ Contam Toxicol 19: 118–123

Bennett WN, Boraas ME (1988) Isolation of a fast-growing strain of the rotifer *Brachionus calyciflorus* Pallas using turbidostat culture. Aquaculture 73: 27–36

Bennett WN, Boraas ME (1989a) A demographic profile of the fastest growing metazoan: a strain of *Brachionus calyciflorus* Pallas (Rotifera). Oikos 55: 365–369

Bennett WN, Boraas ME (1989b) Comparison of population dynamics between slow- and fast-growing strains of the rotifer *Brachionus calyciflorus* Pallas in continuous culture. Oecologia 81: 494–500

Bennett WN, Boraas ME (1989c) An experimental test of the egg-ratio method with instantaneous birth rates as an independent variable. Limnol Oceanogr 34: 1120–1125

Bennett WN, Brooks AS (1989) Measurement of zinc amelioration of cadmium toxicity in *Chlorella pyrenoidosa* using turbidostat culture. Environ Toxicol Chem 8: 877–882

Bennett WN, Brooks AS, Boraas ME (1986) Selenium uptake and transfer in an aquatic food chain and its effects on fathead minnow larvae. Arch Environ Contam Toxicol 15: 513–517

Bergter F (1983) Wachstum von Mikroorganismen. Experimente und Modelle. Verlag Chemie, Weinheim, 161 pp

Bern L (1987) Zooplankton grazing on (methyl-^{3}H)-thymidine-labeled natural particle assemblages: determination of filtering rates and food selectivity. Freshwater Biol 17: 151–159

Berner-Fankhauser H (1986) Populationsdynamik der Planktonrotatorien im Bielersee 1981/82. Thesis Phil Nat Fac, University of Bern, 88 pp

Berquist AM, Carpenter SR, Latino JC (1985) Shifts in phytoplankton size structure and community composition during grazing by contrasting zooplankton assemblages. Limnol Oceanogr 30: 1037–1049

Bērziņš B, Pejler B (1987) Rotifer occurrence in relation to pH. Hydrobiologia 147: 107–116

Bērziņš B, Pejler P (1989a) Rotifer occurrence in relation to temperature. Hydrobiologia 175: 223–231

Bērzinš B, Pejler P (1989b) Rotifer occurrence in relation to oxygen content. Hydrobiologia 183: 165–172

Beverton RJH, Holt SJ (1957) On the dynamics of exploited fish populations. Ministry of Agriculture, Fisheries and food (London), Fish Invest Ser 2, 19: 533 pp

Birch LL (1948) The intrinsic rate of natural increase of an insect population. J Anim Ecol 17: 15–26

Bjørnsen PK (1986) Automatic determination of bacterioplankton biomass by image analysis. Appl Environ Microbiol 51: 1199–1204

Blachére H, Jamart G (1969) A flow cell photometer for bacterial growth monitoring. Biotechnol Bioeng 11: 1005–1010

Blackman FF (1905) Optima and limiting factors. Ann Bot 19: 281–295

Bliss CI, James AT (1966) Fitting the rectangular hyperbola. Biometrics 22: 573–602

Blueweiss LH, Fox H, Kudzma V, Nakashima D, Peters R, Sams S (1978) Relationship between body size and some life history parameters. Oecologia 37: 257–272

Bogdan KG, Gilbert JJ (1982) Seasonal patterns of feeding by natural populations of *Keratella*, *Polyarthra*, and *Bosmina*: Clearance rates, selectivities, and contributions to community grazing. Limnol Oceanogr 27: 918–934

Bogdan KG, Gilbert JJ (1984) Body size and food size in freshwater zooplankton. Proc Natl Acad Sci USA 81: 6427–6431

Bogdan KG, Gilbert JJ (1987) Quantitative comparison of food niches in some freshwater zooplankton. Oecologia 72: 331–340

Bonomi A, Fredrickson AG (1976) Protozoan feeding and bacterial wall growth. Biotechnol Bioeng 18: 239–252

Boraas ME (1974) Behavior of some ecological competition models based on saturation kinetics. Am Zool 14: 1287

Boraas ME (1980) A chemostat system for the study of rotifer-algal-nitrate interactions. In: Kerfoot WC (ed) Evolution and ecology of zooplankton communities. University Press of New England, Hanover, NH, pp 173–182

Boraas ME (1983) Population dynamics of food-limited rotifers in two-stage chemostat culture. Limnol Oceanogr 28: 546–563

Boraas ME, Bennett WN (1988) Steady-state rotifer growth in a two-stage, computer-controlled turbidostat. J Plankton Res 10: 1023–1038

Boraas ME, Seale DB, Horton JB (1990) Resource competition between two rotifer species (*Brachionus rubens* and *B. calyciflorus*): an experimental test of a mechanistic model. J Plankton Res 12: 77–87

Borgmann U, Millard ES, Charlton CC (1988) Dynamics of a stable, large volume, laboratory ecosystem containing *Daphnia* and phytoplankton. J Plankton Res 10: 691–713

Børsheim KY, Andersen S (1987) Grazing and food size selection by crustacean zooplankton compared to production of bacteria and phytoplankton in a shallow Norwegian mountain lake. J Plankton Res 9: 367–379

Bosch FVD, Ringelberg J (1985) Seasonal succession and population dynamics of *Keratella cochlearis* (Ehrbg.) and *Kellicottia longispina* (Kellicot) in Lake Maarsseven (Netherlands). Arch Hydrobiol 103: 273–290

Bottrell HH, Duncan A, Gliwicz ZM, Grygierek E, Herzig A, Hillbricht-Ilkowska A, Kurasawa H, Larson P, Weglenska T (1976) A review of some problems in zooplankton production studies. Norw J Zool 24: 419–456

Boysen-Jensen P (1919) Valuation of the Limfjord 1909–1917. I. Studies on the fish food in Limfjord 1909–1917. Rep Dan Biol Stn 26: 1–44

Brandl Z, Fernando CH (1978) Prey selection by the cyclopoid copepods *Mesocyclops edax* and *Cyclops vicinus*. Verh Int Ver Limnol 20: 2505–2510

Bratbak G (1985) Bacterial biovolume and biomass estimation. Appl Environ Microbiol 49: 1488–1493

Bratbak G, Thingstad TF (1985) Phytoplankton-bacteria interactions: an apparent paradox? Analysis of a model system with both competition and commensalism. Mar Ecol Prog Ser 25: 23–30

Brendelberger H (1985) Filter mesh-size and retention efficiency for small particles: comparitive studies with Cladocera. Arch Hydrobiol Beih (Ergebn Limnol) 21: 135–146

Brendelberger H, Geller W (1985) Variability of filter structures in eight *Daphnia* species: mesh size and filtering areas. J Plankton Res 7: 473–486

Brett JR (1976) Scope for metabolism and growth of sockeye salmon, *Oncorhynchus nerka*, and some related energetics. J Fish Res Board Can 33: 307–313

Brett JR (1979) Environmental factors and growth. In: Hoar WS, Randall DJ, Brett JR (eds) Fish physiology, vol VII. Academic Press, New York, pp 599–675

Brett JR, Groves TDD (1979) Physiological energetics. In: Hoar WS, Randall DJ, Brett JR (eds) Fish physiology, vol VII. Academic Press, New York, pp 279–352

Brett JR, Shelbourn JE, Shoop CT (1969) Growth rate, and body composition of fingerling sockeye salmon (*Oncorhynchus nerka*) in relation to temperature and ration size. J Fish Res Board Can 26: 2363–2394

Brett MT (1989) The rotifer communities of acid-stressed lakes of Maine. Hydrobiologia 186/187: 181–189

Briggs GE, Haldane JB (1925) Note on kinetics of enzyme action. Biochem J 19: 338–339

Brillouin L (1956) Science and information theory. Academic Press, New York, 320 pp

Broderick AJ, Greenshields RN (1982) Semi-continuous and continuous production of *Aspergillus niger* spores in submerged liquid culture. J Gen Microbiol 128: 2639–2645

Brooks JL, Dodson SI (1965) Predation, body size and composition of plankton. Science 150: 28–35

Brzenzinski MA, Nelson DM (1988) Interactions between pulsed nutrient supplies and a photocycle affect phytoplankton competition for limiting nutrients in long-term culture. J Phycol 24: 346–356

Bungay HR, Bungay ML (1986) Microbial interactions in continuous culture. Adv Appl Microbiol 10: 269–290

Burmaster D (1979) The continuous culture of phytoplankton: mathematical equivalence among three steady-state models. Am Nat 113: 123–134

Burns CW (1968) The relationship between body size of filter feeding Cladocera and the maximum size of particles ingested. Limnol Oceanogr 31: 848–858

Burns CW (1988) Starvation resistance among copepod nauplii and adults. Verh Int Ver Limnol 23: 2087–2091

Burns CW, Gilbert JJ (1986a) Effect of daphnid size and density on interference between *Daphnia* and *Keratella cochlearis*. Limnol Oceanogr 31: 848–858

Burns CW, Gilbert JJ (1986b) Direct observation of the mechanism of interference between *Daphnia* and *Keratella cochlearis*. Limnol Oceanogr 31: 858–866

Butler GJ, Wolkowicz GSK (1985) A mathematical model of the chemostat with a general class of functions describing nutrient uptake. SIAM J Appl Math 45: 138–151

Butler GJ, Wolkowicz GSK (1987) Exploitative competition in a chemostat for two complementary, and possibly inhibitory, resources. Math Biosci 83: 1–48

Button KD (1986) Affinity of organisms for substrate. Limnol Oceanogr 31: 453–456

Calcott PH (1981) The construction and operation of continuous culture. In: Calcott PH (ed) Continuous culture of cells, vol I. CRC Press, Boca Raton, pp 13–26

Calow P (1977a) Conversion efficiency in heterotrophic organisms. Biol Rev 52: 385–409

Calow P (1977b) Ecology, evolution, and energetics: a study in metabolic adaptation. Adv Ecol Res 10: 1–62

Calow P (1984) Economics of ontogeny adaptional aspects. In: Shorrocks B (ed) Evolutionary ecology. Blackwell, Oxford, pp 81–104

Canale CR, Lustig TD, Kehrberger PM, Salo JE (1973) Experimental and mathematical modeling studies of protozoan predation on bacteria. Biotechnol Bioeng 15: 707–728

Caperon J (1967) Population growth in microorganisms limited by food supply. Ecology 48: 715–722

Caperon (1969) Time lag in population growth response of *Isochrysis galbana* to a variable nitrate environment. Ecology 50: 188–192

Carpenter SR (ed) (1988) Complex interactions in lake communities. Springer, Berlin Heidelberg New York, 283 pp

Carpenter SR, Kitchell JF, Hodgson JR (1985) Cascading trophic interactions and lake productivity. Fish predation and herbivory can regulate lake ecosystems. BioScience 35: 634–639

Carpenter SR, Kitchell JF, Hodgson JR, Cochran PA, Elser JJ, Elser MM, Lodge DM, Kretchmer D, He X, Von Ende CN (1987) Regulation of lake primary production by food web structure. Ecology 68: 1863–1876

Chalk EA (1981) Cladoceran filter feeding in a Thames valley reservoir. PhD Thesis, Central London Polytechnician/Thames Water Authority (cited by Duncan and Gulati 1983)

Chesson PL (1986) Environmental variation and the coexistence of species. In: Diamond J, Case TJ (eds) Community ecology. Harper and Row, New York, pp 240–256

Christansen FB, Fenchel TM (1977) Theories of populations in biological communities. Ecological studies 20. Springer, Berlin Heidelberg New York, 144 pp

Christiansen ME, Yang WT (1976) Feeding experiments on the larvae of the fiddler crab *Uca pugilator* (Brachyura, Ocypodidae), reared in the laboratory. Aquaculture 8: 91–98

Chu SP (1942) The influence of the mineral composition of the medium on growth of planktonic communities. I. Methods of culture media. J Ecol 30: 284–325

Clark CW (1989) Bioeconomics. In: Roughgarden J, May RM (eds) Perspectives in ecological theory. Princeton University Press, Princeton, NJ, pp 275–286

Cody ML (1966) A general theory of clutch size. Evolution 20: 174–184

Cole LC (1954) The population consequences of life history phenomena. Q Rev Biol 29: 103–137

Condrey RE (1982) The chemostat and Blackman kinetics. Biotechnol Bioeng 24: 1705–1709

Conover RJ (1966) Factors affecting the assimilation of organic matter by zooplankton and the question of superfluous feeding. Limnol Oceanogr 11: 346–354

Conover RJ (1970) Cultivation of plankton populations. Helgol Wiss Meeresunters 21: 401–444

Conrad JM (1986) Bioeconomics and the management of renewable resources. In: Hallam TG, Levin SA (eds) Biomathematics, vol 17. Mathematical ecology. An introduction. Springer, Berlin Heidelberg New York, pp 381–403

Contois DE (1959) Kinetics of bacterial growth: relationship between population density and specific growth rate of continuous cultures. J Gen Microbiol 21: 40–50

Cook JR (1968) A continuous culture device for protozoan cells. J Protozool 15: 452–455

Cooney CL, Wang DIC (1976) Transient response of *Enterobacter aerogenes* under a dual nutrient limitation in a chemostat. Biotechnol Bioeng 18: 189–190

Cooney CL, Koplov M, Häggstrøm M (1981) Transient phenomena in continuous culture. In: Calcott PH (ed) Continuous cultures of cells, vol I. CRC Press, Boca Raton, pp 143–168

Crowley PH (1975) Natural selection and the Michaelis-Menten constant. J Theor Biol 50: 461–475

Cruz EM, James CM (1989) The effect of feeding rotifers (*Brachionus plicatilis typicus*) on the yield and growth of tilapia (*Oreochromis spirulus*) fry. Aquaculture 77: 353–361

Cunningham A, Nisbet RM (1983) Transients and oscillations in continuous culture. In: Bazin M (ed) Mathematics in microbiology. Academic Press, London, pp 77–103

Curds CR (1974) Computer simulation of some complex microbial food chains. Water Res 8: 769–780

Curds CR, Bazin MJ (1977) Protozoan predation in batch and continuous culture. Adv Aquat Microbiol 1: 115–176

Curds CR, Cochburn A (1968) Studies on the growth and feeding of *Tetrahymena pyriformis* in axenic and monoxenic cultures. J Gen Microbiol 54: 343–358

Curds CR, Cockburn A (1971) Continuous monoxenic culture of *Tetrahymena pyriformis*. J Gen Microbiol 66: 95–108

Currie DJ (1982) Estimating Michaelis-Menten parameters: bias, variance and experimental design. Biometrics 38: 907–919

Cushing DH (1981) Fisheries biology. A study in population dynamics, 2nd edn. University of Wisconsin Press, Madison, 295 pp

Czeczuga B (1959) Oviposition in *Eudiaptomus gracilis* G.O. Sars and *E. graciloides* Lilljeborg (Diaptomide, Crustacea) in relation to season and trophic level. Bull Acad Pol Sci Cl II Ser Sci Biol 7: 227–230

Dabes JN, Finn RK, Wilkie CR (1973) Equations for substrate-limited growth: the case for Blackman kinetics. Biotechnol Bioeng 15: 1159–1177

D'Agostino A (1975) Antibiotics in cultures of invertebrates. In: Smith WL, Chanley MH (eds) Cultures of invertebrate animals. Plenum Press, New York, pp 109–133

DeAngelis DL, Waterhouse JC (1987) Equilibrium and nonequilibrium concepts in ecological models. Ecol Monogr 57: 1–21

DeBiase AE, Sanders RW, Porter KG (1990) Relative nutritional value of ciliate protozoa and algae as food for *Daphnia*. Microb Ecol 19: 199–210

DeMott WR (1982) Feeding selectivities and relative ingestion rates of *Daphnia* and *Bosmina*. Limnol Oceanogr 27: 518–527

DeMott WR (1985) Relations between filter mesh-size, feeding mode, and capture efficiency for cladocerans feeding on ultrafine particles. Arch Hydrobiol Beih (Ergebn Limnol) 21: 125–143

DeMott WR (1986): The role of taste in food selection by freshwater zooplankton. Oecologia 69: 334–340

DeMott WR (1989) The role of competition in zooplankton successions. In: Sommer U (ed) Plankton ecology. Succession in plankton communities. Springer, Berlin Heidelberg New York, pp 195–252

DeMott WR, Kerfoot WC (1982) Competition among cladocerans: nature of the interaction between *Bosmina* and *Daphnia*. Ecology 63: 1949–1964

den Boer PJ (1986) The present status of the competitive exclusion principle. Trends Ecol Evol 1: 25–28

Dent AE, Bazin MJ, Saunders PZ (1976) Behavior of *Dictyostelium discoideum* amoebae and *Escherichia coli* grown together in chemostat culture. Arch Microbiol 109: 187–194

Dickman M (1969) Some effects of lake renewal on phytoplankton productivity and species composition. Limnol Oceanogr 14: 660–666

Dieffenbach H, Sachse R (1911) Biologische Untersuchungen an Rädertieren in Teichgewässern. Int Rev Gesamten Hydrobiol Hydrograph Suppl 3: 1–93

Dillon PJ, Rigler FH (1974) The phosphorus-chlorophyll relationship in lakes. Limnol Oceanogr 19: 767–773

Dodson SI (1974) Zooplankton competition and predation: an experimental test of the size-efficiency hypothesis. Ecology 55: 605–613

Doohan M (1973) An energy budget for adult *Brachionus plicatilis* Müller (Rotatoria). Oecologia 13: 351–362

Doohan M (1975) Rotifera. In: Curds CR, Hawkes HA (eds) Ecological aspects of used water treatment, vol I. Academic Press, London, pp 289–304

Doohan M, Rainbow V (1971) Determination of dry weights of small Aschelminthes (< 0.1 μm). Oecologia 6: 380–383

Dorazio RM (1986) Estimating population birthrates of zooplankton when rates egg deposition and hatching are periodic. Oecologia 69: 532–541.

Dougherty EC, Solberg B, Ferral DJ (1961) The first axenic cultivation of a rotifer species. Experientia 17: 131–132

Dowd JE, Riggs DS (1965) A comparison of estimates of Michaelis-Menten kinetic constants from various linear transformations. J Biol Chem 240: 863–869

Doyle RW, Hunte W (1981) Demography of an estuarine and amphipod (*Gammarus lawrencianus*) experimentally selected for high "r": a model of the genetic effects of environmental change. Can J Fish Aquat Sci 38: 1120–1127

Drake JF, Tsuchiya HM (1976) Predation on *Escherichia coli* by *Colpoda steinii*. Appl Environ Microbiol 31: 870–874

Drake JF, Tsuchiya HM (1977) Growth kinetics of *Colpoda steinii* on *Esherichia coli*. Appl Environ Microbiol 34: 18–22

Droop MR (1970) In: Conover RJ (ed) Cultivation of plankton populations. Helgol Wiss Meeresunters 21: 401–444

Droop MR (1974) The nutrient status of algae cells in continuous culture. J Mar Biol Assoc UK 54: 825–855

Droop MR (1976) The chemostat in mariculture. Proc 10th Eur Symp Mar Biol (Ostend) 1: 71–93

Droop MR (1983) 25 years of algal growth kinetics. A personal view. Bot Mar 26: 99–112

Droop MR, Scott JM (1978) Steady-state energetics of a planktonic herbivore. J Mar Biol Assoc UK 58: 749–772

Droop MR, Scott JM (1982) A steady-state approach to some microplankton problems. Ann Inst Oceanogr (Paris) 58: 47–54

Dugdale RC (1967) Nutrient limitation in the sea. Dynamics, identification and significance. Limnol Oceanogr 12: 685–695

Dumont HJ (1977) Biotic factors in the population dynamics of rotifers. Arch Hydrobiol Beih (Ergebn Limnol) 8: 98–122

Dumont HJ, Van de Velde I, Dumont S (1975) The dry weight estimate of biomass in a selection of Cladocera, Copepoda and Rotifers from the plankton, periphyton and benthos of continental waters. Oecologia 19: 75–97

Duncan A (1983) The influence of temperature upon the duration of embryonic development of tropical *Brachionus* species (Rotifera). In: Schiemer F (ed) Limnology of Parakrama Samudra-Sri Lanka. Junk Publ, The Hague, pp 107–115

Duncan A, Gulati RD (1983) Feeding studies with natural food particles on tropical species of planktonic rotifers. In: Schiemer F (ed) Limnology of Parakrama Samudra – Sri Lanka. Junk Publ, The Hague, pp 25–77

Edmondson WT (1960) Reproductive rates of rotifers in natural populations. Mem Ist Ital Idrobiol 12: 21–77

Edmondson WT (1962) Food supply and reproduction of zooplankton in relation to phytoplankton population. Rapp Cons Explor Mer 153: 137–141

Edmondson WT (1965) Reproductive rate of planktonic rotifers as related to food and temperature in nature. Ecol Monogr 35: 61–111

Ehrlich PR, Roughgarden J (1987) The science of ecology. MacMillan, New York, 710 pp

Ejsmont-Karabin J (1974) Studies on the feeding of planktonic polyphage *Asplanchna priodonta* Gosse. Ekol Pol Ser A 22: 311–317

Elliot JM (1976) The energetics of feeding, metabolism and growth of brown trout (*Salmo trutta* L.) in relation to body weight, water temperature and ration size. J Anim Ecol 45: 923–948

Elser JJ, Goldman CR (1991) Zooplankton effects on phytoplankton in lakes of contrasting trophic status. Limnol Oceanogr 36: 64–90

Emmerson WD (1984) Predation and energetics of *Penaeus indicus* (Decapoda: Penaeidae) larvae feeding on *Brachionus plicatilis* and *Artemia* nauplii. Aquaculture 38: 201–209

Endo K, Mochizuki J (1979) Relation between specific growth rate of the rotifer *Brachionus plicatilis* and the concentration of food of a marine species of *Chlorella*. J Ferment Technol 57: 372–374

Engelberg J, Boyarsky LL (1979) The noncybernetic nature of ecosystems. Am Nat 114: 317–324

Epp RW, Lewis WM Jr (1980) Metabolic uniformity over the environmental temperature range in *Brachionus plicatilis* (Rotifera). Hydrobiologia 73: 145–147

Epp RW, Lewis WM Jr (1984) Cost and speed of locomotion for rotifers. Oecologia 61: 289–292

Evans CGT, Herbert D, Tempest DW (1970) The continuous cultivation of micro-organisms. 2. Construction of a chemostat. In: Norris JR, Ribbons DW (eds) Methods in microbiology, vol 2. Academic Press, London, pp 277–348

Felton LD, Dougherty KM (1924) Studies on virulence. II. The increase in virulence in vitro of a strain of pneumococcus. J Exp Med 39: 137–154

Fenchel T (1974) Intrinsic rate of natural increase: the relationship with body size. Oecologia 14: 317–326

Fenchel T (1982) Ecology of heterotrophic microflagellates. II. Bioenergetics and growth. Mar Ecol Prog Ser 8: 225–231

Fenchel T, Finlay BJ (1983) Respiration rates in heterotropic free-living protozoa. Microb Ecol 9: 99–122

Fencl Z (1966) Mathematical theory of homogeneous continuous cultivations of microorganisms. In: Malek I, Fencl Z (eds) Theoretical and methodological basis of continuous culture. I. Academic Press, New York, pp 79–88

Fencl Z, Silinger V, Nusl J, Malek I (1961) Theory of semicontinuous and continuous cultivation applied to the yeast *Torula utilis*. Folia Microbiol 6: 94–103

Filatov VP (1972) The efficiency of natural food by carp larvae (*Cyprinus carpio*). Vopr Ikhtiol (Moskva) 12: 886–892 (in Russian)

Folt C, Goldman CR (1981) Allelopathy between zooplankton: a mechanism for interference competition. Science 213: 1133–1135

Fredrickson AG (1977) Behavior of mixed cultures of organisms. Annu Rev Microbiol 31: 63–87

Fredrickson AG, Stephanopoulos (1981) Microbiol competition. Science 213: 972–979

Fredrickson AG, Tsuchiya HM (1977) Microbiol kinetics and dynamics. In: Lapidus L, Amundson NR (eds) Chemical reactor theory: a review. Prentice-Hall, Englewood Cliffs, pp 405–483

Frost BW (1972) Effect of size and concentration of food particles on the feeding behavior of the marine planktonic copepod *Calanus pacificus*. Limnol Oceanogr 17: 805–815

Frost BW (1975) A threshold feeding behavior in *Calanus pacificus*. Limnol Oceanogr 20: 263–266

Frost BW (1985) Food limitation of the planktonic marine copepods *Calanus finnmarchicus* and *Pseudocalanus* sp. in a temperate fjord. Arch Hydrobiol Beih (Ergebn Limnol) 21: 1–13

Frost TM, DeAngelis DL, Bartell SM, Hall DJ, Hurlbert SH (1988) Scale in the design and interpretation of aquatic community research. In: Carpenter SR (ed) Complex interactions in lake communities. Springer, Berlin Heidelberg New York, pp 229–258

Fukusho K (1989) Biology and mass production of the rotifer *Brachionus plicatilis*. I. Int J Aqua Fish Technol 1: 68–76

Fulks W, Main KL (eds) (1991) Rotifer and microalgae culture systems. Proc US-Asia Workshop, January 28–31, 1991, Honolulu, Hawaii, pp 1–363

Fulton RS III, Pearl HW (1987) Toxic and inhibitory effects of the blue-green alga *Microcystis aeruginosa* on herbivorous zooplankton. J Plankton Res 9: 837–855

Fulton RS III, Pearl HW (1988) Effects of blue-green alga *Microcystis aeruginosa* on zooplankton competitive relations. Oecologia 76: 383–389

Fushimi T (1989) Systematizing large scale culture methods. In: Fukusho K, Hirayama K (eds) The first live feed source—*Brachionus plicatilis*. KoseishaKoseikaku, Tokyo, pp 118–134 (transl from Japanese)

Gaedeke A, Sommer U (1986) The influence of the frequency of periodic disturbances on the maintenance of phytoplankton diversity. Oecologia 71: 25–28

Galkovskaya GA (1963) On the utilization of food for growth and conditions of highest yield in rotatoria *Brachionus calyciflorus*. Zool Zh 42: 506–512 (in Russian, transl in Fish Res Board Can 997 (1969)).

Galkovskaya GA (1987) Planktonic rotifers and temperature. Hydrobiologia 147: 307–317

Gatesoupe FJ, Luquet P (1981) Practical diet for mass culture of the rotifer *Brachionus plicatilis* application to larval rearing of the sea bass, *Dicentrarchus labrax*. Aquaculture 22: 149–163

Gatesoupe FJ, Robin J (1981) Commercial single cell proteins either as sole food or in formulated diets for intensive and continuous production of rotifers (*Brachionus plicatilis*). Aquaculture 25: 1–15

Gause GF (1934) The struggle for existence. Williams and Wilkins, Baltimore

Geller W (1975) Die Nahrungsaufnahme von *Daphnia pulex* in Abhängigkeit von der Futterkonzentration, der Temperatur, der Körpergröße und dem Hungerzustand der Tiere. Arch Hydrobiol Suppl 48: 47–107

Geller W, Müller K (1981) The filtration apparatus of Cladocera: filter mesh-sizes and their implications on food selectivity. Oecologia 49: 316–321

Gerritsen J, Kou J (1985) Food limitation and body size. Arch Hydrobiol Beih (Ergebn Limnol) 21: 173–184

Giani A (1991) Implications of phytoplankton chemical composition for zooplankton production: experimental evidence. Oecologia 87: 409–416

Gilbert JJ (1970) Monoxenic cultivation of the rotifer *Brachionus calyciflorus* in a defined medium. Oecologia 4: 89–101

Gilbert JJ (1974) Dormancy in rotifers. Trans Am Microsc Soc 93: 490–513

Gilbert JJ (1980a) Monoxenic cultivation of the rotifer *Brachionus calyciflorus* in a defined medium. Oecologia 4: 89–101

Gilbert JJ (1980b) Observations on the susceptibility of some protists and rotifers to predation by *Asplanchna girodi*. Hydrobiologia 73: 87–91

Gilbert JJ (1985) Competition between rotifers and *Daphnia*. Ecology 66: 1943–1950

Gilbert JJ (1988a) Suppression of rotifer populations by *Daphnia*: a review of the evidence, the mechanism, and the effects on zooplankton community structure. Limnol Oceanogr 33: 1286–1303

Gilbert JJ (1988b) Susceptibility of ten rotifer species to interference from *Daphnia pulex*. Ecology 69: 1826–1838

Gilbert JJ, Bogdan KG (1981) Selectivity of *Polyartha* and *Keratella* for flagellate and aflagellate cells. Verh Int Ver Limnol 21: 1515–1521

Gilbert JJ, Bogdan KG (1984) Rotifer grazing: in situ studies on selectivity and rates. In: Meyers DG, Strickler JR (eds) Trophic interactions within aquatic ecosystems. Am Assoc Adv Sci Selected Symp 85: 97–133

Gilbert JJ, Starkweather PL (1977) Feeding in the rotifer *Brachionus calyciflorus*. 1. Regulatory mechanisms. Oecologia 28: 125–131

Gilbert JJ, Waage JK (1967) *Asplanchna*, *Asplanchna*-substance and posterolateral spine length variation of the rotifer *Brachionus calyciflorus* in a natural environment. Ecology 48: 1027–1031

Gilbert JJ, Williamson CE (1978) Predator-prey behavior and its effect on rotifer survival in associations of *Mesocyclops edax*, *Asplanchna girodi*, *Polyarthra vulgaris* and *Keratella cochlearis*. Oecologia 37: 13–22

Gilch N (1981) Populationsdynamik des Crustaceenplanktons im Fasaneriesee. Diploma Thesis, Fac Biology University of Munich, 63 pp

Gilley JW, Bungay HR (1967) Oscillatory growth rate responses of *Saccharomyces cerevisiae* in continuous culture. Biotechnol Bioeng 9: 617–622

Girin M (1978) Methods de production de juveniles chez trois poissons marins: le bar (*Dicentrarchus labrax*), le sole (*Solea solea*) et le turbot (*Scophrhalmus maximus*). Thèse de Doctorat d'Etat, Faculté des Sciences de Paris

Gliwicz ZM (1969) Studies on the feeding of pelagic zooplankton in lakes with varying trophy. Ekol Pol 17: 663–708

Gliwicz ZM (1980) Filtering rates, food size selectivity, and filtering rates in cladocerans, another aspect of interspecific competition in filter-feeding zooplankton. In: Kerfoot WC (ed) Evolution and ecology of zooplankton communities. University Press of New England, Hanover, NH, pp 282–291

Gliwicz ZM (1990) Food thresholds and body size in cladocerans. Nature 343: 638–640

Gliwicz ZM, Lampert W (1990) Food thresholds in *Daphnia* species in the absence and presence of blue-green filaments. Ecology 71: 691–702

Gliwicz ZM, Pijanowska J (1989) The role of predation in zooplankton succession. In: Sommer U (ed) Plankton ecology. Succession in plankton communities. Springer, Berlin Heidelberg New York, pp 253–296

Gliwicz ZM, Siedler E (1980) Food size limitation and algae interfering with food collection in *Daphnia*. Arch Hydrobiol 88: 155–177

Gnauck AH (1974) Oberflächengewässer als biochemische Reaktoren. Acta Hydrochim Hydrobiol 2: 433–439

Goel NS, Maitra SC, Montroll EW (1971) On the Volterra and other nonlinear models of interacting populations. Rev Mod Phys 43: 231–276

Gold K (1973) Methods for growing Tintinnida in continuous culture. Am Zool 13: 203–208

Goodrich MS, Lech JJ (1990) A behavioral screening assay for *Daphnia magna*: a method to assess the effects of xenobiotics on spatial orientation. Environ Toxicol Chem 9: 21–30

Gophen M (1977) Food and feeding habits of Mesocyclops leuckarti (Claus) in Lake Kinneret (Israel). Freshwater Biol 7: 513–518

Gordon HS (1954) The economic theory of a common-property resource: the fishery. J Polit Econ 62: 124–142

Gotham IJ, Rhee G-Y (1982) Effects of a hexachlorobiphenyl and pentachlorophenol on growth and photosynthesis of phytoplankton. J Great Lakes Res 8: 328–335

Gould SJ (1977) Ontogeny and physiology. Harvard University Press, Cambridge, 501 pp
Goulden CE, Hornig L, Wilson C (1978) Why do large zooplankton species dominate? Verh Int Ver Limnol 20: 2457–2460
Goulden CE, Henry LL, Tessier AJ (1982) Body size, energy reserves, and competitive ability in three species of Cladocera. Ecology 63: 1780–1789
Goulden CE, Henry LL, Berrigan D (1987) Egg size, postembryonic yolk and survival ability. Oecologia 72: 28–31
Grady CPL Jr, Harlow LJ, Riesing RR (1972) Effect of growth rate and influent substrate concentration on effluent quality from chemostats containing bacteria in pure and mixed culture. Biotechnol Bioeng 14: 391–410
Groeneweg J, Schlüter M (1981) Mass production of freshwater rotifers in liquid wastes. II. Mass production of *Brachionus rubens* Ehrenberg 1838 in the effluent of high rate algal ponds used for the treatment of piggery wastes. Aquaculture 25: 25–34
Gross LJ (1986) Ecology: an idiosyncratic overview. In: Hallam TG, Levin SA (eds) Biomathematics, vol 17, Mathematical ecology. An introduction. Springer, Berlin Heidelberg New York, pp 3–15
Gruhl E (1986) Untersuchung der Zooplanktonentwicklung im Hochwásserückhaltebecken Postmünster von Juli bis Oktober 1985. Diploma Thesis, Fac Biology, University of Munich, 70 pp
Gschloessl T (1985) Die Nahrungsaufnahme von *Brachionus angularis* (Rotatoria) in Abhängigkeit von der Futterart, der Futterkonzentration, der Temperatur und der Fütterungsdauer. Diploma Thesis, Fac Biology, University of Munich, 127 pp
Güde H (1985) Influence of phagotrophic processes on the regeneration of nutrients in two-stage continuous culture systems. Microb Ecol 11: 193–204
Güde H (1989) The role of grazing on bacteria in plankton succession. In: Sommer U (ed) Plankton ecology. Succession in plankton communities. Springer, Berlin Heidelberg New York, pp 337–364
Güde H, Haibel B, Müller H (1985) Development of planktonic bacterial populations in a water column of Lake Constance (Bodensee-Obersee). Arch Hydrobiol 105: 59–77
Guerrero R, Abella C, Miracle MR (1978) Spatial and temporal distribution of bacteria in a meromictic carstic lake basin: relationships with physiochemical parameters and zooplankton. Verh Int Ver Limnol 20: 2264–2271
Guisande C, Mazuelos N (1991) Reproductive pattern of *Brachionus calyciflorus* Pallas at different food concentrations. J Plankton Res 13: 279–286
Guiset A (1977) Stomach contents in *Asplanchna* and *Ploeosoma*. Arch Hydrobiol Beih (Ergebn Limnol) 20: 2264–2271
Gulati RD, Rooth J, Ejsmont-Karabin J (1987) A laboratory study of feeding and assimilation in *Euchlanis dilatata lucksiana*. Hydrobiologia 147: 289–296
Guma'a SA (1978) The food and feeding habits of young perch, *Perca fluviatilis*, in Windermere. Freshwater Biol 8: 177–187
Hagiwara A, Hino A, Hirano R (1988) Effects of temperature and chlorinity on resting egg formation in the rotifer *Brachionus plicatilis*. Nippon Suisan Gakkaishi 54: 569–575
Haken H (1982) Synergetik. Eine Einführung. Springer, Berlin Heidelberg New York, 382 pp
Halbach U (1970a) Einfluß der Temperatur auf die Populationsdynamik des planktischen Rädertieres *Brachionus calyciflorus* Pallas. Oecologia 4: 176–207
Halbach U (1970b) Das Zusammenwirken von Konkurrenz und Räuber-Beute-Beziehungen bei Rädertieren. Zool Anz (Suppl) 33: 72–79
Halbach U (1979) Introductory remarks: strategies in population research exemplified by rotifer population dynamics. Fortschr Zool 25: 1–27
Halbach U (1984) Population dynamics of rotifers and its consequences for ecotoxicology. Hydrobiologia 109: 79–96
Halbach U, Burkhardt HJ (1972) Sind einfache Zeitverzögerungen die Ursachen für periodische Populationsschwankungen? Vergleich experimenteller Untersuchungen an *Brachionus calyciflorus* Pallas (Rotatoria) mit Computersimulationen. Oecologia 9: 215–222
Halbach U, Halbach-Keup G (1974) Quantitative Beziehungen zwischen Phytoplankton und

der Populationsdynamik des Rotators *Brachionus calyciflorus* Pallas. Befunde aus Laboratoriumsexperimenten und Freilandunteruchungen. Arch Hydrobiol 73: 273–309

Halbach U, Siebert M, Wissel C, Klaus J, Beuter K, Delion M (1981) Population dynamics of rotifers as bioassay too for toxic effects of organic pollutants. Verh Int Ver Limnol 21: 1141–1146

Halbach U, Siebert M, Westermayer M, Wissel C (1983) Population ecology of rotifers as a bioassay tool for ecotoxicological tests in aquatic environments. Ecotox Environ Safety 7: 484–513

Hall DJ, Threlkeld ST, Burns CW, Crowley PH (1976) The size efficiency hypothesis and the size structure of zooplankton communities. Annu Rev Ecol Syst 7: 117–208

Hamilton RD, Preslan JE (1970) Observations on the continuous culture of a planktonic phagotropic protozoan. J Exp Mar Ecol 4: 90–99

Hammer C (1985) Feeding behavior of roach (*Rutilus rutilus*) larvae and the fry of perch (*Perca fluviatilis*) in Lake Lankau. Arch Hydrobiol 103: 61–74

Hansen SR, Hubbell SP (1980) Single-nutrient microbial competition: qualitative agreement between experimental and theoretically forecast outcomes. Science 207: 1491–1493

Hanson JM, Leggett WC (1982) Empirical prediction of fish biomass and yield. Can J Fish Aquat Sci 39: 257–263

Hardin G (1960) The competitive exclusion priniciple. Science 131: 1292–1298

Hartman J (1978) Fischwachstum bei Oligo-, Meso- und Eutrophie des Bodensees, Schweiz Z Hydrol 40: 32–39

Hartmann U (1987) Die Populationsdynamik der pelagischen Rotatorien *Brachionus angularis* and *Notholca caudata* in Abhängigkeit von der Futterkonzentration und der Temperatur. Diploma Thesis, Fac Biology University of Munich 91 pp

Harvey A (1972) A semi-continuous culture technique for *Daphnia pulex*. J Appl Ecol 9: 831–834

Hassel MB, Lawton JH, Beddington JR (1977) Sigmoid functional responses by invertebrate predators and parasitoids. J Anim Ecol 46: 249–262

Hassenstein B (1966) Kybernetik and biologische Forschung. In: Gessner F (ed) Handbuch der Biologie, vol I/2. Akademische Verlagsgesellschaft, Frankfurt/M pp 629–719

Havel JE (1987) Predator-induced defenses: a review. In: Kerfoot WC, Sih A (eds) Predation. Direct and indirect impacts on aquatic communities. University of New England Press, Hanover, pp 263–278

Havens KE (1991) The importance of rotiferan and crustacean zooplankton as grazers of algal productivity in a freshwater estuary. Arch Hydrobiol 122: 1–22

Havens KE, DeCosta J (1988) An experimental analysis of the acid sensitivity of the common planktonic rotifer *Keratella cochlearis*. Int Rev Gesamten Hydrobiol 73: 407–416

Healey FP (1980) Slope of the Monod equation as an indicator of advantage in nutrient competition. Microb Ecol 5: 281–286

Helps SK (1982) An examination of prey size selection and its subsequent effect on survival and growth of gilthead sea bream (*Sparus aurata*). Master's Thesis. Plymouth Polytechnic, Plymouth, UK

Hemmingsen AM (1960) Energy metabolism as related to body size and respiratory surfaces, and its evolution. Rep Steno Mem Hosp Copenhagen 9: 1–110

Herbert D (1958) Some priniciples of continuous culture. In: Tunevall G (ed) Recent Progress in Microbiology. Blackwell, Oxford, pp 381–396

Herbert D (1976) Stoichiometric aspects of microbial growth. In: Dean ACR, Ellwood DC, Evans CGT, Melling J (eds) Continuous Culture 6: applications and new field. Ellis Horwood, Chichester, pp 1–28

Herbert D, Elsworth R, Telling RC (1956) The continuous culture of bacteria: a theoretical and experimental study. J Gen Microbiol 14: 601–622

Herzig A (1983) Comparative studies on the relationship between temperature and duration of embryonic development of rotifers. Hydrobiologia 104: 237–246

Herzig A (1987) The analysis of plankton rotifer populations: a plea for long term investigations. Hydrobiologia 147: 163–180

Hill SH, Abott MR, Denman KL (1985) A computer-controlled turbidostat for the culture of planktonic algae. Can J Fish Aquat Sci 42: 744–753

Hino A, Hirano R (1988) Relationship between water chlorinity and bisexual reproduction rate in the rotifer *Brachionus plicatilis*. Nippon Suisan Gakkaishi 54: 1329–1332

Hirata H (1980) Culture methods for the marine rotifer *Brachionus plicatilis*. Min Rev Data File Fish Res 1: 27–46

Hirata H, Yamasaki S, Kawaguchi T, Ogawa M (1983) Continuous culture of the rotifer *Brachionus plicatilis* fed recycled algal diets. Hydrobiologia 104: 71–75

Hirayama K, Funamoto H (1983) Supplementary effect of several nutrients on nutritive deficiency of baker's yeast for population growth of the rotifer *Brachionus plicatilis*. Bull Jpn Soc Sci Fish 49: 505–510

Hirayama K, Kusano T (1972) Fundamental studies on physiology of rotifer for its mass culture. II. Influence of water temperature on population growth of rotifers. Bull Jpn Soc Sci Fish 38: 1357–1363

Hirayama K, Watanabe K, Kusano T (1973) Fundamental studies on physiology of rotifer for its mass culture. III. Influence of pytoplankton density on population growth. Bull Jpn Soc Sci Fish 39: 1123–1127

Hirayama U, Ogawa M (1972) Fundamental studies on physiology of rotifer for its mass culture. I. Filter feeding of rotifers. Bull Jpn Soc Sci Fish 38: 1207–1214

Hoenicke R, Goldman CR (1987) Resource dynamics and seasonal changes in competitive interactions among three cladoceran species. J Plankton Res 9: 397–417

Höffgen B (1988) Abundanz- und Populationsdynamik des Mikrozooplanktons im Brunnsee. Diploma Thesis Fac Biology, University of Munich, 178 pp

Höfle MG (1984) Transient responses of glucose-limited cultures of *Cytophaga johnsonae* to nutrient excess and starvation. Appl Environ Microbiol 47: 356–362

Holling CS (1959a) The components of predation as revealed by a study of small-mammal predation of the European sawfly. Can Entomol 91: 293–320

Holling CS (1959b) Some characteristics of simple types of predation and parasitism. Can Entomol 91: 385–398

Holm NP, Armstrong DE (1981) Role of nutrient limitation and competition in controlling the populations of *Asterionella formosa* and *Microcystis aeruginosa* in semicontinuous culture. Limnol Oceanogr 26: 622–634

Horn W (1985a) Investigation into the food selectivity of the planktic crustaceans *Daphnia hyalina*, *Eudiaptomus gracilis* and *Cyclops vicinus*. Int Rev Gesamten Hydrobiol 70: 603–612

Horn W (1985b) Results regarding the food of the planktic crustaceans *Daphnia hyalina* and *Eudiaptomus gracilis*. Int Rev Gesamten Hydrobiol 70: 703–709

Howell BR (1973) Marine fish culture in Britain. VIII. A marine rotifer, *Brachionus plicatilis* Müller, and the larvae of the mussel, *Mytilus edulis* L. as foods for larval flat fish. J Cons Int Explor Mer 35: 1–6

Hrbáček J, Dvořakova M, Kořínek V, Procházkóva L (1961) Demonstration of the effect of the fish stock on the species composition of zooplankton and the intensity of the whole plankton assemblage. Verh Int Ver Limnol 14: 192–195

Hsu SB, Hubbel S, Waltman P (1977) A mathematical theory for single-nutrient competition in continuous cultures of micro-organisms. SIAM J Appl Math 32: 366–383

Hsu SB, Cheng K-S, Hubbel SP (1981) Exploitative competition of microorganisms for two complementary nutrients in continuous cultures. SIAM J Appl Math 41: 422–444

Huber M (1982) Populationsdynamik der Rädertiere des Fasanerie-Sees im Sommer 1981. Diploma Thesis, Fac Biology, University of Munich, 99 pp

Huffaker CB (1958) Experimental studies on predation: dispersion factors and predator-prey oscillations. Hilgardia 27: 343–383

Hunter JR (1980) The feeding behaviour and ecology of marine fish larvae. In: Bardach JR, Magnusen JJ, May CR, Reinhart JM (eds) Fish behaviour and its use in the capture and culture of fishes. ICLARM Conf Proc 5, Makati, Manila, pp 287–330

Hutchinson GE (1948) Circular causal systems in ecology. Ann NY Acad Sci 50: 221–246

Hutchinson GE (1951) Copepodology for the ornithologist. Ecology 32: 571–577

Hutchinson GE (1961) The paradox of the plankton. Am Nat 95: 137–145

Ito T (1960) On the culture of mixohaline rotifer *Brachionus plicatilis* O. F. Müller in the sea water. Rep Fac Fish Univ Mie 3: 708–736

Jacobs J (1978) Coexistence in similar zooplankton species by differential adaptation to reproduction and escape in an environment with fluctuating food and enemy densities. III. Laboratory experiments. Oecologia 35: 35–54

Jacobs J (1984) Cooperation, optimal density and low density thresholds: yet another modification of the logistic model. Oecologia 64: 389–395

James CM, Abu-Rezeq T (1988) Effect of different cell densities of *Chlorella capsulata* and a marine *Chlorella* sp. for feeding the rotifer *Brachionus plicatilis*. Aquaculture 69: 43–56

James CM, Abu-Rezeq T (1989a) Production and nutritional quality of two small-sized strains of the rotifer *Brachionus plicatilis*. J World Aquac Soc 20: 1–15

James CM, Abu-Rezeq T (1989b) Intensive rotifer cultures using chemostats. Hydrobiologia 186/187: 423–430

James CM, Abu-Rezeq T (1989c) An intensive chemostat culture system for the production of rotifers for aquaculture. Aquaculture 81: 291–301

James CM, Abu-Rezeq T (1990) Efficiency of rotifer chemostats in relation to salinity regimes for producing rotifers for aquaculture. J Aquac Trop 5: 103–116

James CM, Al-Khars AM (1990) An intensive continuous culture system using tubular photobioreactors for producing microalgae. Aquaculture 87: 381–391

James CM, Al-Khars AM, Abbas MB, Al-Ameeri AA (1983a) Nutritional studies on rotifers and *Artemia* for feeding fish larvae. Kuwait Inst Sci Res Annu Res Rep 1983: 61–63

James CM, Bou-Abbas M, Al-Khars AM, Al-Hinty S, Salman AE (1983b) Production of the rotifer *Brachionus plicatilis* for aquaculture in Kuwait. Hydrobiologia 104: 77–84

James CM, Al-Khars AM, Al-Hinty S, Abbas MB (1986) Manual on live food production for aquaculture, 2nd edn. Kuwait Inst Sci Res Annu Res Rep 1986: 1–63

James CM, Dias P, Salman AE (1987) The use of marine yeast (*Candida* sp.) and baker's yeast (*Saccharomyces cerevisiae*) in combination with *Chlorella* sp. for mass culture of the rotifer *Brachionus plicatilis*. Hydrobiologia 147: 263–268

James CM, Al-Khars AM, Chorbani P (1988) pH dependent growth of *Chlorella* in a continuous culture system. J World Aquac Soc 19: 27–35

James CM, Al-Hinty S, Salman AE (1989) Growth and omega 3 fatty acid and amino acid composition of marine microalgae under different temperature regimes. Aquaculture 77: 337–351

Jannasch HW (1967) Enrichments of aquatic bacteria in continuous culture. Arch Mikrobiol 59: 165–173

Jannasch HW (1974) Steady state and the chemostat in ecology. Limnol Oceanogr 19: 716–720

Jannasch HW, Mateles RI (1974) Experimental bacterial ecology studied in continuous culture. In: Rose AH, Tempest DW (eds) Advances in microbial physiology, vol II. Academic Press, London, pp 165–212

Jost JL, Drake JF, Fredrickson AG, Tsuchyia HM (1973a) Interactions of *Tetrahymena pyriformis*, *Escherchia coli*, *Azotobacter vinlandii* and glucose in a minimal medium. J Bacteriol 113: 834–840

Jost JL, Drake JF, Tsuchyia HM, Fredrickson AG (1973b) Microbial food chains and food webs. J Theor Biol 41: 461–484

Jürgens K, Güde H (1990) Incorporation and release of phosphorus by planktonic bacteria and phagotrophic flagellates. Mar Ecol Prog Ser 59: 271–284

Kafuku T, Ikenoue H (1983) Modern methods of aquaculture in Japan. Developments in aquaculture and fisheries science 2. Kodansha, Tokyo and Elsevier, Amsterdam, 216 pp

Karabin A (1978) The pressure of pelagic predators of the genus *Mesocyclops* on small zooplanktons. Ekol Pol 26: 241–257

Karabin A (1985) Pelagic zooplankton (Rotatoria and Crustacea) variations in the process of lake eutrophication. I. Structural and quantitative features. Ekol Pol 33: 567–616

Kareiva P (1989): Renewing the dialogue between theory and experiments in population ecology. In: Roughgarden J, May RM, Levin SA (eds) Perspectives in ecological theory. Princeton University Press, Princeton, pp 68–88

Kerfoot WC (ed) (1980) Evolution and ecology of zooplankton communities. University of New England Press, Hanover, 793 pp

Kerfoot WC (1987) Cascading effects and indirect pathways. In: Kerfoot WC, Sih A (eds)

Predation – direct and indirect impacts on aquatic communities. University of New England Press, Hanover, pp 57–70

Kerfoot WC, Sih A (eds) (1987) Predation. Direct and indirect impacts on aquatic communities. University of New England Press, Hanover, 386 pp

Kersting K (1978) Growth efficiency of *Daphnia magna* 1. The effect of food concentration. Hydrobiol Bull 12: 3–20

Kersting K (1985) Properties of an aquatic micro-ecosystem. V. Ten years of observation of the prototype. Verh Int Ver Limnol 22: 3040–3045

Ketchum BH, Redfield AC (1938) A method for maintaining a continuous supply of marine diatoms by culture. Biol Bull 75: 165–169

Kilham SS (1978) Nutrient kinetics of freshwater planktonic algae using batch and semicontinuous methods. Mitt Int Ver Limnol 21: 147–157

Kilham SS, Kilham P (1978) Natural community bioassays. Predictions of results based on nutrient physiology and competition. Verh Int Ver Limnol 20: 68–74

King CE (1967) Food, age, and the dynamics of a laboratory population of rotifers. Ecology 48: 111–128

King CE, Snell TW (1978) Culture media (natural and synthetic): rotifera. In: Diets, culture media, food supplements. CRC handbook series in nutrition and food. CRC Press, Boca Raton, pp 71–75

Kingsland SE (1985) Modeling nature: episodes in the history of population ecology. University of Chicago Press, Chicago, 267 pp

Koga S, Humphrey AE (1967) Study of the dynamic behavior of the chemostat system. Biotechnol Bioeng 9: 375–386

Kompala DS, Ramkrishna D, Tsao GT (1984) Cybernetic modeling of microbial growth on multiple substrates. Biotechnol Bioeng 26: 1272–1281

Korstad J, Vadstein O, Olsen Y (1989) Feeding kinetics of *Brachionus plicatilis* fed *Isochrysis galbana*. Hydrobiologia 186/187: 51–57

Koste W (1978a) Rotatoria. Die Rädertiere Mitteleuropas. Ein Bestimmungswerk, begründet von M Voigt. Gebr Borntraeger, Berlin, 2nd edn, Textband 673 pp, Tafelband 234 pp

Koste W (1978b) *Synchaeta grandis*, ein in Mitteleuropa vom Aussterben bedrohtes Planktonrädertier. Mikrokosmos 1978: 331–336

Krambeck HJ, Lampert W, Brede H (1981) Messung geringer Mengen von partikulärem Kohlenstoff in natürlichen Gewässern. GIT Fachz Labortech 25: 1009–1012

Krause-Dellin D (1983) Entwicklung des Zooplanktons in einem stark durchströmten See. Diploma Thesis, Fac Biology University of Munich, 112 pp

Lack DL (1954) The natural regulation of animal numbers. Clarendon Press, Oxford, 343 pp

Laing I, Helm MM (1981) Factors affecting the semi-continuous production of *Tetraselmis suecica* (Kylin) Butch. in 200-1 vessels. Aquaculture 22: 137–148

Lam RK, Frost BW (1976) Model of copepod filtering responses to changes in size and concentration of food. Limnol Oceanogr 21: 490–500

Lampert W (1976) A direct coupled, artificial two-step food chain for long term experiments with filter-feeders at constant food concentrations. Mar Biol 37: 349–355

Lampert W (1977) Studies on the carbon balance of *Daphnia pulex* de Geer as related to experimental conditions. IV. Determinations of the "threshold" concentration as a factor controlling the abundance of zooplankton species. Arch Hydrobiol Suppl 48: 361–368

Lampert W (1978) A field study of the dependence of fecundity of *Daphnia* species on food concentration. Oecologia 36: 363–369

Lampert W (1981) Inhibitory and toxic effects of blue-green algae on *Daphnia*. Int Rev Gesamten Hydrobiol 66: 285–298

Lampert W (1984) The measurement of respiration. In: Downing JA, Rigler FH (eds) A manual on methods for the assessment of secondary productivity in fresh waters. Blackwell, Oxford, pp 413–468

Lampert W (ed) (1985) Food limitation and the structure of zooplankton communities. Arch Hydrobiol Beih (Ergebn Limnol) 21, Stuttgart 497 pp

Lampert W (1987) Predictability in lake ecosystems: the role of biotic interactions. In: Schulze

ED, Zwölfer H (eds) Potentials and limitations of ecosystem analysis. Ecological Studies 61. Springer, Berlin Heidelberg New York, pp 333–346

Lampert W (1989) The adaptive significance of diel vertical migration of zooplankton. Funct Ecol 3: 21–27

Lampert W, Muck P (1985) Multiple aspects of food limitation in zooplankton communities: the *Daphnia-Eudiaptomus* example. Arch Hydrobiol Beih (Ergebn Limnol) 21: 311–322

Lampert W, Schober U (1980) The importance of "threshold" food concentrations. In: Kerfoot WC (ed) Evolution and ecology of zooplankton communities. University of New England Press, Hanover, pp 264–267

Lane PA (1975) The dynamics of aquatic systems: a comparative study of the structure of four zooplankton communities. Ecol Monogr 45: 307–336

Lassiter RR (1979) Microcosms as ecosystems for testing ecological models. In: Jørgensen SE (ed) State-of-the-art in ecological modelling, Proc Conf Ecol Modelling, Copenhagen 28 Aug–2 Sept 1978, vol 7. Int Soc Ecol Modelling, Copenhagen, pp 127–161

Laxhuber R, Hartman U (1988) The influence of temperature on the life cycle of the cold-stenothermal rotifer *Notholca caudata*. Verh Int Ver Limnol 23: 2016–2018

Laybourn J (1976) Energy budgets for *Stentor coerulus* Ehrenberg (Ciliophora). Oecologia 22: 431–437

Lederman TC, Rhee G-Y (1981) The influence of a hexachlorobiphenyl on the growth of Great Lakes phytoplankton. Can J Fish Aquat Sci 39: 388–394

Lehman JT (1976) The filter-feeder as an optimal forager and the predicted shapes of feeding curves. Limnol Oceanogr 21: 501–516

Lehman JT (1980) Nutrient recycling as an interface between algae and grazers in freshwater communities. In: Kerfoot WC (ed) Evolution and ecology of zooplankton communities. University Press of New England, Hanover, pp 251–263

Lehman JT (1988) Ecological principles affecting community structure and secondary production by zooplankton in marine and freshwater environments. Limnol Oceanogr 33: 931–945

Lehninger AL (1975) Biochemistry. The molecular basis of cell structure and function, 2nd edn. Worth, New York, 1104 pp

Leibold MA (1989) Resource edibility and the effects of predators and productivity on the outcome of trophic interactions. Am Nat 134: 922–949

Leimeroth N (1980) Respiration of different stages and energy budgets of juvenile *Brachionus calyciflorus*. Hydrobiologia 73: 195–197

Lewis WM Jr (1979) Zooplankton community analysis. Studies on a tropical system. Springer, Berlin Heidelberg New York, 156 pp

Likens GE, Gilbert JJ (1970) Notes on quantitative sampling of natural populations of planktonic rotifers. Limnol Oceanogr 15: 816–820

Lindeman RL (1942) The trophic-dynamic aspect of ecology. Ecology 23: 399–418

Livansky K (1979) Semicontinuous cultivation of autotrophic algae. Folia Microbiol 24: 346–351

Lomnicki (1988) Ecology of individuals. Princeton University Press, Princeton, NJ

Lotka AJ (1922) The stability of the normal age distribution. Proc Natl Acad Sci USA 8: 339–345

Lotka AJ (1925) Elements of physical biology. Williams and Wilkins, Baltimore, 460 pp

Lubzens E (1987) Raising rotifers for use in aquaculture. Hydrobiologia 147: 245–255

Lubzens E, Minkoff G, Marom S (1985) Salinity dependence of sexual and asexual reproduction in the rotifer *Brachionus plicatilis*. Mar Biol 85: 123–126

Lubzens E, Tandler A, Minkoff G (1989) Rotifers as food in aquaculture. Hydrobiologia 186/187: 387–400

Lundgren A (1985) Model ecosystems as a tool in freshwater and marine research. Arch Hydrobiol Suppl (Monogr Beitr) 70: 157–196

Lynch M (1980) The evolution of cladoceran life histories. Q Rev Biol 55: 23–42

Lynch M, Shapiro J (1981) Predation, enrichment, and phytoplankton community structure. Limnol Oceanogr 26: 86–102

Maaløe O, Kjeldgaard NO (1966) Control of macromolecular synthesis. Benjamin Press, New York, 284 pp

MacArthur RH (1972) Geographical ecology: patterns in the distribution of species. Harper and Row, New York

MacArthur RH, Levins R (1964) Competition, habitat selection, and character displacement in patchy environments. Proc Natl Acad Sci USA 51: 1207–1210

MacArthur RH, Wilson EO (1967) The theory of island biogeography. Princeton University Press, Princeton, NJ, 216pp

MacDonald N (1976) Time delay in simple chemostat models. Biotechnol Bioeng 18: 805–812

MacDonald N (1982) Time delays in chemostat models. In: Bazin MJ (ed) Microbiol population dynamic. CRC Press, Boca Raton, pp 33–53

MacIsaac HJ, Gilbert JJ (1989) Competition between rotifers and cladocerans of different body sizes. Oecologia 81: 295–301

MacIsaac HJ, Gilbert JJ (1990) Does exploitative competition from *Daphnia* limit the abundance of *Keratella* in Loch Leven? A reassessment of May and Jones (1989). J Plankton Res 12: 1315–1322

Manning BW, Federle TW, Cerniglia CE (1987) Use of a semicontinuous culture system as a model for determining the role of human intestinal microflora in the metabolism of xenobiotics. J Microbiol Meth 6: 81–94

Marker AFM, Nusch EA, Rai H, Riemann B (1980) The measurement of photosynthetic pigments in freshwaters and standardization of methods: conclusion and recommendations. Arch Hydrobiol Beih (Ergebn Limnol) 14: 91–106

Marr AG, Nilson EH, Clark DJ (1963) The maintenance requirement of *Escherichia coli*. Ann NY Acad Sci 102: 536–548

Mateles RI, Ryu DY, Yasudo T (1965) Measurement of unsteady state growth rates of microorganisms. Nature 208: 263–264

Matveev VF (1985) Competition and population time lags in *Bosmina* (Cladocera, Crustacea). Int Rev Gesamten Hydrobiol 70: 491–508

Matveev VF (1987) Effect of competition on the demography of planktonic cladocerans-*Daphnia* and *Diaphanosoma*. Oecologia 74: 468–477

May L (1983) Rotifer occurrence in relation to water temperature in Loch Leven, Scotland. Hydrobiologia 104: 311–315

May L (1987) Culturing freshwater, planktonic rotifers on *Rhodomonas minuta* var.*nannoplanctica* Skuja and *Stichococcus bacillaris* Nägeli. J Plankton Res 9: 1217–1223

May L, Jones DH (1989) Does interference competition from *Daphnia* affect populations of *Keratella cochlearis* in Loch Leven, Scotland? J Plankton Res 11: 445–461

May RM (1973) Time delay versus stability in population models with two and three trophic levels. Ecology 54: 315–325

Maty RM (1975) Stability and complexity in model ecosystems, 2nd edn. Princeton University Press, Princeton, NJ, 235 pp

May RM (1981) Models for single populations. In: May RM (ed) Theoretical ecology. Blackwell, Oxford, pp 5–29

May RM, Conway GR, Hassel MP, Southwood TRE (1974) Time delay, density dependence, and single species oscillations. J Anim Ecol 43: 747–770

Mazumder A, McQueen DJ, Taylor WD, Lean DRS, Dickman MD (1990) Micro- and mesozooplankton grazing on natural pico- and nanoplankton in contrasting plankton communities produced by planktivore manipulation and fertilization. Arch Hydrobiol 118: 257–282

McCauley E, Kalff J (1981) Empirical relationships between phytoplankton and zooplankton biomass in lakes. Can J Fish Aquat Sci 38: 458–463

McCauley E, Murdoch WW (1987) Cyclic and stable populations: plankton as a paradigm. Am Nat 129: 97–121

McCauley E, Murdoch WW (1990) Predator-prey dynamics in environments rich and poor in nutrients. Nature 343: 455–457

McQueen DJ, Post JR (1988) Cascading tophic interactions: uncoupling at the zooplankton-phytoplankton link. Hydrobiologia 159: 277–296

McQueen DJ, Post JR, Mills EL (1986) Trophic relationships in freshwater pelagic ecosystems. Can J Fish Aquat Sci 43: 1571–1581

Megee RD, Drake JF, Fredrickson AG, Tsuchiya HM (1972) Studies in intermicrobial symbiosis, *Saccharomyces cerevisiae* and *Lactobacillus casei*. Can J Microbiol 18: 1733–1742

Merz L, Jaschek H (1990) Grundkurs der Regelungstechnik, 10th edn. Oldenbourg, München 293 pp

Michaelis L, Menten MM (1913) Die Kinetik der Invertinwirkung. Biochem Z 49: 333–369

Mikschi E (1989) Rotifer distribution in relation to temperature and oxygen content. Hydrobiologia 186/187: 209–214

Mills EL, Forney JL (1988) Trophic dynamics and development of freshwater pelagic food webs. In: Carpenter SR (ed) Complex interactions in lake communities. Springer, Berlin Heidelberg New York, pp 11–30

Mills EL, Forney JL, Wagner KJ (1987) Fish predation and its cascading effect on the Oneida Lake food chain. In: Kerfoot WC, Sih A (eds) Predation: direct and indirect impacts on aquatic communities. University Press of New England, Hanover, NH, pp 118–131

Miracle MR, Serra M (1989) Salinity and temperature influence in rotifer life history characteristics. Hydrobiologia 186/187: 81–102

Mitchell SA (1986) Experiences with outdoor semi-continuous mass culture of *Brachionus calyciflorus* Pallas (Rotifera). Aquaculture 51: 289–297

Monod J (1942) Recherches sur la croissance des cultures bacteriennes. Hermann et Cie, Paris, 210 pp

Monod J (1950) La technique de culture continue: theorie et applications. Ann Inst Pasteur Paris 79: 390–410

Moore MV, Gilbert JJ (1987) Age-specific *Chaoborus* predation on rotifer prey. Freshwater Biol 17: 223–236

Morowitz HJ (1968) Energy flow in biology. Academic Press, New York, 179 pp

Mosel C (1991) Entwicklung des Zooplanktons in einem abwasserbelasteten Flußstausee (Ismaninger Speichersee). Diploma Thesis, Fac Biology, University of Munich, 89 pp

Moser A (1981) Bioprozeßtechnik, Berechnungsgrunlagen der Reaktionstechnik biokatalytischer Prozesse. Springer, Berlin Heidelberg New York, 198 pp

Moser H (1957) Contributions to the theory of the continuous bacterial growth apparatus. I. Kinetics of growth of homogenous populations. Proc Natl Acad Sci USA 43: 222–225

Müller H (1972) Wachstum und Phosphatbedarf von *Nitzschia actinastroides* (Lemm.) v. Goor in statischer und homokontinuierlicher Kultur unter Phosphatlimitierung. Arch Hydrobiol Suppl 38: 399–484

Mullin MM, Stewart EF, Fuglister FJ (1975) Ingestion by planktonic grazers as a function of concentration of food. Limnol Oceanogr 20: 259–262

Munson RJ (1970) Turbidostats. In: Norris JR, Ribbons DW (eds) Methods in microbiology, vol 2. Academic Press, New York, pp 349–376

Myers J, Clark LB (1944) Culture conditions and the development of the photosynthetic mechanism. II. An apparatus for the continuous culture of *Chlorella*. J Gen Physiol 28: 103–112

Myers JH, Krebs CJ (1974) Population cycles in rodents. Sci Am 230 (6): 38–46

Nachtigall W (1977) Zur Bedeutung der Reynoldszahl und der damit zusammenhängenden strömungsmechanischen Phänomene in der Schwimmphysiologie und Flugbiophysik. Fortschr Zool 24: 13–56

Nathan HA, Laderman AD (1959) Rotifers as biological tools. Ann NY Acad Sci 77: 96–101

Naumann E (1923) Spezielle Untersuchungen über die Ernährungsbiologie des tierischen Limnoplanktons II (Copepoden, Rotatoria). Lunds Univ Arskr NF 19: 3–17

Nauwerck A (1963) Die Beziehungen zwischen Zooplankton und Phytoplankton im See Erken. Symp Bot Upps 17: 1–153

Neijssel OM, Tempest DW (1976) The role of energy-spilling reactions in the growth of *Klebsiella aerogenes* NCTC 418 in aerobic chemostat culture. Arch Microbiol 110: 305–315

Neill WE (1974) The community matrix and interdependence of the competition coefficients. Am Nat 108: 399–408

Neill WE (1975) Experimental studies of microcrustacean competition, community composition, and efficiency of resource utilization. Ecology 56: 809–826

Neill WE (1984) Regulation of rotifer densities by crustecean zooplankton in an oligotrophic mountain lake in British Columbia. Oecologia 61: 175–181

Neill WE (1985) The effects of herbivore competition upon the dynamics of *Chaoborus* predation. Arch Hydrobiol Beih (Ergebn Limnol) 21: 483–491

Neill WE (1988) Complex interactions in oligotrophic lake food webs: response to nutrient enrichment. In: Carpenter SR (ed) Complex interactions in lake communities. Springer, Berlin Heidelberg New York, pp 31–44

Nichols JD, Conley W, Batt B, Tipton AR (1976) Temporally dynamic reproductive strategies and the concept of r- and K-selection. Am Nat 110: 995–1005

Nicholson AJ (1954) An outline of the dynamics of animal populations. Aust J Zool 2: 9–65

Nisbet RM, Gurney WSC (1976) A simple mechanism for population cycles. Nature 263: 319–320

Novick A, Szilard L (1950) Description of the chemostat. Science 112: 715–716

Nümann W (1973) Versuch einer Begründung für den Wandel in der qualitativen und quantitativen Zusammensetzung des Fischbestandes im Bodensee während der letzten 60 Jahre und eine Bewertung der Besatzmassnahmen. Schweiz Z Hydrol 35: 206–238

Nusch EA (1978) Development of planktonic algae in the Ruhr river dependent on nutrient supply, water flow, irradiance and temperature. Verh Int Ver Limnol 20: 1837–1843

O'Connel MF, Andrews C (1987) Plankton ecology in relation to flushing rate in four Newfoundland ponds. Int Rev Gesamten Hydrobiol 72: 487–515

Oka A, Suzuki N, Watanabe T (1980) Effect of fatty acids in rotifers on growth and fatty acid composition of larval Ayu *Plecoglossus altivelis*. Bull Jpn Soc Sci Fish 46: 1413–1418

Oksanen L (1988) Ecosystem organization: mutualism and cybernetics or plain Darwinian struggle for existence? Am Nat 131: 424–444

O'Neill RV (1989) Perspectives in hierarchy and scale. In: Roughgarden J, May RM, Levin SA (eds) Perspectives in ecological theory. Princeton University Press, Princeton, pp 140–156

Orcutt JD Jr, Pace ML (1984) Seasonal dynamics of rotifer and crustacean zooplankton populations in a eutrophic monomictic lake with a note on rotifer sampling techniques. Hydrobiologia 119: 73–80

Paasche E (1973) Silicon and the ecology of marine plankton diatoms. II. Silicate uptake kinetics in five diatom species. Mar Biol 19: 262–269

Pace ML, Porter KG, Feig YS (1983) Species- and age-specific differences in bacterial resource utilization by two co-occurring cladocerans. Ecology 64: 1145–1156

Pace ML, McManus GB, Findlay SEG (1990) Planktonic community structure determines the fate of bacterial production in a temperate lake. Limnol Oceanogr 35: 795–808

Paine RT (1980) Food webs: linkage, interaction strength and community infrastructure. J Anim Ecol 49: 667–685

Paloheimo JE (1974) Calculation of instantaneous birth rates. Limnol Oceanogr 19: 692–694

Pappers U (1985) Untersuchung des Zooplanktons im Rottstausee Postmünster im Jahre 1984. Diploma Thesis, Fac Biology, University of Regensburg, 73 pp

Pascual E, Yúfera M (1983) Crecimento en cultivo de una cepa de *Brachionus plicatilis* O. F. Müller en funcion de la temperature y la salinidad. Invest Pesq 47: 151–159

Patridge D, Lopez PD, Johnston VS (1984) Computer programs as theories in biology. J Theor Biol 108: 539–564

Patten BC (1959) An introduction to the cybernetics of the ecosystem: the trophic dynamic aspect. Ecology 40: 221–231

Patten BC Odum EP (1981) The cybernetic nature of ecosystems. Am Nat 118: 886–895

Pauli HR (1990) Seasonal succession of rotifers in large lakes. In: Tilzer MM, Serruya C (eds) Large lakes. Ecological structure and function. Springer, Berlin Heidelberg New York, pp 459–474

Pearl R (1925) The biology of population growth. Knopf, New York, 260 pp

Pearl R, Reed LS (1920) On the rate of growth of the population of the United States since 1790 and its mathematical representation. Proc Natl Acad Sci USA 6: 275–288

Pejler B (1977) Experience with rotifer cultures based on *Rhodomonas*. Arch Hydrobiol Beih (Ergebn Limnol) 8: 264–266

Perram JW (1973) Relaxation times in bacterial culture and the approach to steady state. J Theor Biol 38: 571–578

Persson L, Andersson G, Hamrin SF, Johansson L (1988) Predator regulation and primary production along the productivity gradient of temperature lake ecosystems. In: Carpenter SR (ed) Complex interactions in lake communities. Springer, Berlin Heidelberg New York, pp 45–65

Peters RH (1983) The ecological implications of body size. Cambridge University Press, New York, 329 pp

Peters RH (1984) Methods for the study of feeding, grazing and assimilation by zooplankton. In: Downing JA, Rigler FH (eds) A manual on methods for the assessment of secondary productivity in fresh waters. Blackwell, Oxford, pp 336–412

Phillips JN, Myers J (1954) Measurement of algal growth under controlled steady-state conditions. Plant Physiol 29: 148–161

Pianka FR (1970) on r and K selection. Am Nat 104: 592–597

Pianka FR (1972) r and K selection or b and d selection. Am Nat 106: 581–588

Pianka FR (1978) Evolutionary ecology. Harper & Row, New York, 397 pp

Pickett AM (1982) Growth in a changing environment. In: Bazin MJ (ed) Microbial population dynamics. CRC Press, Boca Raton, pp 91–124

Pielou EC (1974) Population and community ecology: principles and methods. Gordon and Breach, New York, 424 pp

Pijanowska J (1990) Cyclomorphosis in *Daphnia*: an adaptation to avoid invertebrate predation. Hydrobiologia 198: 41–50

Pilarska J (1977a) Eco-physiological studies on *Brachionus rubens* Ehrbg (Rotatoria). I. Food selectivity and feeding rate. Pol Arch Hydrobiol 24: 319–328

Pilarska J (1977b) Eco-physiological studies on *Brachionus rubens* Ehrbg (Rotatoria). II. Production and respiration. Pol Arch Hydrobiol 24: 329–341

Pilarska J (1977c) Eco-physiological studies on *Brachionus rubens* Ehrbg (Rotatoria). III. Energy balances. Pol Arch Hydrobiol 24: 343–354

Pirt SJ (1965) The maintenance energy of bacteria in growing cultures. Proc R Soc B 163: 224–231

Pirt SJ (1975) Principles of microbe and cell cultivation. Blackwell, Oxford, 274 pp

Platzek J (1986) Die Nahrungsaufnahme planktischer Crustaceen: Grazing-Experimente in einem eutrophen Baggersee. Diploma Thesis, Fac Biology, University of Munich, 217 pp

Porter KG (1977) The plant-animal interface in freshwater ecosystems. Am Sci 65: 159–170

Porter KG, Feig YS (1980) The use of DAPI for identifying and counting aquatic microflora. Limnol Oceanogr 25: 943–948

Porter KG, Gerritsen J, Orcutt JD Jr (1982) The effect of food concentration on swimming patterns, feeding behavior, ingestion, assimilation, and respiration by *Daphnia*. Limnol Oceanogr 27: 935–949

Porter KG, Feig YS, Vetter EF (1983a) Morphology, flow regimes and filtering rate of *Daphnia*, *Ceriodaphnia* and *Bosmina* fed natural bacteria. Oecologia 58: 156–163

Porter KG, Orcutt JD, Gerritsen J (1983b) Functional response and fitness in a generalist filter feeder, *Daphnia magna* (Cladocera, Crustacea). Ecology 64: 735–742

Porter KG, Paerl H, Hodson R, Pace M, Priscu J, Riemann B, Scavia D, Stockner J (1988) Microbial interactions in Lake food webs. In: Carpenter SR (ed) Complex interactions in lake communities. Springer, Berlin Heidelberg New York, pp 209–227

Pourriot R (1965) Recherches sur l'ecologie des rotifères. Vie Milieu Suppl 21, 224 pp

Pourriot R (1970) Quelques *Trichocera* (Rotifères) et leurs regimes alimetaires. Ann Hydrobiol 1: 155–171

Pourriot R (1973a) Rapports entre la température, la taille des adults, la longueur des ouefs et le taux de développement embryonnaire chez *Brachionus calyciflorus* Pallas (Rotifères). Ann Hydrobiol 4: 103–115

Pourriot R (1973b) Influence de la teneur en protéines, de la température et du jeune sur la respiration de rotifères héléoplanctoniques. Verh Int Ver Limnol 18: 1429–1433

Pourriot R (1977) Food and feeding habit of rotifera. Arch Hydrobiol Beih (Ergebn Limnol) 8: 243–260

Pourriot R (1980) Workshop on culture techniques of rotifers. Hydrobiologia 73: 33–35

Pourriot R, Deluzarches M (1971) Recherches sur la biologie des Rotifères. II. Influence de la température sur la durée du développement embryonnaire et post-embryonnaire. Ann Limnol 7: 25–52

Pourriot R, Rougier C (1975) Dynamique d'une population expérimentale de *Brachionus dimidiatus* (Bryce) (Rotifère) en fonction de la nourriture et de la température. Ann Limnol 11: 125–143

Powell EO (1956) Growth rate and generation time of bacteria with special reference to continuous culture. J Gen Microbiol 15: 492–511

Powell EO (1958) Criteria for the growth of contaminants and mutants in continuous culture. J Gen Microbiol 18: 259–268

Powell T, Richerson PJ (1985) Temporal variation, spatial heterogeneity, and competition for resources in plankton systems: a theoretical model. Am Nat 125: 431–464

Pratt DM (1943) Analysis of population development in *Daphnia* at different temperatures. Biol Bull 85: 116–140

Precht E (1983) Beziehungen der Populationsdynamik des Crustaceenplanktons zu zwei Größenkategorien des Phytoplanktons und zu den Bakterien in einem eutrophen Baggersee. Diploma Thesis, Fac Biology, University of Munich, 88 pp

Price PW, Gaud WS, Slobodchikoff CN (1984) Introduction: is there a new ecology? In: Price PW, Slobodchikoff CN, Gaud WS (eds) A new ecology. Novel approach to interactive systems. Wiley, New York, pp 1–11

Prigogine I (1978) Time, structure and fluctuations. Science 201: 777–785

Proper G, Garver JM (1966) Mass culture of the protozoa *Colpoda steinii*. Biotechnol Bioeng 8: 279–296

Provasoli L, Pintner IJ (1960) Artificial media for freshwater algae: problems and suggestions. In: Tyron CA Jr, Hartmann RT (eds) The ecology of algae. Special Publ 2, Pymatuning Laboratory of Field Biology, University of Pittsburg, pp 84–96

Provasoli L, McLaughlin JJA, Droop MR (1957) The development of artificial media for marine algae. Arch Mikrobiol 25: 392–428

Purcell EM (1977) Life at low Reynold numbers. Am J Phys 45: 3–11

Quinian AV (1986) A semicontinuous culture model that links cell growth to extracellular nutrient concentration. Biotechnol Bioeng 28: 1455–1461

Rao TR, Sarma SSS (1986) Demographic parameters of *Brachionus patulus* Müller (Rotifera) exposed to sublethal DDT concentrations at low and high food levels. Hydrobiologia 139: 193–200

Ratkowsky DA (1983) Non-linear regression modeling: a unified practical approach. Dekker, New York, 276 pp

Reed JV (1979) A general technique for the culture of rotifers. Microsc J Queckett Microsc Club 33: 448–459

Regan DL, Roper GH (1971) Response of continuous cultures to stimuli in glucose feed rate and dilution rate. Biotechnol Bioeng 13: 815–824

Reich S (1989) Die Nahrungsaufnahme planktischer Rotatorien: Grazing-Experimente in einem eutrophen Baggersee. Diploma Thesis, Fac Biology, University of Munich, 159 pp

Remmert H (1989) Ökologie. Ein Lehrbuch, 4th edn. Springer, Berlin Heidelberg New York, 374 pp

Rezeq TA, James CM (1987) Production and nutritional quality of the rotifer *Brachionus plicatilis* fed marine *Chlorella* sp. at different cell densities. Hydrobiologia 147: 257–261

Reznick DN (1985) Costs of reproduction: an evaluation of the empirical evidence. Oikos 44: 257–267

Reznick DN, Perry R, Travis J (1988) Measuring the cost of reproduction: a comment on papers by Bell. Evolution 40: 1338–1344

Rhee GY (1980) Continuous culture in phytoplankton ecology. In: Droop MR, Jannasch HW (eds) Advances in aquatic microbiology, vol 2. Academic Press, London, pp 151–203

Richman S (1958) The transformation of energy by *Daphnia pulex*. Ecol Monogr 28: 273–291

Richman S, Dodson SI (1983) The effect of food quality on feeding and respiration by *Daphnia* and *Diaptomus*. Limnol Oceanogr 28: 984–956

Ričica J (1966) Technique of continuous laboratory cultivations. In: Malek I, Fencl Z (eds) Theoretical and methodological basis of continuous culture of microorganisms. Publ House of the Czechoslovak Academy of sciences, Prague, and Academic Press, New York, pp 157–313

Rigler FM (1961) The relation between concentration of food and feeding rate of *Daphnia magna* Straus. Can J Zool 39: 857–868

Robertson JR, Salt GW (1981) Responses in growth, mortality and reproduction to variable food levels by the rotifer, *Asplanchna girodi*. Ecology 62: 1585–1596

Röhler R (1973) Biologische Kybernetik. Teubner, Stuttgart, 180 pp

Romanovsky YE (1984) Food limitation and life history strategies in cladoceran crustaceans. Arch Hydrobiol Beiheft (Ergebn Limnol) 21: 363–372

Romanovsky YE, Feniova IY (1985) Competition among Cladocera: effect of different levels of food supply. Oikos 44: 243–252

Romanovsky YE, Polishchuk LV (1982) A theoretical approach to calculation of secondary production at the population level. Int Rev Gesamten Hydrobiol 67: 341–359

Ross PE, Munawar M (1981) Preference for nannoplankton in Lake Ontario zooplankton grazing. J Great Lakes Res 7: 65–67

Rothbucher F (1988) Einfluß der Futterkonzentration auf Eivolumen, Körpervolumen, Embryonal- und Juvenilentwicklungszeit bei *Brachionus angularis* (Rotatoria). Diploma Thesis Fac Biology University of Munich, 57 pp

Rothhaupt KO (1985) A model approach to the population dynamics of the rotifer *Brachionus rubens* in two-stage chemostat culture. Oecologia 65: 252–259

Rothhaupt KO (1988a) Ökophysiologische Grundlagen der Populationsdynamik und der zwischenartlichen Konkurrenz bei Rädertieren der Gattung *Brachionus*. Thesis Math Nat Fac University of Kiel, 123 pp

Rothhaupt KO (1988b) Mechanistic resource competition theory applied to laboratory experiments with zooplankton. Nature 333: 660–662

Rothhaupt KO (1990a) Population growth rates of two closely related rotifer species: effects of food quantity, particle size, and nutritional quality. Freshwater Biol 23: 561–570

Rothhaupt KO (1990b) Differences in particle size-dependent feeding efficiencies of closely related rotifer species. Limnol Oceanogr 35: 16–23

Rothhaupt KO (1990c) Changes of the functional responses of the rotifers *Brachionus rubens* and *Brachionus calyciflorus* with particle sizes. Limnol Oceanogr 35: 24–32

Rothhaupt KO (1990d) Resource competition of herbivorous zooplankton: a review of approaches and perspectives. Arch Hydrobiol 118: 1–29

Rothhaupt KO (1991) The influence of toxic and filamentous blue-green algae on feeding and population growth of the rotifer *Brachionus rubens*. Int Revue Gesamten Hydrobiol 76: 67–72

Rothhaupt KO and Lampert W (1992) Growth-rate dependent feeding rates in *Daphnia pulicaria* and *Brachionus rubens:* adaptation to intermediate time-scale variations in food abundance. J Plankton Res 14: 737–751

Rothschild BJ (1991) Food-signal theory: population regulation and the functional response. J Plankton Res 13: 1123–1135

Rubenstein DI, Koehl MAR (1977) The mechanism of filter-feeding: some theoretical considerations. Am Nat 111: 981–994

Ruhrverband (1980) Ruhrwassergüte im Wasserwirtschaftsjahr 1979. Jahresbericht, Essen, 86 pp

Ruttner F (1962) Grundriss der Limnologie, 3rd edn. de Gruyter, Berlin, 332 pp

Ruttner-Kolisko A (1972a) Rotatoria. In: Elster HJ, Ohle W (eds) Die Binnengewässer 26: Das Zooplankton der Binnengewässer, 1. Teil. E. Schweizerbart, Stuttgart, pp 99–324

Ruttner-Kolisko A (1972b) Der Einfluß von Temperatur und Salzgehalt des Mediums auf Stoffwechsel- und Vermehrungsintensität von *Brachionus plicatilis* (Rotatoria). Dtsch Zool Ges 65: 89–95

Sachs L (1992) Angewandte Statistik Anwendung statistischer Methoden, 7. Aufl. Springer, Berlin Heidelberg New York, 846 pp

Salt GW, Sabbadini GF, Commins ML (1978) Trophi morphology relative to food habits in six species of rotifers (Asplanchnidae). Trans Am Micros Soc 97: 496–485

Sanders RW, Porter KG (1990) Bacterivorous flagellates as food resources for the freshwater crustacean zooplankter *Daphnia ambigua*. Limnol Oceanogr 35: 188–191

Sarma SSS, Rao TR (1987) Effect of food level on body size and egg size in a growing population of the rotifer *Brachionus patulus* Müller. Arch Hydrobiol 111: 245–253

Scheda SM, Cowell BC (1988) Rotifer grazers and phytoplankton: seasonal experiments and natural communities. Arch Hydrobiol 114: 31–44

Scheffer M (1991) Should we expect strange attractors behind plankton dynamics—and if so, should we bother? J Plankton Res 13: 1291–1305

Schiemer F (1985) Bioenergetic niche differentiation of aquatic invertebrates. Verh Int Ver Limnol 22: 3014–3018

Schlosser HJ, Anger K (1982) The significance of some methodological effects on filtration and ingestion rates of the rotifer *Brachionus plicatilis*. Helgol Meeresunters 35: 215–225

Schlüter M (1980) Mass culture experiments with *Brachionus rubens*. Hydrobiologia 73: 45–50

Schlüter M (1984) Kontinuierliche Massenkultur des planktischen Rotators *Brachionus rubens*. Thesis, University of Essen, 194 pp

Schlüter M, Soeder CJ, Groeneweg J (1987) Growth and food conversion of *Brachionus rubens* in continuous culture. J Plankton Res 9: 761–783

Schmid-Arya JM (1992) The biochemical composition and calorific content of a rotifer and its algal food: comparison of a two stage chemostat and batch culture. Oecologia 92: 327–338

Schober U (1980) Kausalanalytische Untersuchungen der Abundanzschwankungen des Crustaceen-Planktons im Bodensee. Thesis, Fac Biology, University of Freiburg/Brsg, 160 pp

Schroeder LA (1981) Consumer growth efficiencies: their limits and relationships to ecological energetics. J Theor Biol 93: 805–828

Schulze KL, Lipe RS (1964) Relationship between substrate concentration, growth rate, and respiration rate of *Escherichia coli* in continuous culture. Arch Mikrobiol 48: 1–20

Schwerdtfeger F (1978) Lehrbuch der Tierökologie. Paul Parey, Hamburg, 384 pp

Scott AP, Middleton C (1979) Unicellular algae as food for turbot (*Sophthalmus maximus* L.) larvae—the importance of dietary long-chain polyunsaturated fatty acids. Aquaculture 18: 227–240

Scott JM (1977) Rotifer reproduction under controlled experimental conditions. Arch Hydrobiol Beih (Ergebn Limnol) 8: 169–171

Scott JM (1980) Effect of growth rate of the food alga on the growth/ingestion efficiency of a marine herbivore. J Mar Biol Assoc UK 60: 681–702

Scott JM (1981) The vitamin B_{12}-requirement of the marine rotifer *Brachionus plicatilis*. J Mar Biol Assoc UK 61: 983–994

Scott JM (1988) Effects of growth rate on the physiological rates of a chemostat-grown rotifer *Encentrum linnhei*. J Mar Biol Assoc UK 68: 165–177

Seale DB (1980) Influence of amphibian larvae on primary production, nutrient flux, and competition in a pond ecosystem. Ecology 61: 1531–1550

Seale DB (1987) Amphibia. In: Pandian TJ, Vernberg FJ (eds) Animal energetics. Academic Press, New York, pp 467–552

Seale DB, Beckvar N (1980) The comparative ability of anuran larvae (genera: *Hyla*, *Bufo*, and *Rana*) to ingest suspended blue-green algae. Copeia 1980: 495–503

Seale DB, Boraas ME, Warren G (1987) Effects of sodium and phosphate on growth of cyanobacteria. Water Res 21: 625–631

Segel IH (1975) Enzyme kinetics. Behavior and analysis of rapid equilibrium and steady-state enzyme systems. Wiley, New York, 957 pp

Serra M, Miracle MR (1985) Enzyme polymorphism in *Brachionus plicatilis* populations from several Spanish lagoons. Vert Int Ver Limnol 22: 2991

Shapiro J (1980) The importance of trophic-level interactions to the abundance and species composition of algae in lakes. In: Barica J, Mur LR (eds) Hypertrophic ecosystems. Junk, The Hague, pp 105–116

Siebeck O (1986) Elemente der Steuerung und Regulation in der Pelagialbiozönose. Eine

Einführung in das Thema des Symposiums. In: Siebeck O (ed) Elemente der Steuerung und der Regulation in der Pelagialbiozönose. Akademie für Naturschutz und Landschaftspflege, Laufen/Salzach. Laufener Seminarbeiträge 2/86, pp 15–21
Siegfried CA (1991) The pelagic rotifer community of an acidic clearwater lake in the Adirondack Mountains of New York State. Arch Hydrobiol 122: 441–462
Siegfried CA, Sutherland JW, Quinn SO, Bloomfield JA (1984) Lake acidification and the biology of Adirondack lakes. I. Rotifer communities. Verh Int Ver Limnol 22: 549–558
Silliman PR, Gutsell (1958) Experimental exploitation of fish populations. Fish Bull Fish Wildl Serv US 58: 215–241
Sleigh MA, Blake JR (1977) Methods of ciliary propulsion and their size limitation. In: Pedley TJ (ed) Scale effects in animal locomotion. Academic Press, London, pp 243–256
Slobodkin LB (1962) Energy in animal ecology. Adv Ecol Res 1: 69–101
Slobodkin LB, Richman S (1956) The effect of removal of fixed percentages of newborn on size and variability in populations of *Daphnia pulicaria*. Limnol Oceanogr 1: 209–237
Smith FE (1952) Experimental methods in population dynamics. A critique. Ecology 33: 441–471
Smith FE (1963) Population dynamics in *Daphnia magna* and a new model for population growth. Ecology 44: 651–663
Smith HL (1981) Competitive coexistence in an oscillating chemostat. SIAM J Appl Math 40: 498–522
Smith SW, Cooper SD (1982) Competitive among cladocerans. Ecology 63: 1004–1015
Smith VH (1991) Competition between consumers. Limnol Oceanogr 36: 820–823
Smith-Gill SJ, Gill DE (1978) Curvilinearities in the competition equations: an experiment with ranid tadpoles. Am Nat 112: 557–570
Smouse PE (1981) Mathematical models for continuous culture growth dynamics subsisting on a heterogeneous resource base. II. Predation and trophic structure. Theor Popul Biol 20: 127–149
Smyly WJD (1978) The crustacean zooplankton of Grasmere before and after a change in sewage effluent treatment. Int Rev Gesamten Hydrobiol 63: 389–403
Snell TW (1980) Blue-green algae and selection in rotifer populations. Oecologia 46: 343–346
Snell TW, Boyer EM (1988) Thresholds for mictic female production in the rotifer *Brachionus plicatilis* (Müller). J Exp Mar Biol Ecol 124: 73–85
Snell TW, King CE (1977) Life span and fecundity patterns in rotifers: the cost of reproduction. Evolution 31: 882–890
Snell TW, Biebrich CJ, Fuerst R (1983) The effect of green and blue-green algal diets on the reproductive rate of the rotifer *Brachionus plicatilis*. Aquaculture 31: 21–30
Snell TW, Childress MJ, Boyer EM, Hoff FH (1987) Assessing the status of rotifer mass cultures. J World Aquacult Soc 18: 270–277
Solomon ME (1958) Meaning of density dependence and related terms in population dynamics. Nature 181: 1778–1780
Solomons GL (1972) Improvements in the design and operation of the chemostat. J Appl Chem Biotechnol 22: 217–228
Sommer U (1984) The paradox of the plankton: fluctuation of phosphorus availability maintain diversity of phytoplankton in flow-through cultures. Limnol Oceanogr 29: 633–636
Sommer U (1985) Comparison between steady state and non-steady state competition: experiments with natural zooplankton. Limnol Oceanogr 30: 335–346
Sommer U (1988a) Phytoplankton succession in microcosm experiments under simultaneous grazing pressure and resource limitation. Limnol Oceanogr 33: 1037–1054
Sommer U (1988b) Does nutrient competition among phytoplankton occur in situ? Verh Int Ver Limnol 23: 707–712
Sommer U (ed) (1989a) Plankton ecology. Succession in plankton communities, Springer, Berlin Heidelberg New York (Brock/Springer Series in Contemporary Bioscience), 369 pp
Sommer U (1989b) The role of competition for resources in phytoplankton succession. In: Sommer U (ed) Plankton ecology. Succession in plankton communities. Springer, Berlin Heidelberg New York, pp 57–106
Sommer U (1989c) Nutrient status and nutrient competition of phytoplankton in a shallow, hypertrophic lake. Limnol Oceanogr 34: 1162–1173

Sommer U (1989d) Toward a Darwinian ecology of plankton. In: Sommer U (ed) Plankton ecology. Succession in plankton communities. Springer, Berlin Heidelberg New York, pp 1–8

Sommer U, Gliwicz ZM, Lampert W, Duncan A (1986) The PEG-model of seasonal succession of planktonic events in fresh waters. Arch Hydrobiol 106: 433–471

Sorgeloos P, Persoone G (1973) A culture system for *Artemia, Daphnia*, and other invertebrates, with continuous separation of the larvae. Arch Hydrobiol 72: 133–138

Sorokin C, Krauss RW (1958) The effects of light intensity on the growth rates of green algae. Plant Physiol 33: 109–113

Stachowiak H (1973) Allgemeine Modelltheorie. Springer, Berlin Heidelberg New York, 494 pp

Starkweather PL (1980) Aspects of the feeding behavior and trophic ecology of suspension-feeding rotifers. Hydrobiologia 73: 63–72

Starkweather PL (1981) Trophic relationships between the rotifer *Brachionus calyciflorus* and the blue-green alga *Anabaena flos-aquae*. Verh Int Ver Limnol 21: 1507–1514

Starkweather PL (1987) Rotifera. In: Pandian TJ, Vernberg SJ (eds) Animal energetics. Academic Press, New York, pp 159–183

Starkweather PL, Gilbert JJ (1977a) Feeding in the rotifer *Brachionus calyciflorus*. 1. The importance of gut passage time. Arch Hydrobiol Beih (Ergebn Limnol) 8: 261–263

Starkweather PL, Gilbert JJ (1977b) Feeding in the rotifer *Brachionus calyciflorus*. 2. Effect of food density on feeding rates using *Euglena gracilis* and *Rhodotorula glutinis*. Oecologia 28: 133–139

Starkweather PL, Keller PE (1983) Utilization of cyanobacteria by *Brachionus calyciflorus: Anabaena flos-aquae* (NRC-44-1) as a sole or complementary food source. Hydrobiologia 104: 373–377

Starkweather PL, Gilbert JJ, Frost TM (1979) Bacterial feeding by the rotifer *Brachionus calyciflorus*: clearance and ingestion rate, behavior and population dynamics. Oecologia 44: 26–30

Steele JH (1974) The structure of marine ecosystems. Harvard University Press, Cambridge

Steiff B (1981) Das Zooplankton des Fasaneriesees im Winterhalbjahr 1979/1980. Diploma Thesis, Fac Biology, University of Munich, 58 pp

Stemberger RS (1981) A general approach to the culture of planktonic rotifers. Can J Fish Aquat Sci 38: 721–724

Stemberger RS (1987) The potential growth of *Ascomorpha ecaudis*. Hydrobiologia 147: 297–301

Stemberger RS, Evans MS (1984) Rotifer seasonal succession and copepod predation in Lake Michigan (USA). J Great Lakes Res 10: 417–428

Stemberger RS, Gilbert JJ (1984) Body size, ration level, and population growth in *Asplanchna*. Oecologia 64: 355–359

Stemberger RS, Gilbert JJ (1985) Body size, food concentration, and population growth in planktonic rotifers. Ecology 66: 1151–1159

Stemberger RS, Gilbert JJ (1987a) Rotifer threshold food concentrations and the size-efficiency hypothesis. Ecology 68: 181–187

Stemberger RS, Gilbert JJ (1987b) Defenses of planktonic rotifers against predators. In: Kerfoot WC, Sih A (eds) Predation, direct and indirect impacts on aquatic communities. University Press of New England, Hanover, pp 227–239

Stenson JAE (1982) Fish impact on rotifer community structure. Hydrobiologia 87: 57–64

Stenson JAE (1983) Changes in the relative abundance of *Polyarthra vulgaris* and *P. dolichoptera*, following the elimination of fish. Hydrobiologia 104: 269–273

Stenson JAE, Bohlin T, Henrikson L, Nilsson BI, Nyman HG, Ocarson HG, Larsson P (1978) Effects of fish removal from a small lake. Verh Int Ver Limnol 20: 794–801

Stephanopoulos G, Fredrickson AG, Aris R (1979) The growth of competing microbial populations in a CSTR with periodically varying inputs. AIChE J 25: 864–873

Sterner RW (1989) The role of grazers in phytoplankton succession. In: Sommer U (ed) Plankton ecology, succession in plankton communities. Springer, Berlin Heidelberg New York, pp 107–170

Stewart FM, Levin BR (1973) Partitioning of resources and the outcome of interspecific competition: a model and some general considerations. Am Nat 107: 171–198

Stich HB (1985) Untersuchungen zur tagesperiodischen Vertikalwanderung planktischer Crustaceen im Bodensee. Thesis, Fac Biology, University of Freiburg/Br, 282 pp

Stoecker DK, Capuzzo JM (1990) Predation on protozoa: its importance to zooplankton. J Plankton Res 12: 891–908

Stubbs M (1977) Density dependence in the life cycles of animals and its importance in K- and r-strategies. J Anim Ecol 46: 677–688

Stugren B (1986) Grundlagen der Allgemeinen Ökologie, 4th edn. Gustav Fischer, Stuttgart, 356 pp

Sudo R, Kobayashi K, Aiba S (1975) Some experiments and analysis of a predator-prey model: interaction between *Colpidium campylum* and *Alcaligenes faecalis* in continuous and mixed culture. Biotechnol Bioeng 17: 167–184

Sulkin SD (1975) The significance of diet in the growth and development of larvae of the blue crab *Callinectes sapidus* under laboratory conditions. J Exp Mar Biol Ecol 20: 119–135

Suttle CA, Harrison PJ (1986) Phosphate uptake rates of phytoplankton assemblages grown at different dilution rates in semicontinuous culture. Can J Fish Aquat Sci 43: 1474–1481

Swift MC, Fedorenko AY (1975) Some aspects of prey capture by *Chaoborus* larvae. Limnol Oceanogr 20: 418–425

Taub FB (1980) Use of continuous culture technique to control nutritional quality. In: Shelef G, Soeder CJ (eds) Algae biomass. Elsevier/North Holland Biomedical Press, Amsterdam, pp 708–721

Taub FB, McKenzie DH (1973) Continuous cultures of an alga and its grazer. Bull Ecol Res Commun Stockholm 17: 371–377

Taylor BE (1985) Effects of food limitation on growth and reproduction of *Daphnia*. Arch Hydrobiol Beih (Ergebn Limnol) 21: 285–296

Taylor PA, Williams PJLeB (1975) Theoretical studies on the coexistence of competing species under continuous-flow conditions. Can J Microbiol 21: 90–98

Taylor RJ (1984) Predation. Chapman and Hall, New York 166 pp

Taylor WD (1978) Growth responses of ciliate protozoa to the abundance of their bacterial prey. Microb Ecol 4: 204–216

Teissier G (1936) Les lois quantitatives de la croissance. Ann Physiol Physicochim Biol 12: 527–573

Tempest DW, Herbert D (1965) Effect of dilution rate and growth-limiting substrate on the metabolic activity of *Torula utilis* cultures. J Gen Microbiol 41: 143–150

Tempest DW, Hunter JR, Sykes J (1965) Magnesium limited growth of *Aerobacter aerogenes* in a chemostat. J Gen Microbiol 39: 355–366

Terry O, Edmunds LN Jr (1969) Semi-continuous culture and monitoring system for temperature-synchronized *Euglena*. Biotechnol Bioeng 11: 745–756

Tessier AJ, Henry LL, Goulden CE, Durand MW (1983) Starvation in *Daphnia*: energy reserves and reproductive allocation. Limnol Oceanogr 28: 667–676

Theilacker GH, McMaster MF (1971) Mass culture of the rotifer *Brachionus plicatilis* and its evaluation as a food for larval anchovies. Mar Biol 10: 183–188

Thienemann A (1918) Lebensgemeinschaft und Lebensraum. Naturwiss Wochenschr 33 (NF 17): 297–303

Thompson RJ, Bayne BL (1974) Some relationship between growth, metabolism and food in the mussel *Mytilus edulis*. Mar Biol 27: 317–326

Threlkeld ST (1976) Starvation and the size structure of zooplankton communities. Freshwater Biol 6: 489–497

Threlkeld S, Choinski EM (1987) Rotifers, cladocerans and planktivorous fish: what are the major interactions? Hydrobiologia 147: 239–243

Tilman D (1981) Experimental tests of resource competition theory using four species of Lake Michigan algae. Ecology 62: 802–815

Tilman D (1982) Resource competition and community structure. Princeton University Press, Princeton, 269 pp

Tilman D (1987) The importance of the mechanisms of interspecific competition. Am Nat 129: 769–774

Tilman D, Kilham SS, Kilham P (1982) Phytoplankton community ecology: the role of limiting nutrients. Annŭ Rev Ecol Syst 13: 349–372

Toman M, Rejic M (1985) The influence of certain types of waste water upon the rotifer biocoenosis in activated sludge. Z Wasser Abwasser Forsch 18: 169–177

Toth LG, Zankai NP (1985) Feeding of *Cyclops vicinus* (Uljanin) (Copepoda: Cyclopoida) in Lake Balaton on the basis of gut content analysis. Hydrobiologia 122: 251–260

Toth LG, Zankai NP, Messner OM (1987) Alga consumption of four dominant planktonic crustaceans in Lake Balaton (Hungary). Hydrobiologia 145: 323–332

Townsend CR, Calow P (1981) Physiological ecology: an evolutionary approach to resource use. Sinauer Assoc, Sunderland, 393 pp

Trotta P (1980) A simple and inexpensive system for continuous culture of *Brachionus plicatilis* Müller as a basis for mass production. In: Shelef G, Soeder CJ (eds) Algae biomass. Elsevier/North Holland Biomedical Press, Amsterdam, pp 307–313

Tsuchiya HM, Drake JF, Jost JL, Fredrickson AG (1972) Predator-prey interactions of *Dictystelium discoideum* and *Escherichia coli* in continuous culture. J Bacteriol 110: 1147–1153

Turner AM, Mittelbach GG (1990) Predator avoidance and community structure: interactions among piscivores, planktivores, and plankton. Ecology 71: 2241–2254

Uhlmann D (1971) Influence of dilution, sinking and grazing rate on phytoplankton populations of hyperfertilized ponds and micro-ecosystems. Mitt Int Ver Limnol 19: 100–124

Uhlmann D (1972) Das Staugewässer als offenes System und als Reaktor. Verh Int Ver Limnol 18: 761–778

Uhlmann D (1985) Scaling of microcosms and the dimensional analysis of lakes. Int Rev Gesamten Hydrobiol 7: 47–62

Uhlmann D, Mihan H, Kuchta M (1980) Experimentation with plankton ecosystems on the basis of semicontinuous or continuous cultivation methods. Contin Cultiv Microorganisms, Proc 7th Symp Prague, July 10–14, 1978, pp 383–394

USEPA (1979) Methods for chemical analysis of water and wastes (EPA-600-4-79-020). Environmental monitoring and support laboratory. Office of research and Development, Cincinnati, Ohio

Vancil JE (1983) A method for the laboratory culture of the planktonic rotifer *Keratella cochlearis* (Gosse). Hydrobiologia 107: 47–50

Vanni MJ (1986a) Competition in zooplankton communities: suppression of small species by *Daphnia pulex*. Limnol Oceanogr 31: 1039–1056

Vanni MJ (1986b) Fish predation and zooplankton demography: indirect effects. Ecology 67: 337–354

Vanni MJ (1987) Effects of food availability and fish predation on a zooplankton community. Ecol Monogr 57: 61–88

Vanni MJ, Findlay DL (1990) Trophic cascades and phytoplankton community structure. Ecology 71: 921–937

Varesci E, Jacobs J (1984) The ecology of Lake Nakuru (Kenya). V. Production and consumption of consumer organisms. Oecologia 61: 83–98

Veldkamp H (1976) Mixed culture studies with the chemostat. In: Dean ACR, Ellwood DC, Evans CGT, Melling J (eds) Continuous culture 6: applications and new fields. Ellis Horwood, Chichester, pp 315–328

Veldkamp H (1977) Ecological studies with the chemostat. In: Alexander M (ed) Advances in microbial ecology. Plenum Press, New York, pp 59–94

Veldkamp H, Jannasch HW (1972) Mixed culture studies with the chemostat. In: Dean ACR, Pirt SJ, Tempest DW (eds) Environmental control of cell synthesis and function. Academic Press, London, pp 105–123

Veldkamp H, Kuenen JG (1973) The chemostat as a model system for ecological studies. Bull Ecol Res Commun 17: 347–355

Verhulst PF (1838) Notice sur le loi que la population suit dans son accroissement. Corresp Math Phys 10: 113–121

Vijverberg K (1989) Culture techniques for studies on the growth, development and reproduction of copepods and cladocerans under laboratory and *in situ* conditions: a review. Freshwater Biol 21: 317–373

Volterra V (1926) Variazione e fluttuazini del numero d'individui in specie animali conviventi. Mem Acad Naz Lincei (Ser VI) 2: 31–113

Volterra V (1931) Variations and fluctuations of the number of individuals in animal species living together. Translated from the Italian in: Chapman RN (ed) Animal ecology, McGraw-Hill, New York, pp 409–448

Vucković M (1981) Einfluß der Temperatur und der Nahrungsquantität auf die Populationsdynamik des planktischen Rädertiers *Brachionus rubens* Ehrenberg. Diploma Thesis, Fac Biology, University of Frankfurt/M, 75 pp

Waldon MG (1975) Competition models. Am Nat 109: 487–490

Walker KF (1981) A synopsis of ecological information on the saline lake rotifer *Brachionus plicatilis* Müller 1786. Hydrobiologia 81: 159–167

Wallace R, Snell TW (1991) *Rotifera*. In: Thorp JH, Covich AP (eds) Ecology and classification of North American freshwater invertibrates. Academic Press, New York, pp 187–248

Walz N (1978a) The energy balance of the freshwater mussel *Dreissena polymorpha* Pallas in Laboratory experiments and in Lake Constance. III. Growth under standard conditions. Arch Hydrobiol Suppl 55: 121–141

Walz N (1978b) The energy balance of the freshwater mussel *Dreissena polymorpha* Pallas in laboratory experiments and in Lake Constance. IV. Growth in Lake Constance. Arch Hydrobiol Suppl 55: 142–156

Walz N (1983a) Individual culture and experimental population dynamics of *Keratella cochlearis* (Rotatoria). Hydrobiologia 107: 35–43

Walz N (1983b) Continuous culture of the pelagic rotifer *Keratella cochlearis* and *Brachionus angularis*. Arch Hydrobiol 98: 70–92

Walz N (1986a) Wie werden Rotatorienpopulationen reguliert? Beziehungen zwischen Rotatorien, Crustaceen und Phytoplankton. In: Siebeck O (ed) Elemente der Steuerung und der Regulation in der Pelagialbiozönose. Akademie für Naturschutz und Landschaftpflege, Laufener Seminarbeiträge 2/86, pp 80–90

Walz N (1986b) Funktionelle Abhängigkeit von Größen der Populationsdynamik und Stoffbilanz bei Rotatorien von der Algenkonzentration. Verh Dtsch Zool Ges 79: 323–324

Walz N (1987a) Comparative population dynamics of *Brachionus angularis* and *Keratella cochlearis*. Hydrobiologia 147: 209–213

Walz N (1987b) Stoffumsatz und Kinetik von Regulationsprozessen bei Zooplankton-Populationen. Analysen und Modelle in Rotatorien-Chemostaten und im Plankton eines Sees. Habilitation Thesis, Fac Biology, University of Munich, 225 pp

Walz N (1992) Model simulations of continuous rotifer cultures. Hydrobiologia (in press)

Walz N, Gschloessl T (1988) Functional response of ingestion and filtration rate of the rotifer *Brachionus angularis* to the food concentration. Verh Int Ver Limnol 23: 1993–2000

Walz N, Rothbucher F (1991) Effect of food concentration on body size, egg size, and population dynamics of *Brachionus angularis* (Rotatoria). Verh Int Ver Limnol 24: 2750–2753

Walz N, Elster HJ, Mezger M (1987) The development of the rotifer community structure in Lake Constance during its eutrophication. Arch Hydrobiol Suppl (Monogr Beitr) 74: 457–487

Walz N, Gschloessl T, Hartmann U (1989) Temperature aspects of ecological bioenergetics in *Brachionus angularis* (Rotatoria). Hydrobiologia 186/187: 363–369

Wangersky PJ (1978) Lotka-Volterra population models. Annu Rev Ecol Syst 9: 189–218

Ware DM (1975) Relation between egg size, growth and natural mortality of larval fish. J Fish Res Board Can 32: 2503–2512

Watanabe T, Kitajima C, Fujita S (1983) Nutritional values of live organisms used in Japan for mass propagation of fish. A review. Aquaculture 34: 115–143

Weber DN, Russo A; Seale DB, Spieler RE (1991) Waterborne lead effects feeding abilities and neurotransmitter levels of juvenile fathead minnows (Pimephales promelas). Aquat Toxicol 21: 71–80

Webster IA (1983) The continuously fed batch reactor for measuring microbial growth rates. Biotechnol Bioeng 25: 2981–2990

Weglenska T (1971) The influence of various concentrations of natural food on the development, fecundity and production of planktonic filtrators. Ekol Pol Ser A 19: 427–473

Wesenberg-Lund C (1904) Plankton investigations of the Danish lakes. Special Part, Gyldendalske Boghandel, Copenhagen, cited by Hutchinson (1951)

Widigo B (1988) Experimentelle Untersuchungen zur Eignung des Rotators *Brachionus calyciflorus* als Erstfutter für Karpfenlarven (*Cyprinus carpio* L.) Thesis, Fac Biology, University of Munich, 119 pp

Wiens JA (1984) Resource systems, populations, and communities. In: Price PW, Slobodchikoff CN, Gaud WS (eds) A new ecology. Novel approach to interactive systems. Wiley, New York, pp 397–436

Wieser W (1986) Bioenergetik. Energietransformation bei Organismen. Thieme, Stuttgart, 248 pp

Wilbert M (1962) Über die Festlegung und Einhaltung der mittleren Dichte von Insektenpopulationen. Z Morphol Oekol Tiere 50: 576–615

Wilbert H (1968) Die Regulation tierischer Populationen nach kybernetischem Prinzip. Umschau 1968: 746–750

Williams FM (1971) Dynamics of Microbial populations. In: Patten BC (ed) Systems analysis and simulation in ecology. Academic press, London, pp 197–267

Williams FM (1972) Mathematics of microbial populations, with emphasis on open systems. Trans Conn Acad Arts Sci 44: 396–426

Williams FM (1973) Mathematical modelling of microbial populations. Bull Ecol Res Commun (Stockholm) 17: 417–426

Williams FM (1980) On understanding predator-prey interactions In: Ellwood DC, Hedger JN, Latham MJ, Lynch JM, Slater JH (eds) contemporary microbial ecology. Academic Press, London, pp 349–375

Williamson CE (1983) Invertebrate predation on planktonic rotifers. Hydrobiologia 104: 385–396

Williamson CE, Butler NM (1986) Predation on rotifers by the suspension feeding copepod *Diaptomus pallidus*. Limnol Oceanogr 31: 393–402

Winberg GG (1971) Methods for the estimation of production of aquatic animals. Academic Press, London New York, 175 pp

Wood ED, Armstrong FJ, Richards FA (1967) Determination of nitrate in sea water by cadmium-copper reduction to nitrate. J Mar Biol Assoc UK 47: 23–31

Wu TY (1977) Introduction to the study of the scaling of aquatic animal locomotion. In: Pedley TJ (ed) Scale effects in animal locomotion. Academic Press, London, pp 203–232

Wynne D, Gophen M (1986) Phytoplankton-zooplankton feedback effects on the food chain. Arch Hydrobiol 106: 145–154

Wynne-Edward VC (1965) Self-regulating systems in populations of animals. Science 147: 1543–1548

Yee J, Broekhoven L, Colgan P, Harmsen R (1981) Determination of maximal growth rate using continuous culture. J Protozool 28: 146–148

Yúfera M (1987) Effect of algal diet and temperature on the embryonic development time of the rotifer *Brachionus plicatilis*. Hydrobiologia 147: 319–322

Yúfera M, Pascual E (1985) Effects of algal food concentration on feeding and ingestion rates of *Brachionus plicatilis* in mass culture. Hydrobiologia 122: 181–187

Yúfera M, Lubian LM, Pascual E (1983) Efecto do cuatro algas marinas sobre el crecimiento poblacional de dos cepas de *Brachionus plicatilis* (Rotifer: Brachionidae) en cultivo. Invest Pesq 47: 325–337

Zankai NP (1984) Predation of *Cyclops vicinus* (Copepoda: Cyclopoida) on small zooplankton animals in Lake Balaton (Hungary). Arch Hydrobiol 99: 360–378

Zaret TM (1980) Predation and freshwater communities. Yale University Press, New Haven, 187 pp

Zimmermann C (1974) Die pelagischen Rotatorien des Sempachersees, mit spezieller Berücksichtigung der Brachioniden und der Ernährungsfrage. Schweiz Z Hydrol 36: 205–300

Subject Index

extinction 146, 162, 168, 174, 232, 242

· Mercedesdruck, Berlin
·ıchbinderei Lüderitz & Bauer, Berlin